ARCO

# Master the
# Firefighter Exam

### 15th Edition

Fred M. Rafilson, Ph.D.
President, I/O Solutions,
Publisher of the
National Firefighters
Selection Inventory

**THOMSON**

**PETERSON'S**

Australia • Canada • Mexico • Singapore • Spain • United Kingdom • United States

**Petersons.com/publishing**
Check out our Web site at www.petersons.com/publishing
to see if there is any new information regarding the test
and any revisions or corrections to the content of this book.
We've made sure the information in this book is accurate
and up-to-date; however, the test format or content may
have changed since the time of publication.

# Contents

# Contents

# Before You Begin

## HOW TO USE THIS BOOK

*ARCO Master the Firefighter Exam* will show you what to expect and will give you a speedy brushup on the subjects tested in your exam. Some of these are subjects not taught in schools at all. Even if your study time is very limited, you should:

- Become familiar with the type of examination you will take

- Improve your general test-taking skills

- Improve your skills in analyzing and answering questions involving reasoning, judgment, comparison, and evaluation

- Improve your speed and skills in reading and in understanding what you read (an important part of your ability to learn and an important part of most tests)

This book will help you prepare by presenting many different types of questions that have appeared on actual firefighter exams. It also:

**Finds your weaknesses.** Once you know what subjects you're weak in, you can get right to work and concentrate on those areas. This kind of selective study yields maximum test results.

**Gives you the *feel* of the exam.** All the questions in the practice exams and in the sections that give instructions in answering reading comprehension questions, reasoning and judgment questions, and in dealing with spatial orientation questions were taken from actual previous firefighter exams given at various times in various places. This book will give you experience in answering real questions.

**Gives you confidence as you prepare for the test.** It will build your self-confidence as you proceed and will prevent the kind of test anxiety that causes low test scores.

Every city creates its own firefighter exam, so the exams are not all alike. The number of questions varies; timing varies; and the actual content of the exam varies. Nearly all firefighter exams, however, use the multiple-choice question format and a few basic subject areas such as reading-based questions, reasoning and judgment, use of maps and diagrams, and observation and memory. This book will give you lots of practice with questions in all these areas.

A special feature of ARCO books is an explanation for every answer. You can learn immediately why a correct answer is correct and why a wrong answer is wrong. There's a great deal of information in the answer explanations. Studying answer explanations for all questions, even those you got right, will serve as a course in itself.

# PART I

## FIREFIGHTING BASICS

# What Firefighters Do

## chapter 1

## OVERVIEW

- The nature of the work
- Employment outlook
- Places of employment
- Essential job tasks performed by firefighters
- Training and advancement
- Related occupations
- Sources of additional information

## THE NATURE OF THE WORK

Every year, fires destroy thousands of lives and property worth millions of dollars. Firefighters help protect the public against this danger. This book is concerned with paid career firefighters; it does not cover the many thousands of volunteer firefighters in communities across the country.

During duty hours, firefighters must be prepared to respond to a fire and to handle any emergency that arises. Because firefighting is dangerous and complicated, it requires organization and teamwork. At every fire, firefighters perform specific duties assigned by a company officer such as a lieutenant, a captain, or another department officer. Firefighters may connect hoselines to hydrants, operate a pump, or position ladders. Since their duties can change several times while the company is in action, they must be skilled in many different firefighting activities, such as rescue, ventilation, and salvage. Some firefighters operate fire apparatus, emergency rescue vehicles, and fireboats. In addition, they help people to safety and administer first aid.

Most fire departments also are responsible for fire-prevention activities. They provide specially trained personnel to inspect public buildings for conditions that might cause a fire. They may check building plans, the number and working condition of fire escapes and fire doors, the storage of flammable materials, and other potential hazards. In addition, firefighters educate the public about fire prevention and safety measures. They frequently speak on this subject before school assemblies and civic groups; in some communities, they inspect private homes for fire hazards.

Between alarms, they have practice drills and classroom training and are responsible for cleaning and maintaining the equipment they use.

## EMPLOYMENT OUTLOOK

The tragic events of September 11, 2001, highlighted both the dangers of the firefighting occupation and the courage of the men and women who are called to the profession. According to the U.S. Fire Administration, 343 New York firefighters lost their lives that day. However, because of their selfless actions, many more lives were saved. Firefighters have become America's new heroes since the attacks, and rightly so.

Even though the hazards of the profession have become all too clear to us, thousands of men and women strongly desire to join the fire service and serve their communities. It is an attractive career for many because the educational requirement is usually a high school diploma, salaries are relatively high, and a pension is guaranteed upon retirement. Therefore, expect some stiff competition for available job opportunities. The number of qualified applicants typically exceeds the number of available jobs despite the fact that the written examination and physical requirements eliminate many applicants. The increasing competition for firefighter positions is expected to persist over the next few years.

Firefighting jobs are expected to grow at a rate of about 3 to 9 percent between 2000 and 2010. Compared to the growth rate of 10 to 20 percent for all other jobs, job opportunities for firefighters are likely to be slim. This trend is expected to continue through the year 2010 as fire departments compete with other public service providers for funding.

The increased level of specialized training in this occupation has made it increasingly difficult for volunteer firefighters to remain qualified. Therefore, most job growth will occur as volunteer firefighting positions are converted to paid positions. Large urban fire departments are expected to experience the least amount of growth. The turnover rate of firefighting jobs is particularly low at this time. This is somewhat unusual considering that the job of a firefighter is hazardous and requires a relatively limited investment in formal education. Other job openings will occur as firefighters retire, stop working for other reasons, or transfer to other occupations.

In recent years, firefighters have become involved in much more than preventing fires. In response to the expanding role of firefighters, some municipalities have combined fire prevention, public fire education, safety, and emergency medical services into a single organization. Some local and regional fire departments are being consolidated into countywide public safety establishments. This has allowed for the reduction of overhead, reduction of administrative staff, and the establishment of consistent training standards and work procedures.

Layoffs of firefighters continue to be uncommon. Fire protection is an essential service for all citizens. As a result, city officials are likely to feel considerable pressure to expand or at least preserve the level of fire-protection coverage. When budget cuts do occur, local fire departments usually cut expenses by postponing equipment purchases or by not hiring new firefighters rather than by a lay-off of staff.

## PLACES OF EMPLOYMENT

According to the U.S. Department of Labor's Bureau of Labor Statistics, about 258,000 men and women worked as paid career firefighters in the year 2000. Nine out of ten firefighters work in municipal fire departments. These fire departments vary widely in size; some very large cities have several thousand workers on the payroll, whereas many small towns have fewer than 25. Federal installations often have their own fire departments; firefighting opportunities can also be found at airports and in large manufacturing plants.

## ESSENTIAL JOB TASKS PERFORMED BY FIREFIGHTERS

- Put on and wear protective equipment.
- Step off an engine or truck wearing full equipment.
- Pull uncharged hose off an engine to a hydrant.
- Advance preconnected, uncharged $2\frac{1}{2}$-inch hose until it is fully extended.
- Hold a charged $1\frac{3}{4}$- and/or $1\frac{1}{2}$-inch hose and open the nozzle.
- While on a ladder, direct water at fire.
- Remove a K-12 saw from the truck and start it.
- Carry a 35-foot extension ladder from the truck to the fire scene, with assistance, or carry a 24-foot extension ladder from the truck to the fire scene without assistance.
- Carry a 50-foot section of rolled hose unassisted.
- Make and unmake coupling connections.

- Carry a jump bag where needed.

- Raise a 24-foot and/or 35-foot extension ladder.

- Extend the fly section of a 24-foot and/or a 35-foot ladder.

- Operate a charged line from confined spaces.

- Advance preconnected, uncharged $1\frac{3}{4}$-inch hose until it is fully extended.

- Extend the small line to a fire.

- Operate a fire extinguisher.

- Remove a smoke ejector (a fan, for example) from the truck; transport and place it in operation.

- Climb an aerial ladder to a height of 75 to 100 feet wearing full equipment.

- Remove and transport electrical extension cords from the truck as needed.

- Wrap a hose around the hydrant, unscrew the hydrant cap, screw the hose connection to the hydrant, and turn on the hydrant.

- Be aware of electrical lines when setting up ladders and directing water streams.

- Operate a chain or circular saw during firefighting and rescue operations.

- Enter smoking buildings/rooms with a hose in hand while wearing SCBA (self-contained breathing apparatus).

- Fight fires in smoky buildings when visibility is poor or nonexistent.

- Seek the source of a fire and extinguish.

- Prevent the extension of a fire.

- Select the correct tool or equipment for various firefighting maneuvers.

- Crawl on a floor and if you cannot see, feel for the heat of the fire source.

- Safely shut off utility services to buildings in emergency situations.

- Determine the stability of supporting surfaces.

- Systematically search for trapped persons.

- Operate a line from elevated positions.

- Carry equipment up stairs while wearing full turnout gear and SCBA.

- Use equipment (ax, sledge hammer, etc.) to make forcible entries.

- Determine when to open roofs, walls, windows, and doors.

- Use a ceiling hook to pull down a ceiling.

- Carry a standpipe rack up numerous flights of stairs while wearing full equipment.

- Chop/cut ventilation holes in a roof with a fire ax or power saw to let out smoke and heat.

- Evacuate people from a fire area.

- Remove people from entrapments.

- Carry or drag a victim out of a building while wearing full turnout gear.

- Carry or drag a victim up or down stairs while wearing full turnout gear.

- Carry people down ladders while wearing full turnout gear.

- Remove burned and charred waste.

- Re-bed hoses and put them back onto the engine/truck.

- Check for smoldering fire inside walls and ceilings after a fire is extinguished.

- Lower ground ladders and re-bed them onto the truck.

- Extricate people from automobiles.

- Respond to hazards related to electrical emergencies.

## First Aid

New firefighters are typically eager to extinguish large blazes and perform daring rescues in dangerous situations. However, firefighters have taken on many more duties over the years because the number of fire incidents is down. As a result, aspiring firefighters may be surprised to learn that fire departments respond to more medical calls or cardiac emergencies than any other type of call, including fires. Firefighters will certainly be called upon to respond to all sorts of fires, but they will also provide assistance at car accidents and emergency treatment to those who may be seriously ill or injured.

Providing emergency medical care often is a part of the firefighter's role. Firefighters deal with burns, broken bones, serious cuts and abrasions, and breathing stoppages. All of these medical emergencies occur with some frequency at the sites of major fires. Every new recruit to a fire department receives training in dealing with these medical emergencies as part of firefighter training. In fact, in the current climate of cutting costs by combining administrative operations and by avoiding duplication of services, many municipalities are merging firefighting forces with emergency medical services. This means that some firefighters are being trained to serve as emergency medical technicians, and some emergency medical technicians and paramedics are being trained as firefighters.

No applicant about to take an open competitive exam for a firefighter position is expected to know how to drive a fire engine, raise an extension ladder, charge a hose,

**NOTE**

The training of firefighters is by no means restricted to training in putting out fires.

or immobilize a broken limb. Firefighter training academies offer training in all aspects of the firefighter's job. All recruits undergo rigorous and thorough training and must pass mastery examinations before being appointed to the force. Testing in knowledge and performance comes after instruction and practice. The initial open competitive exam, the one for which you are preparing right now in the hope of being accepted into the firefighter training program, will test your ability to learn, common sense and judgment, and general awareness of safety measures and first aid.

Before providing anyone with first-aid treatment, firefighters must ensure that they take proper body substance isolation (BSI) precautions in order to protect themselves against infectious diseases or other hazardous conditions. Firefighters put themselves at great risk when treating patients in the prehospital setting because they may be exposed to the patient's blood and other body fluids, and a patient's infection status is often unknown.

An overzealous firefighter may not want to take the time for these precautions, such as wearing personal protective equipment or cleaning surfaces stained with body fluids, especially when victims are in urgent need, but a firefighter who exposes him- or herself unnecessarily to infectious diseases will soon be no help to anyone.

Firefighters must be trained to assess patients rapidly at the scene of an emergency. This assessment involves quickly determining the following:

- Overall condition of the patient

- Extent of injury or illness

- Type of intervention and first aid to be used

- Prioritization of patients according to the seriousness of their injuries or illness

- Immediate transportation of seriously ill patients

The most important aspects of the initial assessment can be easily remembered with the acronym ABCDE, which stands for the following:

**A**—Airway

**B**—Breathing

**C**—Circulation

**D**—Disability (Level of Consciousness)

**E**—Exposure for examinations/protection from the environment

Firefighters must be familiar with procedures to treat cardiac emergencies because 50 percent of deaths due to cardiac disease occur outside of the hospital. Therefore, the first responder plays a critical role in the treatment of these patients. The firefighter or EMT who encounters a cardiac patient in the field will be asked to perform vital lifesaving procedures. CPR skills are vital. In addition, defibrillation with automatic external defibrillator (AED) devices have become a new (and in some cases, legally mandated) responsibility of the first-aid provider. Research has shown that CPR alone (although a critical component of the lifesaving process) is not enough to resuscitate a patient. The computer systems of the AED help the first responder diagnose and treat a cardiac patient with electricity that stops the irregular heart rhythm. AEDs are now simpler to use and maintain, and more and more public safety professionals are capable of learning how to use this vital lifesaving device with only a few hours of training.

The following is a brief list of additional first-aid procedures that firefighters will typically use at an emergency scene:

- Treating and transporting patients with burns and smoke inhalation injuries

- Recognizing the symptoms of heat cramps, heat exhaustion, and heat stroke

- Treating patients with wounds, fractures, and other traumatic injuries

- In some areas, treating patients who have nearly drowned or have suffered from hypothermia

- Treating patients who appear to be suffering from a respiratory ailment, such as asthma

- Recognizing the symptoms of carbon monoxide poisoning and providing treatment and transport

- Recognizing the symptoms of an adverse reaction to a hazardous material and providing treatment and transport

Firefighters must make themselves aware of the possibility of litigation when providing emergency medical treatment. Lawsuits against people in the medical field, even those who provide first aid or pre-hospital care, are becoming more and more common, and public safety personnel are no longer immune from criminal and civil liability. All fire department personnel should be familiar with the concepts of civil liability, tort liability, and negligence as well as applicable laws and statutes regarding medical care. In order for firefighters to protect themselves from being the focus of a criminal or civil lawsuit, it is critical they know their responsibilities at the scene of a medical emergency. Firefighters must follow all procedures carefully, document every action they take, and know a patient's rights.

**NOTE**

Knowing what can and cannot be done during a medical emergency is the first line of defense.

## Heartbeat

Your first thought about injuries connected with fires is undoubtedly that fires cause burns. A firefighter is much less likely to have to deal with the burns, however, than to deal with breathing stoppages, heart stoppages, and shock. If a fire victim's heart has stopped beating, survival is limited to only a few minutes. A victim whose heart has stopped beating will have no pulse. Your training will include cardiopulmonary resuscitation (CPR), and you must initiate CPR immediately—with one exception. If the fire is gaining on you and the victim, you both must get out. By remaining in the path of a fire to apply CPR to a victim, you simply create two victims. Pick up the victim and get out. This is a matter of common sense. You might encounter a question setting up a scenario such as this on your exam. The question might ask you which of a number of steps to take first in such a situation. The examiners are not really testing your knowledge of first-aid practices and procedures. They are looking for signs of logical reasoning and good judgment on your part. Removing the victim from the fire is ALWAYS the first step.

## Breathing

If your fire victim's heart is beating, the next priority is the victim's breathing, regardless of any other injuries he or she might have sustained. Check for absence of breathing by observing the victim's chest or by putting your ear to his or her nose to listen or feel for signs of breathing. An individual can live for only a very short time without breathing. A person who manages to survive after a period without breathing is apt to suffer severe brain damage.

The preferred method of attempting to restore breathing is mouth-to-mouth resuscitation. This is a relatively simple procedure:

1. Position the victim on his or her back.

2. Clear the upper airway by turning the head to one side and running your finger through the mouth to remove the tongue from the throat and to wipe out debris.

3. Tilt the head back so the neck is stretched and the head is in a chin-up position.

4. If the victim is an adult, pinch the nose closed. If it's a child, you should cover both the nose and mouth with your own mouth.

5. Take a deep breath, open your mouth wide, and breathe forcefully into the victim's mouth. Watch to see if the victim's chest rises. If not, breathe harder. If your breath is still not reaching the victim's lungs, there might be an obstruction in the airway. Turn the victim and strike him or her between the shoulders to dislodge the obstruction.

⑥ Take another deep breath and again breathe into the victim's mouth. Repeat rhythmically 12 to 20 times per minute until the victim breathes alone or until help arrives.

⑦ With infants and small children, breathe more shallowly from your cheeks rather than from deep in your lungs.

Your examination will not ask you how to perform mouth-to-mouth resuscitation, but it might ask you a question about when such a procedure is appropriate or how long you should continue. Your answer will indicate awareness and judgment.

## Bleeding

Severe bleeding is not a direct result of exposure to fire. Nonetheless, severe bleeding among fire victims is not uncommon. The bleeding is, of course, the result of an injury. A person attempting to escape a fire might rush through a smoke-filled area with poor visibility and crash through jagged glass or impale him- or herself on fire-damaged wood or metal. The escape attempt might entail a fall or a jump that leads to an injury with severe bleeding. Bleeding often looks worse than it is. Spurting bright-red blood from a severed artery, however, leads to massive loss of blood and often is fatal. Uncontrolled, long-term, heavy bleeding (hemorrhage) from a large vein also can prove fatal.

The person bleeding heavily should first lie down. If the bleeding is from a limb, the limb should be elevated to slow the blood flow. The firefighter, EMT, or bystander pressed into service should then cover the wound with the cleanest cloth readily available and press hard. Pressure over the wound should slow the bleeding, thereby limiting the loss of blood and allowing formation of a clot. You might have to maintain this pressure over the wound for a considerable period of time as you wait for a self-limiting clot to form. You will know that a clot has formed when bleeding has noticeably slowed; the bloody area of the covering cloth will stop expanding, and perhaps the cloth will feel a bit drier. You then can secure the cloth over the wound by tying it snugly but not tightly. If pressure over the wound stops the bleeding, it definitely is the treatment of choice. Because maintaining pressure on a wound can entail a considerable length of time, you should look around for an onlooker who can take over this task so that you, the professional firefighter, can return to the fire scene to search for more victims or to help fight the fire. Recognizing when and how to utilize nonprofessional help also is a part of judgment.

Arterial bleeding (the bright-red spurting kind) sometimes can be stopped, at least temporarily or in conjunction with direct pressure over the wound, by applying pressure on pressure points. Pressure points are areas on the body where the artery lies close to bone; external pressure can narrow the artery against resistance from the bone. There are pressure points at the temple just in front of the ear, below the jaw, behind the collarbone, in the upper and lower arms, at the wrist, in the groin, at the

# NOTE

Your firefighter exam will not ask you to name pressure points, though it surely can do you no harm to know where they are. Instead, it might ask a general question about control of bleeding to check your grasp of priorities.

knee, and in the ankle. Apply pressure at the pressure point between the wound and the heart that is closest to the wound. If you apply pressure at both the appropriate pressure point and over the wound itself, you can slow the flow of blood and hasten clot formation. The accompanying diagram shows the location of pressure points on the body.

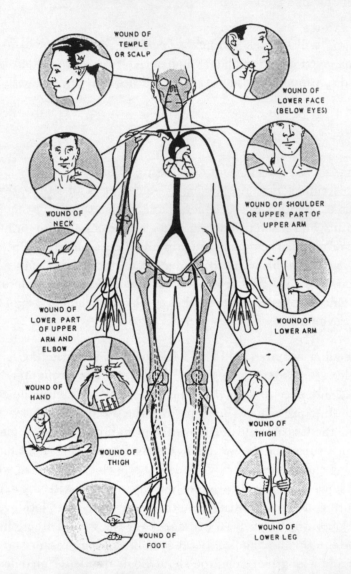

**Pressure points for temporary control of arterial bleeding.**

A tourniquet is a tight, constricting band placed entirely around an extremity (arm or leg) to totally stop the flow of blood. A tourniquet is truly the method of last resort. A tourniquet shuts off the entire flow of blood to all tissue beyond it. Lack of blood causes tissue to die. After a tourniquet has been put into place and tightened, it must not be loosened. A tourniquet should be removed only by a doctor. Before placing a tourniquet, consider how soon the victim can reach qualified medical attention. On the other hand, if you find it impossible to stop arterial bleeding in any other way, it is surely better to lose the limb and save the life. This is a judgment call. Delaying a tourniquet and letting the victim bleed to death would not be prudent.

## Other Emergencies

Absence of heartbeat, breathing stoppage, and hemorrhage represent three emergency situations in which firefighters must take immediate, positive action. In these situations, time is of the essence. If you find yourself facing any one of these medical emergencies, you cannot wait for an EMT, a paramedic, an ambulance, or a doctor. You must act. And you must remember this for answering questions on your firefighter exam.

Other emergency medical situations require immediate attention from firefighters and EMTs at the scene but do not require quite such instantaneous or proactive response. These situations, while not immediately life threatening, can lead too easily to shock and subsequent death. Two of these medical events, in particular, are fractures and burns.

### *Fractures*

A fracture is a break in a bone. The broken bone can be in an extremity, such as an arm, leg, hand, or foot; in the head, such as a skull fracture; or in the torso, such as a broken collarbone, rib, or pelvis. Fractures must be diagnosed and treated by a physician, and they are extremely painful. A fracture victim in extreme pain may well go into shock. (Shock is discussed in detail later.) The role of the firefighter at the scene is to lessen the chance or severity of shock by alleviating pain. The pain of a fracture is best alleviated by immobilizing the affected area. If it is safe to leave the victim wherever he or she is sitting or lying, the person should be told to stay in one place and move as little as possible. If the person must be moved, movement must be done as gently as possible with minimal jarring to the broken bone(s). Where possible, a makeshift splint should be fashioned, extending beyond joints on either side of the break. Such a splint—made of a board or stick, padded with cloths or grass or moss, and secured with string or belts—should keep pieces of bone from moving and thereby spare flesh from friction. The less motion, the less damage to tissue, the less bleeding, the less pain, and the less likelihood of shock.

In cases of fire, bone breakage often comes from jumps from high windows and from falls through floors weakened by the fire. Unfortunately, these jumps and falls frequently cause back injuries, and the fracture with the greatest potential for grave damage is a fracture of the spine. Enclosed within the bony structure of the spinal column is the spinal cord. The spinal cord carries nerve messages from the brain to the nervous system and hence to the muscles of the entire body. If the spinal column is severed, the messages cannot get through. Broken bones have sharp, ragged edges. A broken bone in the spinal column, if moved, can injure the spinal cord. A

**TIP**

If the choice is between possible paralysis and certain death, is there a choice? Remember this as you take your firefighter exam.

spinal cord injury in the lower spine affects the lower extremities. It can leave the victim a paraplegic, unable to walk. A spinal cord injury farther up on the spine can affect all parts of the body below the injury. A broken neck can leave the victim paralyzed from the neck down, a quadriplegic. The general rule in first aid is that a person with a suspected or even possible back or neck injury should never be moved until totally immobilized on a backboard. In firefighting, however, there is always the one consideration that takes precedence—getting away from the fire.

### Burns

The most obvious effects of exposure to fire are smoke inhalation and burns. Smoke inhalation does not lend itself to first aid. The firefighter can only administer oxygen as instructed at the training academy and wait for professional medical assistance. The burns of the victim also cannot be treated at the scene, but the burn victim can and must receive attention. Except for the most superficial burns, all burns are serious. Body fluids are lost very rapidly through the large expanses of exposed flesh of burns, and this can lead to dehydration, lowered blood pressure, and shock. The intense pain of burns also is a contributor to shock. After removing the burned person from the fire, the firefighter must make that person lie down. The firefighter then should loosely cover burned areas with the cleanest cloth available. If the person is conscious and is not extremely nauseated, the firefighter must encourage him or her to begin fluid replacement as soon as possible. Lightly salted cool or cold water is best, but if salt is unavailable, plain water will do. The burn victim should start with small sips and then work up to full glasses.

### Shock

Shock is the body's involuntary response to major injury. Shock can occur immediately upon the injury to the body or might develop as a reaction later. Shock is characterized by a swift and severe loss of blood pressure, rapid lowering of body temperature, rapid heartbeat, bodily weakness, and fainting. With lowered blood pressure comes decreased blood circulation and oxygen starvation of cells, especially cells of the vital organs. In particular, the brain, liver, heart, and kidneys can be rapidly damaged by shock. If shock deepens and continues, the result is liver and kidney failure.

The injuries that fire victims sustain are precisely the types of injuries that tend to throw a person into shock. For a fragile personality, fear, fatigue, and pain are adequate to initiate the shock syndrome. In combination with loss of blood from hemorrhage or loss of body fluids from burns, the fear, fatigue, and pain almost guarantee shock. In addition, fire victims often face legitimate apprehension about the fate of their loved ones who might have been involved in the same fire and insecurity about their future health, homes, and lifestyles. The physical and mental pressures on a fire victim are almost overwhelming. Shock is the response.

Everyone who deals with fire victims must be aware of symptoms of shock and of the best ways to forestall, lessen, or reverse these symptoms. The medical treatment of choice is intravenous replacement of lost fluids—blood or plasma by transfusion and other fluids by infusion of saline, sugar, or electrolyte solutions. Obviously, this medical treatment must follow the first-aid treatment by the firefighter at the scene. The firefighter's role is to:

❶ Remove the victim to a safe location for the safety of the victim as well as to allay the victim's fear of danger and additional pain that might be caused by the fire.

❷ You must calm the victim. Convince the victim to lie down. Unless the injuries themselves keep the victim off his or her feet, a person who has just been traumatized by fire is likely to be highly agitated and to walk about frantically. This type of activity will hasten shock.

❸ Control bleeding. Control of bleeding is a lifesaving measure in itself. Blood is absolutely necessary to sustain life, and massive loss of blood is seriously life threatening. More moderate loss of blood also is contributory to shock. In combination with loss of blood pressure and diminished circulation, lower blood volume can lead to a critical situation. As a firefighter, you can contribute to shock prevention by helping to stop the bleeding and by encouraging the victim to drink water to replace lost fluids and prevent dehydration.

❹ Make the victim comfortable. A common reaction to trauma is shivering. If the victim's clothing is wet, remove it. Loosen constricting clothing. Cover the victim loosely. Keep him or her warm but avoid overheating. Extreme chilling and extreme overheating both adversely affect blood circulation. You must watch for erratic changes in temperature and adjust the coverings accordingly. If a cold victim is not nauseated, warm liquids can be soothing and can help with the warming process. A nauseated victim, on the other hand, should be served cool or cold liquids to avoid vomiting.

A few words about liquids:

- Never try to force liquids into an unconscious person.

- If the victim is unable to sit up, administer liquids with a straw or spoon. A person who cannot lift his or her head to drink might be able to get liquids by sucking on a cloth repeatedly dipped into the liquid. This method also might work for a person who is nauseated. It is less likely to induce vomiting.

- Never administer alcoholic drinks.

❺ Relieve pain. The firefighter at the scene has very limited resources for relieving pain. Immobilizing broken bones can help. Covering open wounds to protect them from wind, water, or falling debris also can be useful. Giving reassurance and moral support can help, too. Believe it or not, your calming words can help slow the victim's breathing and heart rate. You can make the victim feel better and can slow the progress of shock just by talking gently.

In short, as a firefighter, you are the first at the scene of a fire. As such, you face the primary mandate to save lives. You save those lives by removing people from danger caused by fire itself, smoke, and crumbling structures. If any people you remove from danger require immediate lifesaving attention, it is your responsibility to initiate that care until some other person arrives to take over. Your task then is to extinguish the fire.

These are your priorities. First save lives and *then* save property. First help keep victims alive and then do what you can to reduce physical damage to those victims. If you remember these simple rules, you should be able to answer most of the commonsense and judgment questions on your firefighter entrance exam.

## Serving as an Emergency Medical Technician (EMT)

Fire departments typically either have dedicated Emergency Medical Technicians (EMTs) to respond to medical calls or their firefighters will handle these incidents themselves. Cost-cutting measures have forced more and more departments to certify firefighters as EMTs to handle the large volume of these types of calls. Therefore, it is a good idea for you to know the requirements for these certifications. When applying for firefighter positions, you should look carefully at all requirements and prerequisites for each department because the specific EMT certification requirements will vary from department to department.

### Levels of EMT Certification

There are three levels of certification: **EMT-Basic** (the first level of certification), **EMT-Intermediate,** and **EMT-Paramedic.** All three types of certifications require the completion of an appropriate course based on the *National Standard Curriculum*. In order to become an EMT-Paramedic, prior certification as an EMT-Basic or EMT-Intermediate is required.

**EMT-Basic Certification** involves training in basic life support (BLS) procedures and other tasks required to successfully fulfill the duties of the job. EMT-Basic responsibilities include:

- Vehicle operation

- Emergency medical care to adult, pediatric, and trauma patients

- Cardiopulmonary resuscitation (CPR)

- Using an automated external defibrillator (AED)

- Treating wounds

**NOTE**

It should be noted that some departments require all applicants to have valid certification; other departments award preference points to those applicants who are already certified.

- Assisting with childbirth
- Administering some medications
- Responding to environmental emergencies

**EMT-Intermediate Certification** requires certification in EMT-Basic and additional training in advanced life support (ALS) procedures, including:

- Intravenous therapy
- Treatment of respiratory emergencies
- Advanced airway procedures
- Interpretation of cardiac rhythms
- Emergency pharmacology
- Management of obstetrical emergencies

**EMT-Paramedic Certification** requires the completion of all of the BLS and ALS skills that EMT-Basics and EMT-Intermediates have obtained. However, training in the following areas is also required:

- Patient assessment
- Drug therapy
- Airway management and ventilation
- Trauma and hemorrhage
- Defibrillation

Some applicants who lack necessary EMT credentials required by a department believe that they can obtain these licensures in a short period of time. They soon discover, however, that these certifications require extensive training. The EMT-Basic and EMT-Intermediate levels generally require between 300 to 400 hours of instruction, while the EMT-Paramedic level requires 1,000 to 1,200 hours of instruction. These certifications are usually good for two or three years, depending on the state. Many states also require EMTs to participate in continuing education courses.

According to the U.S. Department of Labor's *Dictionary of Occupational Titles* and the Occupational Information Network (O*NET OnLine), there are some additional tasks performed by EMTs and paramedics, including:

**NOTE**

All three levels of certification require knowledge of the various ethical and legal issues that arise during medical emergencies.

- Treating sick and injured individuals in a prehospital setting as authorized by a physician

- Determining the type of treatment(s) that must be performed for a patient, the priority of those treatments, and the need for additional assistance

- Recording patients' vital signs

- Performing endotracheal intubation as part of airway management and ventilation

- Inflating an antishock garment on a patient as a means to improve blood circulation

- Assisting in extricating trapped individuals

- Transporting patients to an emergency medical facility

- Driving a mobile intensive care unit to the emergency scene

- Acting as team leader for emergency medical technicians

- Using communication devices to relay important information to physicians and other medical personnel

- Maintaining all occupational equipment

- Replenishing supplies when needed

- Comforting and reassuring patients and their families

- Working in conjunction with other emergency medical technicians, medical personnel, and police and fire department personnel

- Communicating with dispatchers

- Arranging for the reception of victims at the emergency medical facility

- Lifting and carrying patients

Aspiring firefighters should keep the last bullet point in mind. Performing many EMT duties requires considerable strength. However, since firefighting is physically strenuous to begin with, qualified applicants will have the necessary physical fitness to perform the essential tasks of all aspects of the job. It is important for applicants to spend time training, exercising, and maintaining their physical fitness. Due to the high volume of medical calls, it is likely that firefighters will be performing physically strenuous tasks on a regular basis. (Refer to Chapter 3 for more information regarding physical fitness.)

Firefighters who perform EMT duties also need considerable stress tolerance and an ability to manage stress safely and effectively. EMTs and paramedics often witness gruesome trauma scenes that sometimes involve infants and children. EMTs and paramedics may also encounter mentally disturbed individuals or find themselves in violent surroundings. Again, firefighting is also a dangerous and stressful occupation, so well-qualified applicants should be able to adapt well to their EMT responsibilities.

Although many people may not automatically think of emergency medical care as a primary duty of a firefighter, it is clearly just as important as putting out fires and cleaning hazardous materials spills. It can be a very rewarding extension of a firefighter's lifesaving vocation.

## Working Conditions

Firefighters spend much of their time at fire stations, which usually have facilities for dining and sleeping. When an alarm comes in, firefighters must respond rapidly, regardless of the weather or hour. They may spend long periods outdoors fighting fires in adverse weather.

Firefighting is among the most hazardous occupations. The job of a firefighter involves the risk of death or injury from sudden cave-ins of floors or toppling walls and danger from exposure to flames and smoke. Firefighters also might come in contact with poisonous, flammable, and explosive gases and chemicals.

In some cities, firefighters are on duty for 24 hours, are then off for 48 hours, and receive an extra day off at intervals. In other cities, they work a day shift of 10 hours for three or four days, a night shift of 14 hours for three or four nights, have three or four days off, and then repeat the cycle. Although in many large cities, particularly in the eastern United States, firefighters work a standard 40-hour week, many firefighters average as many as 56 hours per week. In addition to scheduled hours, firefighters often must work extra hours when they are bringing a fire under control. Fire lieutenants and fire captains work the same hours as the firefighters they supervise. Duty hours may include some time when firefighters are free to read, study, or pursue other personal interests.

## TRAINING AND ADVANCEMENT

Applicants for municipal firefighting jobs must pass a written test, a medical examination, and tests of strength, physical stamina, and agility, as specified by local regulations. These examinations are open to people who are at least 18 years of age and have a high school education or the equivalent. Those who receive the highest scores on the examinations have the best chances for appointment. Extra credit usually is given for military service. Experience gained as a volunteer firefighter might also improve an applicant's chances for appointment.

As a rule, beginners in large fire departments are trained for several weeks at the department's training center. Through classroom instruction and practical training, the recruits study firefighting techniques, fire prevention, hazardous materials, local building codes, and emergency medical procedures. They also learn how to use axes, saws, chemical extinguishers, ladders, and other firefighting and rescue equipment. After completing this training, they are assigned to a fire company, where they are evaluated during a probationary period.

A small but growing number of fire departments have accredited apprenticeship programs that last for three to four years. These programs combine formal, technical instruction with on-the-job training under the supervision of experienced firefighters. Technical instruction covers subjects such as firefighting techniques and equipment, chemical hazards associated with various combustible building materials, emergency medical procedures, and fire prevention and safety.

Many fire departments offer continuing in-service training to members of the regular force. This training may be offered during regular on-duty hours at an easily accessible site connected to the station house alarm system. Such courses help firefighters maintain certain seldom-used skills and permit thorough training on any new equipment the department may acquire.

Most experienced firefighters continue to study to improve their job performance and to prepare for promotion examinations. To progress to higher-level positions, firefighters must acquire expertise in the most advanced firefighting equipment and techniques and in building construction, emergency medical procedures, writing, public speaking, management and budgeting procedures, and labor relations. Fire departments frequently conduct training programs, and some firefighters attend training sessions sponsored by the National Fire Academy on topics such as executive development, anti-arson techniques, and public fire safety and education. Most states also have extensive firefighter training programs.

Many colleges and universities offer courses such as fire engineering and fire science that are helpful to firefighters, and fire departments often offer firefighters incentives such as tuition reimbursement or higher pay to pursue advanced training. Many fire captains and other supervisory personnel have college training.

Among the personal qualities firefighters need are mental alertness, courage, mechanical aptitude, endurance, and a sense of public service. Initiative and good judgment are extremely important because firefighters often must make quick decisions in emergencies. Because members of a crew eat, sleep, and work closely together under conditions of stress and danger, they should be dependable and be able to get along well with others in a group. Leadership qualities are assets for officers, who must establish and maintain discipline and efficiency as well as direct the activities of firefighters in their companies.

Opportunities for promotion are good in most fire departments. As firefighters gain experience, they can advance to a higher rank. After three to five years of service, they might become eligible for promotion to the grade of lieutenant. The line of further promotion usually is to captain, battalion chief, assistant chief, deputy chief, and finally to chief. Advancement generally depends on scores from a written examination, performance on the job, and seniority. Increasingly, fire departments are using assessment centers—which simulate a variety of actual job performance tasks—to screen for the best candidates for promotion. However, many fire departments require a master's degree—preferably in public administration or a related field—for promotion to positions higher than battalion chief.

## Firefighting in the Twenty-First Century

The terrorist attacks of September 11, 2001 underscored the need for a strong incident command system and effective communications. As a result, new firefighter responsibilities have been brought to the fore, such as investigating possible hazardous materials and suspicious activity. In many cases, fire departments must adapt to these new responsibilities without additional manpower or funding. Consequently, fire department personnel must work even harder to learn new disaster-management techniques and streamline the department's emergency operations.

Prior to the events of 9/11, many departments and the general public were not prepared for the possibility of a large-scale disaster that would result in the deaths of hundreds of firefighters and thousands of civilians. Today fire departments must develop plans to manage all sorts of worst-case scenarios: suicide bombers, additional airplane hijackings, and/or attacks with weapons of mass destruction.

To combat these plots and respond most effectively to future catastrophes, fire departments have had to make some drastic changes in a short period of time, placing a new emphasis on rescue and improved communications. Firefighting personnel must also be able to take advantage of available state and federal resources and the assistance of neighboring departments. Departments must be willing to adopt the policies and procedures of new post-9/11 governmental agencies, such as the Department of Homeland Security. Other necessary changes include coordinating efforts with other agencies, adopting a common terminology at an emergency scene, seeking additional funding when needed, and exploring new technology and procedures that can make operations more efficient and save more lives. Firefighters must learn to be especially vigilant at all times and be prepared for unexpected emergencies.

These changes cannot be implemented successfully without continuous, up-to-date training courses and materials. Both large and small cities now devote time and energy to preparing for large-scale emergencies. This training is not only limited to rookie firefighters. All personnel are required to refresh their knowledge on these topics and absorb new information throughout their careers.

Although we will probably never be able to eliminate the threat of terrorist attacks completely, the willingness of firefighting organizations to change their traditional operations and examine the problem from new perspectives has led to the development of disaster response plans and increased large-scale disaster training for firefighters.

It should also be noted that fire departments must not neglect their responsibility for providing the community with basic fire safety education. Although it is necessary to plan for such disasters as biological, chemical, or even nuclear attacks in our communities, it is just as important to remind the public to maintain their smoke detectors, sprinklers, and other fire-prevention devices. All families should have a fire evacuation plan, and the owners and managers of buildings must be held accountable for abiding by local fire codes. In other words, the new demands of the post-9/11 world must be balanced with the department's traditional fire-prevention responsibilities.

## Earnings

Many career firefighters and company officers are unionized and belong to the International Association of Firefighters; chief officers often belong to the International Association of Fire Chiefs. The unions have been highly successful in securing wages and benefits for their members to compensate the efforts they make, the responsibilities they take on, and the dangers they face.

According to the U.S. Department of Labor's Bureau of Labor Statistics, the median hourly earnings of firefighters was $17.99 in 2003. Fire lieutenants and fire captains may earn considerably more. The mean annual salary for sworn full-time firefighters was $38,810 in 2003.

Firefighters who work more than a specified number of hours per week are required to be paid for overtime. Firefighters often earn overtime for working extra shifts to maintain minimum staffing levels or in the event of special emergencies.

Some of the benefits that firefighters usually receive include the following:

- Medical insurance
- Liability insurance
- Sick leave
- Vacation
- Some paid holidays

Almost all fire departments provide protective clothing, such as helmets, boots, and coats, as well as breathing apparatus. Many departments also provide dress uniforms. Firefighters are generally covered by pension plans that provide retirement at half pay after twenty five years of service or if disabled in the line of duty.

## RELATED OCCUPATIONS

Firefighters work to prevent fires and to save lives and property when fires and other emergencies occur, such as explosions and chemical spills. Related fire-protection occupations include wildland firefighters and fire-protection engineers, who identify fire hazards in homes and workplaces and design prevention programs and automatic fire detection and extinguishing systems. Other occupations in which workers respond to emergencies include police officers and emergency medical technicians.

## SOURCES OF ADDITIONAL INFORMATION

Contact your local civil service office or local fire department for information about obtaining a job as a firefighter. Check the classified ads in your local newspaper for classified ads regarding current available firefighting positions. You may also find listings for fire departments across the country if you subscribe to the member services on the homepage for Public Safety Recruitment (www.ifpra.com).

Information about a career as a firefighter can be obtained from the following:

**International Association of Firefighters**
1750 New York Avenue, NW
Washington, DC 20006
Phone: 202-737-8484
Fax: 202-737-8418
www.iaff.org

**International Association of Fire Chiefs**
4025 Fair Ridge Drive, Suite 300
Fairfax, VA 22033-2868
Phone: 703-273-0911
Fax: 703-273-9363
www.iafc.org

**U.S. Fire Administration**
16825 S. Seton Avenue
Emmitsburg, MD 21727
Phone: 301-447-1000
Fax: 301-447-1346
www.usfa.fema.gov

**Women in the Fire Service, Inc.**
P.O. Box 5446
Madison, WI 53705
Phone: 608-233-4768
www.wfsi.org

Information about firefighter professional qualifications can be obtained from the
following:

**National Fire Protection Association**
1 Batterymarch Park
Quincy, MA 02269
Phone: 617-770-3000
Fax: 617-770-0700
www.nfpa.org

# Test-Taking Techniques

## chapter 2

## OVERVIEW

- Preparing for the exam
- How to take an exam
- Test-day strategy

## PREPARING FOR THE EXAM

You want to become a firefighter. That's great. Every city, every town, and every village needs qualified and enthusiastic new recruits to maintain staffing at a full level. You have made a good start toward joining your firefighting force by buying this book.

This book has been carefully researched and written to help you through the qualifying process. The information in this book will prepare you for the written exam and will give you valuable tips toward preparing yourself for the physical performance exam.

It is important for you to allow the book to help you. Read it carefully. Do not skip any information. You must know what to expect and then prepare yourself for it. If you are prepared, you can feel self-confident. If you feel confident, you can answer questions quickly and decisively, finish the exam, and earn a high score. If you feel confident, you can enter the physical performance course without hesitation and can prove that you are fit for the job. Here are some tips to help start your studying:

❶ **Make a study schedule.** Assign yourself a period of time each day to devote to preparation for your firefighter exam. A regular time is best, but the important thing is daily study.

❷ **Study alone.** You will concentrate better when you work by yourself. Keep a list of questions you find puzzling and points of which you are unsure. Later, talk it over with a friend preparing for the same exam. Plan to exchange ideas at a joint review session just before the test.

③ **Eliminate distractions.** Choose a quiet, well-lit spot as far away as possible from the telephone, television, and family activities. Try to arrange not to be interrupted.

④ **Begin at the beginning and read.** Underline points that you consider to be significant. Make margin notes. Flag the pages you think are especially important with little Post-it™ Notes.

⑤ **Concentrate on the information and instruction chapters.** Learn the vocabulary of the job. Get yourself psyched for the whole world of firefighting. Learn how to handle reading-based questions. Focus on the approach to eliminate wrong answers. This information is important for answering all multiple-choice questions, but it is especially vital to questions of reasoning and judgment.

⑥ **Answer the practice questions in each chapter.** Study the answer explanations. You can learn a great deal from answer explanations. Even when you have answered correctly, the explanation might bring out points that had not occurred to you. This same suggestion—read all the explanations—applies throughout this book to the exams as well as the instructional chapters.

⑦ **Try the practice exams.** When you believe you are well prepared, move on to the exams. If possible, answer an entire exam in one sitting. If you must divide your time, divide it into no more than two sessions per exam.

When you take the practice exams, treat them with respect. Consider each as a dress rehearsal for the real thing. Time yourself accurately and do not peek at the correct answers. Remember, you are taking these for practice. They will not be scored and do not count, so learn from them. Learn to think; learn to reason like an effective firefighter; learn to pace yourself so you can answer all the questions. Then learn from the explanations.

## HOW TO TAKE AN EXAM

① **Get to the examination room about 10 minutes ahead of time.** You'll get a better start when you are accustomed to the room. If the room is too cold, too warm, or not well ventilated, call these conditions to the attention of the person in charge.

② **Make sure you read the instructions carefully.** In many cases, test-takers lose points because they misread an important part of the directions. (An example would be selecting the incorrect choice instead of the correct choice.)

③ **Mark the answer sheet clearly.** Unless you have applied to a fire force in a very small locality, your exam will be machine scored. You will mark your answers on a separate answer sheet. Each exam in this book has its own answer sheet so

**TIP**

Do not memorize questions and answers. Any question that has been released will not be used again. You might run into questions that are very similar, but you will not be tested with any of these exact questions.

you can practice marking all your answers in the correct way. Tear out the answer sheet before you begin each exam. Do not try to flip pages back and forth. Because your answer sheet will be machine scored, you must fill it out clearly and correctly. You cannot give any explanations to the machine. This means:

- You must blacken your answer space firmly and completely. ● is the only correct way to mark the answer sheet. ◗,⊗,⊘, and ⊘ are all unacceptable. The machine might not read them at all.

- You must mark only one answer for each question. If you mark more than one answer, you will be considered wrong even if one of the answers is correct.

- If you change your mind, you must erase your mark. Attempting to cross out an incorrect answer like this ● will not work. You must erase any incorrect answer completely. An incomplete erasure might be read as a second answer.

- All of your answers should be in the form of blackened spaces. The machine cannot read English. Do not write any notes in the margins. If you have done any figuring in the margins of the test booklet itself, be sure to mark the letter of your answer on the answer sheet. Correct answers in the test booklet do not count. Only the answer sheet is scored.

- MOST IMPORTANT: Answer each question in the correct place. Question 1 must be answered in space 1, question 52 in space 52. If you skip an answer space and mark a series of answers in the wrong places, you must erase all the answers and do the questions over, marking your answers in the proper places. You cannot afford to use the limited time in this way. Therefore, as you answer each question, look at its number and make sure you are marking your answer in the space with the same number. The risk involved in slipping out of line on the answer sheet is the reason we recommend that you answer every question in order, even if you have to guess. Make an educated guess if you can. If not, make a wild guess. Just mark the question number in the test booklet so you can try again if there is time.

❹ **Don't be afraid to guess.** The best policy, of course, is to pace yourself so you can read and consider each question. Sometimes this does not work. Most civil service exam scores are based on the number of questions answered correctly. This means that a wild guess is better than a blank space. There is no penalty for a wrong answer, and you just might guess right. If you see that time is about to run out, mark all the remaining spaces with the same answer. According to the law of averages, some will be right.

You bought this book, however, with entire exams for practice answering firefighter questions. Part of your preparation is learning to pace yourself so you need not answer randomly at the end. Far better than a wild guess is an educated guess. You make this kind of guess not when you are pressed for time but when you are not sure of the correct answer. Usually, one or two of the choices are obviously wrong.

Eliminate the obviously wrong answers and try to reason among those remaining. Then, if necessary, guess from the smaller field. The odds of choosing a right answer increase if you guess from a field of two rather than from a field of four. When you make an educated guess or a wild guess in the course of the exam, make a note next to the question number in the test booklet. If there is time, you can go back for a second look.

**Reason your way through multiple-choice questions carefully and methodically.** Here are a few samples we can walk through together:

**Q** On the job, your supervisor gives you a hurried set of directions. As you start your assigned task, you realize you are not quite clear on the directions given to you. The best action to take would be to

(A) continue with your work, hoping to remember the directions and to do the best you can.

(B) ask a coworker in a similar position what he or she would do.

(C) ask your supervisor to repeat or clarify certain directions.

(D) go on to another assignment.

In this question, you are given four possible answers to the problem described. Though the four choices are all possible actions, it is up to you to choose the best course of action in this particular situation.

Choice (A) will likely lead to a poor result. Given that you do not recall or understand the directions, you would not be able to perform the assigned task properly. Keep choice (A) in abeyance until you have examined the other alternatives. It could be the best of the four choices given.

Choice (B) also is a possible course of action, but is it the best? Consider that the coworker you consult has not heard the directions. How could he or she know? Perhaps his or her degree of incompetence is greater than yours in this area. Of choices (A) and (B), the better of the two is still choice (A).

Choice (C) is an acceptable course of action. Your supervisor will welcome your questions and will not lose respect for you. At this point, you should hold choice (C) as the best answer and eliminate choice (A).

The course of action in choice (D) is decidedly incorrect because the job at hand would not be completed. Going on to something else does not clear up the problem; it simply postpones having to make a necessary decision.

**A** After careful consideration of all choices given, choice (C) stands out as the best possible choice of action. You should select choice (C) as your answer.

Every question is written about a fact or an accepted concept. This question indicates the concept that, in general, most supervisory personnel appreciate subordinates questioning directions that might not have been fully understood. This type of clarification precludes subsequent errors on the part of the subordinates. On the other hand, many subordinates are reluctant to ask questions for fear that their lack of understanding will detract from their supervisor's evaluation of their abilities.

The supervisor, therefore, has the responsibility to issue orders and directions in such a way that subordinates will not be discouraged from asking questions. This is the concept on which the sample question was based.

If you are familiar with this concept, you should have no trouble answering the question. If you are not familiar with it, however, the method outlined here of eliminating incorrect choices and selecting the correct one should prove successful for you.

You have now seen how important it is to identify the concept and the key phrase of the question. Equally important, or perhaps even more so, is identifying and analyzing the key word—the qualifying word—in a question. This word is usually an adjective or adverb. Some of the most common key words are *most, least, best, highest, lowest, always, never, sometimes, most likely, greatest, smallest, tallest, average, easiest, most nearly, maximum, minimum, only, chiefly, mainly, but,* and *or*. Identifying these key words usually is half the battle in understanding and, consequently, answering all types of exam questions.

Let's use the elimination method on some additional questions:

**Q** On the first day you report for work after being appointed as a firefighter, you are assigned to routine duties that seem to you to be very petty in scope. You should

**(A)** perform your assignment perfunctorily while conserving your energies for more important work in the future.

**(B)** explain to your superior that you are capable of greater responsibility.

**(C)** consider these duties an opportunity to become thoroughly familiar with the firehouse.

**(D)** try to get someone to take care of your assignment until you have become thoroughly acquainted with your new associates.

Once again, you are confronted with four possible answers from which you are to select the best one.

Choice (A) will not lead to getting your assigned work done in the best possible manner in the shortest possible time. This would be your responsibility as a newly appointed firefighter, and the likelihood of getting to do more important work in the future following the approach stated in this choice is remote. However, since this is only the first choice you have read so far, you must hold it aside because it might turn out to be the best of the four choices given.

Choice (B) is better than choice (A) because your superior might not be familiar with your capabilities at this point. You therefore should drop choice (A) and retain choice (B) because, once again, it might be the best of the four choices.

The question clearly states that you are newly appointed. Therefore, wouldn't it be wise to perform whatever duties you are assigned in the best possible manner? This way, you not only use the opportunity to become acquainted with firehouse procedures but also to demonstrate your abilities. Choice (C) contains a course of action that will benefit you and the firehouse in which you are working because it gets needed work done. At this point, you should drop choice (B) and retain choice (C) because it is by far the better of the two.

**A**    The course of action in choice (D) is not likely to get the assignment completed, and it will not enhance your image to your fellow firefighters. Choice (C), when compared to choice (D), is far better and should be selected as the best choice.

Now let's examine a question that appeared on a police officer examination:

1. An off-duty police officer in civilian clothes riding in the rear of a bus notices two teenage boys tampering with the rear emergency door. The most appropriate action for the officer to take is to

   **(A)** tell the boys to discontinue their tampering, pointing out the dangers to life that their actions might create.

   **(B)** report the boys' actions to the bus operator and let the bus operator take whatever action is deemed best.

   **(C)** signal the bus operator to stop, show the boys his officer's badge, and then order them off the bus.

   **(D)** show the boys his officer's badge, order them to stop their actions, and take down their names and addresses.

Before considering the answers to this question, you must accept the well-known fact that a police officer is always on duty to uphold the law even though he or she might be technically off duty.

In choice (A), the course of action taken by the police officer will probably serve to educate the boys and will get them to stop their unlawful activity. Because this is only the first choice, you will hold it aside.

In choice (B), you must realize that the authority of the bus operator in this instance is limited. He can ask the boys to stop tampering with the door, but that is all. The police officer can go beyond that point. Therefore, you should drop choice (B) and continue to hold choice (A).

Choice (C), as a course of action, will not have a lasting effect. What is to stop the boys from boarding the next bus and continuing their unlawful action? Therefore, drop choice (C) and continue to hold choice (A).

**A** Choice (D) might have some beneficial effects, but it would not deter the boys from continuing their actions in the future. When you compare choice (A) with choice (D), (A) is the better choice overall and should therefore be your selection.

The next example illustrates a type of question that has gained popularity in recent examinations and that requires a two-step evaluation. First, you must evaluate the condition in the question as being "desirable" or "undesirable." After the determination has been made, you then are left with making a selection from two choices instead of the usual four.

2. A visitor to an office in a city agency tells one of the office aides that he has an appointment with the supervisor of the office who is expected shortly. The visitor asks for permission to wait in the supervisor's private office, which is unoccupied at the moment. For the office aide to allow the visitor to do so would be

**(A)** desirable; the visitor would be less likely to disturb the other employees or to be disturbed by them.

**(B)** undesirable; it is not courteous to permit a visitor to be left alone in an office.

**(C)** desirable; the supervisor might want to speak to the visitor in private.

**(D)** undesirable; the supervisor might have left confidential papers on the desk.

You first must evaluate the course of action on the part of the office aide. Permitting the visitor to wait in the supervisor's office is very undesirable. There is nothing said of the nature of the visit; it might be for a purpose that is not friendly or congenial. There might be papers on the supervisor's desk that he or she does not want the visitor to see or to have knowledge of. Therefore, at this point, you have to make a choice between (B) and (D).

This is definitely not a question of courtesy. Although all visitors should be treated with courtesy, permitting the visitor to wait in the supervisor's office is not the only possible act of courtesy. Another comfortable place could be found for the visitor to wait.

**TIP**

One word of caution: After you have made a determination that a course of action is desirable or undesirable, confine your thoughts to the situations illustrating your choice. Do not revert to again considering the other extreme, or you will flounder and will surely not select the correct answer.

**A** Choice (D) contains the exact reason for evaluating this course of action as being undesirable. When compared with choice (B), choice (D) is far better.

## TEST-DAY STRATEGY

**❶ Allow the test itself to be the main attraction of the day.** Do not squeeze it in between other activities.

**❷ Arrive rested, relaxed, and on time.** In fact, plan to arrive a little bit early. Leave plenty of time for traffic tie-ups or other complications that might upset you and interfere with your test performance.

**❸ If you do not understand any of the examiner's instructions, ASK QUESTIONS.** Make sure that you know exactly what to do. In the test room, the examiner will provide forms for you to fill out. He or she will give you the instructions you must follow in taking the examination. The examiner will tell you how to fill in the grids on the forms. Time limits and timing signals will be explained.

**❹ Follow instructions exactly during the examination.** Fill in the grids on the forms carefully and accurately. Filling in the wrong grid might lead to a loss of veterans' credits to which you might be entitled or to an incorrect address for your test results. Do not begin until you are told to do so. Stop as soon as the examiner tells you to stop. Do not turn pages until you are told to do so. Do not go back to parts you have already completed. Any infraction of the rules is considered cheating. If you cheat, your test paper will not be scored, and you will not be eligible for appointment.

**❺ READ every word of every question.** Once the signal has been given and you begin the exam, be alert for exclusionary words that might affect your answer—words like *not, most, least, all, every,* and *except.*

**❻ READ all the choices before you mark your answer.** It is statistically true that most errors are made when the last choice is the correct answer. Too many people mark the first answer that seems correct without reading through all the choices to find out which answer is best.

The following list summarizes the suggestions we already have given for taking this exam. Read the suggestions now and again before you attempt the practice exams in this book. Review them before you take the actual exam. You will find them all useful.

- Mark your answers by completely blackening the answer space of your choice.

- Mark only ONE answer for each question, even if you think more than one answer is correct. You must choose only one. If you mark more than one answer, the scoring machine will consider you wrong even if one of your answers is correct.

- If you change your mind, erase your first response completely. Leave no doubt as to which answer you have chosen.

- If you do any figuring in the test booklet or on scratch paper, make sure to still mark your answer on the answer sheet.

- Check often to make sure that the question number matches the answer space number and that you have not skipped a space by mistake. If you do skip a space, you must erase all the answers after the skip and answer all the questions again in the right places.

- Answer every question in order but do not spend too much time on any one question. If a question seems to be "impossible," do not take it as a personal challenge. Guess and move on. Remember that your task is to answer correctly as many questions as possible. You must apportion your time to give yourself a fair chance to read and answer all the questions. If you guess at an answer, mark the question in the test booklet so you can find it easily if time permits.

- Guess intelligently if you can. If you do not know the answer to a question, eliminate the answers you know are wrong and guess from the remaining choices. If you have no idea whatsoever as to the answer to a question, guess anyway. Choose an answer other than the first. The first choice generally is the correct answer less often than the other choices. If your answer is a guess, either an educated guess or a wild one, mark the question in the question booklet so you can give it a second try if time permits. If time is about to run out and you have not had a chance to answer all the questions, mark all the remaining questions with the same answer. According to the law of averages, you should get 25 percent of them right.

- If you happen to finish before time is up, check to make sure that each question is answered in the right space and that there is only one answer for each question. Return to the difficult questions you marked in the booklet and try them again. There is no bonus for finishing early, so use all your time to perfect your exam paper.

# Firefighter Screening Process

## OVERVIEW

- A typical Notice of Examination
- The application
- Medical standards
- Fitness and physical ability tests
- Staying fit
- Personality testing and psychological evaluations
- The interview

## A TYPICAL NOTICE OF EXAMINATION

The general provisions of the notice of examination and the general examination regulations of the New York City Department of Citywide Administrative Services are part of this notice of examination. They are available at the NYC Department of Citywide Administrative Services (www.nyc.gov).

The following pages show a copy of a typical Notice of Examination.

chapter 3

THE CITY OF NEW YORK
DEPARTMENT OF CITYWIDE
ADMINISTRATIVE SERVICES
APPLICATIONS CENTER
18 WASHINGTON STREET
NEW YORK, NY 10004

| | |
|---|---|
| **REQUIRED FORMS** | |
| APPLICATION FORM | |

MICHAEL R. BLOOMBERG
Mayor

MARTHA K. HIRST
Commissioner

# NOTICE
# OF
# EXAMINATION

---

### PROMOTION TO FIREFIGHTER

Exam. No.  4532

---

**WHEN TO APPLY:**   From: November 3, 2004     **APPLICATION FEE: $40.00**
              To:      November 23, 2004     *Payable only by money order to D.C.A.S. (EXAMS)*

**THE TEST DATE:** The multiple-choice test is expected to be held on **Saturday, March 5, 2005.**

---

**WHAT THE JOB INVOLVES:** Under supervision, Firefighters assist in the control and extinguishment of fires, in providing pre-hospital emergency medical care, and other emergency response duties, including hazmat incidents, and in the enforcement of laws, ordinances, rules and regulations regarding the prevention, control and extinguishment of fires, as well as perform Fire Safety Education activities; perform inspections and related enforcement duties to assure compliance with provisions of the Fire Prevention Code and applicable sections of the Building Code, Multiple Dwelling Law, Housing Maintenance Code, Labor Law and other laws, rules, regulations, within enforcement purviews of the New York City Fire Department; perform inspection of equipment and schedule, as necessary, the maintenance of various tools and equipment, including, but not limited to, S.C.B.A. power tools, company apparatus, and personal safety equipment; and perform related work.

Some of the physical activities performed by Firefighters and environmental conditions experienced are: wearing protective clothing, such as bunker suit, helmet, boots and breathing apparatus; crawling, crouching and standing, often for prolonged periods, while extinguishing fires; driving fire apparatus and other Department vehicles; climbing stairs, ladders and fire escapes; raising portable ladders; using forcible entry tools, such as axes, sledge hammers, power saws and hydraulic tools; searching for victims in smoke-filled hostile environments; carrying or dragging victims from dangerous locations; connecting, stretching and operating hose lines; locating hidden fire by feel and smell; providing medical assistance to injured or ill citizens; and providing control and mitigation of hazardous materials incidents while wearing chemical protective clothing.

(This is a brief description of what you might do in this position and does not include all the duties of this position.)

**THE SALARY:** The current minimum salary is $36,878 per annum. This rate is subject to change. In addition, employees will receive holiday, night shift and overtime pay.

**HOW TO APPLY:** If you believe you are eligible to take this examination, refer to the "Required Form" section below for the form that you must fill out. Return the completed form and the application fee to DCAS Applications Section, 1 Centre Street, 14th floor, New York, NY 10007 **by mail only**. DCAS will not accept applications in person from candidates.

**ELIGIBILITY TO TAKE EXAMINATION:** This examination is open to each employee of the New York City Fire Department who **on the date of the multiple-choice test:**

     (1)     is permanently (not provisionally) employed in or appears on a Preferred List (see Note, below) for the title of Emergency Medical Specialist - EMT or Emergency Medical Specialist - Paramedic; and

     (2)     is not otherwise ineligible.

(Note: A "Preferred List" is a civil service list which is only for certain former permanent incumbents of the eligible title who have rehiring rights.)

---

**READ CAREFULLY AND SAVE FOR FUTURE REFERENCE**

---

Exam. No. 4532 - Page 2

If you do not know your permanent title or whether you are on a Preferred List, check with **your agency's personnel office**.

You may be given the test before we verify your eligibility. You are responsible for determining whether or not you meet the eligibility requirements for this examination prior to submitting your application. If you are marked "Not Eligible," your application fee will not be refunded and your test paper(s) will not be rated.

**Age Requirement:** Pursuant to Section 54 of the New York Civil Service Law and Section 15-103 of the Administrative Code, you must be at least 17½ years of age by the end of the application period and you must not have reached your twenty-ninth birthday by the beginning of the application period to be eligible to take this examination.

**Exception to Maximum Age Requirement:** All persons who were engaged in military duty as defined in Section 243 of the New York State Military Law may deduct from their actual age the length of time spent in such military duty provided the total deduction for military duty does not exceed six years.

**ELIGIBILITY TO BE PROMOTED:** In order to be eligible for promotion, you must have completed your probationary period in an eligible title as indicated in the above "Eligibility To Take Examination" section, and you must be permanently employed and in the eligible title or your name must appear on a Preferred List for the eligible title at the time of promotion. Additionally, you must have served permanently in the eligible title for at least one year, unless your probationary period in that eligible title has been waived pursuant to Rule 5.2.4 of the Personnel Rules and Regulations of the City of New York. Time served prior to a break in service of more than one year will not be credited toward meeting this requirement.

**Education and Experience Requirements:** In order to be promoted, you must have successfully completed 30 semester credits from an accredited college or university or have completed two years of honorable full-time U.S. military service.

**Certification Requirement:** In order to be promoted, you must have as a minimum a Certified First Responder Certification with Defibrillation (CFR-D). This certification must be maintained for the duration of employment.

**REQUIRED FORM:**

**Application for Examination:** Make sure that you follow all instructions included with your application form, including payment of fee. Save a copy of the instructions for future reference.

**REQUIREMENTS TO BE PROMOTED:**

**Driver License Requirement:** By the time you are promoted to this position, you must have a motor vehicle driver license valid in the State of New York. This license must be maintained for the duration of your employment.

**Minimum Age Requirement:** You must have reached your twenty-first birthday to be eligible for promotion.

**Medical and Psychological Requirements:** Medical and psychological guidelines have been established for the position of Firefighter. Candidates will be examined to determine whether they can perform the essential functions of the position of Firefighter. Additionally, employees will be expected to continue to perform the essential functions of the position of Firefighter throughout their careers, and may, therefore, be medically tested periodically throughout their careers. Where appropriate, a reasonable accommodation will be provided for a person with a disability to enable him or her to take these medical and psychological examinations, and/or to perform the essential functions of the job.

**Drug Screening Requirement:** You must pass a drug screening in order to be promoted. Drug tests will also be administered to all probationary Firefighters as part of the medical examination prior to the completion of probation.

**Character and Background:** Proof of good character and satisfactory background will be an absolute prerequisite to promotion. In accordance with provisions of law, persons convicted of a felony or who have received a dishonorable discharge from the Armed Forces are not eligible for promotion to this position.

**Residency Requirement:** The New York Public Officers Law requires that any person employed as a Firefighter in the New York City Fire Department be a resident of the City of New York or of Nassau, Westchester, Suffolk, Orange, Rockland or Putnam Counties.

**Citizenship Requirement:** United States citizenship is required at the time of promotion.

Exam. No. 4532 - Page 3

**THE TEST**: There will be a written multiple-choice test, weight 50, and a physical test, weight 50. The pass mark for the written test will be determined after an analysis of the test results.

The written test is designed to test the candidate's ability to learn and to perform the work of a Firefighter. It may include questions involving the understanding of written language and information, using language to communicate information or ideas to other people, memorizing information, recognizing or identifying the existence of problems, applying general rules to specific situations, applying prioritized rules to specific situations, determining position or spatial orientation within a larger area, visualizing how objects or structures might appear from different perspectives or after changes, finding a rule or concept which fits or describes a situation; standards of proper employee ethical conduct, including the provisions of Mayor's Executive Order No. 16 of 1978 as amended; and other related areas.

The physical test will consist of a series of events designed to test the candidate's capacity to perform the physical aspects of a Firefighter's job. A more detailed description of the physical test will be made available at a later date. Candidates called to participate in the physical test will be required to pay an additional fee prior to taking the physical test and will be notified of the method of payment prior to the physical test. Failure to pay the additional application fee, in a timely manner, will result in disqualification from further participation in the examination. The additional application fee for the physical test will be waived for a New York City resident receiving public assistance who submits a clear photocopy of a current Benefit Card at the time of the physical test.

Candidates must pass the written test to be summoned for the physical test. Medical evidence to allow participation in the physical test may be required.

For candidates who pass <u>both</u> the written and physical test, scores on the written and physical tests will be converted to standard scores. The standard score on each test will then be multiplied by the weight of that test, and these products will be added resulting in a combined weighted standardized score. Ranking of candidates will be based on this combined weighted standardized score. This combined weighted standardized score will then be transformed to scores between 70 and 100. Only those candidates who receive a score between 70 and 100 will be credited with Veterans' credit, if applicable.

**ADMISSION CARD**: You should receive an Admission Card in the mail about 10 days before the date of the test. If you do not receive an Admission Card at least 4 days before the test date, you must go to the Examining Service Section, 1 Centre Street, 14th floor, Manhattan, to obtain a duplicate card.

**THE TEST RESULTS:** If you pass both the multiple-choice test and physical test, your name will be placed in final score order on an eligible list and you will be given a list number. You will be notified by mail of your test results. If you meet all requirements and conditions, you will be considered for promotion when your name is reached on the eligible list.

**ADDITIONAL INFORMATION:**

**Study Guide:** A study guide to assist candidates in preparing for the physical test will be available, at no cost, to applicants for this examination on the day of the written test. To obtain this study guide, follow the directions given to you on the day of the written test.

**Investigation:** This position is subject to investigation before promotion. At the time of investigation, eligibles will be required to pay a $75.00 fee for fingerprint screening.

At the time of investigation and at the time of promotion, eligibles must present originals or certified copies of all required documents and proof, including, but not limited to, proof of date and place of birth by transcript of record of the Bureau of Vital Statistics or other satisfactory evidence, naturalization papers if necessary, proof of any military service, and proof of meeting educational requirements.

Any willful misstatement or failure to present any documents required for investigation will be cause for disqualification.

**Probationary Period**: The probationary period is 12 months. As part of the probationary period, probationers will be required to successfully complete a prescribed training course. Probationers who fail to complete successfully such training course, at the close of such training course, may be demoted by the agency head.

**SPECIAL ARRANGEMENTS:**

**Late Filing:** Consult **your agency's personnel office** to determine the procedure for filing a late application if you meet one or more of the following conditions:

(1)    You are absent from work for at least one-half of the application period and cannot apply for reasons such as vacation, sick leave or military duty; or

(2)    You are appointed to an eligible title after the above application period but on or before the date of the multiple-choice test.

**Special Test Accommodations**: If you plan to request special testing accommodations due to disability or an alternate test date due to your religious belief, follow the instructions included with the "Application for Examination."

Exam. No. 4532 - Page 4

**Make-up Test:** You may apply for a make-up test if you cannot take the test on the regular test date for any of the following reasons:

(1)  compulsory attendance before a public body;
(2)  on-the-job injury or illness caused by municipal employment;
(3)  absence for one week following the death of a spouse, domestic partner, parent, sibling, child or child of a domestic partner;
(4)  absence due to ordered military duty; or
(5)  a clear error for which the Department of Citywide Administrative Services or the examining agency is responsible.

To request a make-up test, contact the Examining Service Section, 1 Centre Street, 14th floor, New York, NY 10007, in person or by certified mail as soon as possible and provide documentation of the special circumstances.

## THE APPLICATION

The application form you must file could be as simple as the New York City application form reproduced here or could be considerably more complex and more detailed like the Experience Paper that follows it. What is important is that you fill out your application form neatly, accurately, completely, and truthfully.

Obviously, neatness counts. If your application is misread, it is likely to be misfiled or misinterpreted to your detriment. If your application is extremely sloppy, it brands you as a slob. No one wants to hire a slob.

Accuracy is important. A single wrong digit in your Social Security number or exam number can misroute your application forever. Completeness is a requirement. An incomplete application gets tossed into the wastebasket. No one will take the time to chase you down to fill in the blanks.

The need to tell the truth should be self-evident. Certainly, you should describe your duties and responsibilities in the best possible light, but do not exaggerate. Do not list a degree you did not receive. Do not give a graduation date if you did not graduate. If your application form asks for reasons for leaving previous employment, give a very brief but truthful reason. Any statement you make on an application form is subject to verification. There are staff members at the civil service office whose job it is to check with schools and former employers. These investigators ask questions about your job performance, attendance record, duties, and responsibilities.

Any misstatements that are not picked up by background investigators are likely to be unmasked during an interview. The last step of the screening process is likely to be a personal interview before one person or before a panel of interviewers. The interviewer(s) will sit with a copy of your application form and will ask you some questions based upon it. This will be your opportunity to explain unfavorable items such as dismissal from a job or frequency of job changes. If you have made false claims or have stretched the truth too far, this is the point at which you are likely to trip yourself up. It is difficult to maintain a lie under the pressure of pointed questioning. The interviewer(s) are prepared and at ease. You are in an unfamiliar situation and are nervous about qualifying for a job. Do not make it harder for yourself.

**DEPARTMENT OF CITYWIDE ADMINISTRATIVE SERVICES**
**DIVISION OF CITYWIDE PERSONNEL SERVICES**
1 Centre Street, 14th floor          New York, NY 10007

## APPLICATION FOR EXAMINATION

(Directions for completing this application are on the *back* of this form. Additional information is on the Special Circumstances Sheet)

Download this form on-line: nyc.gov/html/dcas

**FOLLOW DIRECTIONS ON BACK**
Fill in all requested information clearly, accurately, and completely.

*The City will only process applications with complete, correct, legible information which are accompanied by correct payment or waiver documentation.*

*All unprocessed applications will be returned to the applicant.*

Check One:
☐ Open Competitive
☐ Promotion

1. EXAM #:

2. EXAM TITLE:

3. SOCIAL SECURITY NUMBER:

4. LAST NAME:

5. FIRST NAME:

6. MIDDLE INITIAL:

7. MAILING ADDRESS:

8. APT. #:

9. CITY OR TOWN:

10. STATE:

11. ZIP CODE:

12. PHONE:

13. OTHER NAMES USED IN CITY SERVICE:

Questions 14 & 15:
Discrimination on the basis of sex, sexual orientation, age, creed, color, age, disability status, veteran status or religious observance is prohibited by law. The City of New York is an equal opportunity employer. The identifying information requested on this form is to be used to determine the representation of protected groups among applicants. This information is voluntary and will not be made available to individual decision making hiring decisions.

14. RACE/ETHNICITY (Check One):
☐ White          ☐ American Indian/Alaskan Native
☐ Black          ☐ Asian/Pacific Islander
☐ Hispanic

15. SEX (Check One):
☐ Male
☐ Female

16. ARE YOU EMPLOYED BY HEALTH AND HOSPITALS CORPORATION? (Check One):
☐ YES   ☐ NO

17. CHECK ALL BOXES THAT APPLY TO YOU: (Directions for this section are found on the "Special Circumstances" Sheet)
☐ I AM A SABBATH OBSERVER AND WILL REQUEST AN ALTERNATE TEST DATE: (Verification required. See Item A on Special Circumstances Sheet)
☐ I HAVE A DISABILITY AND WILL REQUEST SPECIAL ACCOMMODATIONS (Verification required. See Item B on Special Circumstances Sheet)
☐ I CLAIM VETERANS' CREDIT (For qualifications see Item C on Special Circumstances Sheet)
☐ I CLAIM DISABLED VETERANS' CREDIT (For qualifications see Item C on Special Circumstances Sheet)
☐ I CLAIM PARENT LEGACY CREDIT (For qualifications see Item D on Special Circumstances Sheet)
☐ I CLAIM SIBLING LEGACY CREDIT (For qualifications see Item D on Special Circumstances Sheet)

18. **Your Signature:**                    **Date:**

**MICHAEL R. BLOOMBERG**
Mayor

**MARTHA K. HIRST**
Commissioner

*NOTE: You should apply for an examination **only** if you meet the qualification requirements set forth in the Notice of Examination. Read the Notice of Examination carefully before completing the application form.*

*Fill in all requested information clearly, accurately, and completely. **The City will only process applications with complete, correct, legible information which are accompanied by correct payment or waiver documentation. All unprocessed applications will be returned to the applicant.***

*Included in this material is a voter registration form. If you take this opportunity to register to vote, please mail the postage-paid form directly to the Board of Elections. The provision of government services is not conditioned on being registered to vote.*

***When appropriate the City will issue a refund for unprocessed applications after the close of the filing period.***

---

## DIRECTIONS FOR SUBMITTING APPLICATION FOR EXAMINATION

**FORMS**

All required forms which are listed in the upper-right-hand corner of the Notice of Examination must accompany your application. Failure to include these forms may result in your disqualification and you <u>will not</u> receive test scores.

**FEE**

The amount of the fee is stated in the Notice of Examination. Only a *MONEY ORDER* made out to *D.C.A.S. (EXAMS)* is acceptable payment (check or cash **are not** accepted). On the front of the money order you must clearly write your *full name*, *social security number* and *the exam number*. Keep your Money Order receipt as proof of filing.

**FEE WAIVER**

A filing fee is not charged if you are a New York City resident receiving public/cash assistance from the New York City Department of Social Services. To have the fee waived, you **must** be receiving full benefits and not partial benefits. If you are eligible, you must enclose a legible photocopy of your current Benefit Card (**formerly known as the Medicaid Card**) with your application. **The Food Coupon Photo Identification Card is** <u>unacceptable</u>. You must write your *social security number* and the *exam number* on the front of the photocopy of the Benefit Card. The name on your application must exactly match the name printed on your Benefit Card.

<u>**Fee Waivers are limited to persons who are recipients of Public Assistance at the time of submission of the application.**</u> Any person who falsifies information concerning current receipt of Public Assistance in order to obtain a fee waiver may be banned from appointment to any position within the City of New York, and may be subject to criminal prosecution. All such violations will be referred to the Department of Investigation.

**APPLICATION SUBMISSION**

Your application must be postmarked no later than the last day of the application period indicated on the Notice of Examination. Mail the completed application, supporting documents, and required filing fee or fee waiver to:
DCAS Application Section
1 Centre Street, 14th Floor, New York, NY 10007
*C/O Exam #, Exam Title*

---

## INSTRUCTIONS FOR COMPLETING APPLICATION FORM PROPERLY
To ensure proper processing of Application print all information *CLEARLY*. Failure to do so will delay or disqualify your application.

**1. EXAM NO. / EXAM TITLE**

See the Notice of Examination prior to filling in the exact exam number and exam title. Check either the Open Competitive (OC) or Promotion (PRO) box to indicate the type of examination you are applying for.

**2. - 12. GENERAL INFORMATION**

$ The address you give will be used as your mailing address for all official correspondence.
S Only one (1) address for each person is maintained in the files of this Department.
S If you change your mailing address after applying, see the Change Of Address section on Special Circumstances Sheet.

**13. OTHER NAMES USED**

If you have worked for a New York City agency under another name, write the other name in this section. If you have not used other names, skip this section.

**14. - 15. ETHNICITY / SEX**

Completing this information is voluntary. This information will <u>not</u> be made available to individuals making hiring decisions.

**16. HHC EMPLOYEE**

If you are employed by the Health and Hospitals Corporation, check the **YES** box in this section.

**17. SPECIAL CIRCUMSTANCES**

(Sabbath/Religious Observers, Special Accommodations because of a Disability, Veterans' or Disabled Veterans' Credit, Parent or Sibling Legacy Credit)
Please see the "Special Circumstances" direction sheet for qualifications and definitions associated with this section.

**18. SIGNATURE**

Signing the application indicates that all statements you provided on this form and all other forms required for this examination are true and subject to the penalties of perjury.

Applicants who do not receive an admission card at least **4 days** prior to the tentative test date must obtain an admission card by coming to the Examining Service Section of the New York City Department of Citywide Administrative Services, 1 Centre Street, 14th Floor, Room 1448.

Rev. 05/2004

**DEPARTMENT OF CITYWIDE ADMINISTRATIVE SERVICES**
**DIVISION OF CITYWIDE PERSONNEL SERVICES**

**Exam Support Group – Application Section**
One Centre Street, 14th Floor
New York, NY 10007
Automated Telephone: (212) 669-1357 • Fax: (212) 669-4734

---

### APPLICATION SUPPLEMENT

---

Exam Title: _____     Exam No: _____

---

Section 50-b of the New York State Civil Service Law requires that all applicants for Civil Service examinations be asked the following questions:

1.  Do you have any loans made or guaranteed by the New York State Higher Education Services Corporation which are currently outstanding?

    **CHECK ONLY ONE:**     YES ☐          NO ☐

**RETURN THIS SUPPLEMENT WITH YOUR APPLICATION FOR CIVIL SERVICE EXAMINATION ONLY IF YOU HAVE CHECKED THE YES BOX.**

---

2.  If you checked the YES box in *Question 1,* are you presently in default on such loan?

    **CHECK ONLY ONE:**     YES ☐          NO ☐

---

**SOCIAL SECURITY NUMBER:**     ☐☐☐ - ☐☐ - ☐☐☐☐

---

PLEASE PRINT CLEARLY:
FULL NAME:     _____
                        (Last Name,   First Name,   Middle Initial)

ADDRESS:     _____
                        (Include the Apartment Number, Floor, and/or In Care of- C/O,   if applicable)

CITY, STATE, ZIP:     _____

---

**COMPLETE THIS AFFIRMATION:**     I affirm under penalties of perjury that all statements made on this application and all supplementary information are true.

Signature: _____     Date: ___/___/___

DP-2512A (Rev. 05/2003)

The Official New York City Web Site
www.nyc.gov

## MEDICAL STANDARDS

Firefighting is a physically stressful occupation. Firefighters who spend much of their time doing relatively undemanding physical tasks in the firehouse and who spend a good deal of the time in sedentary activity, such as card playing and watching television, are suddenly called to put forth major physical exertion. The rapid and extreme shifts in demands on the body of a firefighter result in real punishment. This is why the selection process places such emphasis on medical condition and on physical fitness and capacity.

The actual firefighting activity involves heavy lifting and carrying while encumbered by heavy protective gear under conditions of extreme temperature, water exposure, and poor air quality. This work must by conducted at great speeds and with constant awareness of personal danger. No wonder there is such concern for hiring potential firefighters who are physically and emotionally suited for the job.

Federal law prohibits discrimination against job applicants on the basis of age. However, when youth is a bona fide prerequisite for effective performance of the duties of the job, employers are permitted to prescribe age limits. There is no question that the age limitation on firefighter applicants is justified. The work requires young, agile bodies. There is a big investment of time, effort, and money in the training of firefighters. The goal is to hire and train healthy, capable young men and women who will serve many productive years as effective firefighters.

In fact, the following three medical standards are a matter of official policy:

❶ This position is physically demanding and affects public health and safety. Therefore, the department requires candidates to meet appropriately high standards of physical and mental fitness. The object of the pre-employment examination is to obtain personnel fit to perform in a reasonable manner all the activities of the position. Appointees might be subject to reexamination at any time during their probationary periods.

❷ A candidate will be medically disqualified upon a finding of physical or mental disability that renders the candidate unfit to perform in a reasonable manner the duties of this position or that might reasonably be expected to render the candidate unfit to continue to perform in a reasonable manner the duties of this position. (For example, a latent impairment or a progressively debilitating impairment may, in the judgment of the designated medical officer, reasonably be expected to render the candidate unfit to continue to perform the duties of the position.)

❸ The fitness of each candidate is determined on an individual basis in relation to this position. The designated medical officer may utilize diagnostic procedures including the use of scientific instruments or other laboratory methods that, in his or her discretion, would determine the true condition of the candidate before he or she is accepted. The judgment of the designated medical officer, based on his or her knowledge of the activities involved in the duties of this position and the candidate's condition, is determinative.

The following definitions apply:

- **Physical or Mental Disability:** A physical, mental, or medical impairment resulting from an anatomical, physiological, or neurological condition that prevents the exercise of a normal bodily function or that is demonstrable by medically accepted clinical or laboratory diagnostic techniques. Such impairment may be latent or manifest.

- **Accepted Medical Principles:** Fundamental deduction consistent with medical facts and based on the observation of a large number of cases. To constitute an accepted medical principle, the deduction must be based on the observation of a large number of cases over a significant period of time and be so reasonable and logical as to create a moral certainty that it is correct.

- **Impairment of Function:** Any anatomic or functional loss, lessening, or weakening of the capacity of the body or any of its parts to perform that which is considered by accepted medical principles to be normal.

- **Latent Impairment:** Impairment of function not accompanied by signs and/or symptoms but of such a nature that there is reasonable and moral certainty, according to accepted medical principles, that signs and/or symptoms will appear within a reasonable period of time or upon a change of environment.

- **Manifest Impairment:** Impairment of function accompanied by signs and/or symptoms.

- **Medical Capability:** General ability, fitness, or efficiency (to perform duty) based on accepted medical principles.

As you can see, the medical standards for prospective firefighters are extremely strict and rigid. Conditions that would not deter a person's performance in a less strenuous occupation are absolutely disqualifying for firefighters. Some of these conditions are temporary, such as pregnancy; others, such as the use of drugs or obesity (being severely overweight), might be correctable. Temporary or correctable conditions can cause the candidate to be barred from the physical performance test at a given time and thus might effectively keep the person from competition and employment. Once the temporary or correctable conditions are overcome, however, the candidate can, if still within the age limit, compete at a later date.

A number of medical conditions, depending on their severity, can bar a candidate from employment as a firefighter even if the candidate is able to score high on the physical performance test and is otherwise highly qualified. In fact, a person with any of these conditions might cause him- or herself injury or permanent complications of the condition as a result of undergoing strenuous physical exercise, participating in conditioning programs, or competing in the physical performance test. Some of these conditions include:

- Alcoholism
- Anemia
- Asthma
- Chronic gastrointestinal disorders
- Diabetes
- Drug dependence
- Epilepsy or other seizure disorder
- Heart disease
- High blood pressure
- Joint disease
- Kidney disease
- Liver disease
- Lung disease
- Muscular disorders
- Obesity (greater than simply being overweight)
- Proneness to heat illness such as heat stroke or heat exhaustion
- Sickle cell disease
- Ulcers

The preceding list represents conditions that are likely to be permanently disqualifying. Other conditions might militate against taking part in physical training or taking the physical performance test until they are removed by passage of time or by active curative measures. Examples of such conditions include:

- Acute gastrointestinal disorder
- Dehydration
- Drug use
- Hernia
- Infections
- Overweight
- Pregnancy
- Severe underweight

The cooperation and assistance of a physician can help a candidate attain the health and medical status to permit participation in fitness programs and in the test itself. A person who has suffered from any of these conditions or similar ones should remain under the supervision of a doctor while undergoing vigorous physical training and should not attempt to participate in a physical performance test until given approval by the doctor.

Still other physical/medical conditions can disqualify a firefighter candidate simply by virtue of their presence and without regard to physical fitness. Considering the requirements of firefighting, it should be obvious that the following conditions disqualify candidates:

- Severe hearing deficit
- Dependency on glasses or contact lenses or on hearing aids that can be lost, broken, or rendered dysfunctional under heat stress
- Severe allergies
- Cancer

People with the above conditions might be in excellent health and be physically strong and able, but they cannot serve as firefighters.

You probably are very much aware of any disabilities or major problems that would keep you from serving as a firefighter. If you are considering yourself as a firefighting candidate, you evidently think of yourself as a healthy, physically fit person. Even so, consult with your doctor before proceeding to invest time and money in the application process and in preparing for the written exam. It also is wise to get the approval of your doctor before you embark on a vigorous physical conditioning program. Tell your doctor about the type of work you have in mind, describe the physical demands, and ask for an assessment of your potential to withstand these rigors. If your doctor foresees any potential problems, either in passing the exams or in facing the demands of the job, discuss corrective measures and remedial programs right now. Follow the medical advice you receive. While speaking with your doctor, describe the physical performance test. You might be able to pick up special tips to prepare yourself to earn the top score on the performance test itself. Your doctor might even have a physical conditioning program to recommend.

The exam for which you are preparing yourself is the first step toward your career in firefighting. If all goes well, you are planning to spend the bulk of your working life in this career. You cannot expect to prepare in a few weeks for a career that will last a lifetime. Plan ahead. Start now.

The book you have before you is a good starting place for preparation for the written test. Now consider preparing for the physical test at the same time.

**TIP**

Because so much of the hiring decision is based on the physical status of the applicant, it makes sense to devote at least as much attention to preparing your body for the physical test as to preparing your mind for the written exam.

## NOTE

In some states, smoking is grounds for disqualification.

**Step One:** Go to your doctor for a full medical exam. If you get the go-ahead from your doctor, you can begin an aggressive physical fitness training program. If your doctor has any reservations, take care of your health problems. Follow your doctor's advice for improving your general health, for curing any specific problems, and for building strength and stamina at a rate consistent with your overall condition.

**Step Two:** Stop polluting your body. If you smoke, stop smoking now. Medical authorities do not agree on all the details, but they all acknowledge that some harm is caused by tobacco smoke. Smoking has been linked to lung cancer, breathing disorders, heart disease, and circulatory diseases. Smoking begins to affect the body long before actual diseases develop. Smokers generally do not have the stamina of nonsmokers. Smoking reduces the power of the lungs to distribute oxygen. With less oxygen going to the muscles of the body, smokers cannot run or climb as long, and cannot carry as much and cannot sustain effort to the degree that nonsmokers can. For a top score on the physical performance test and for a long career in firefighting, stop smoking now.

If you use drugs—inhaled, injected, ingested, or smoked—stop now. Different drugs affect the body in different ways, but all alter functioning in some way. Regardless of whether the drug you use affects your vision, your hearing, your emotions, or your thinking, it creates an unnatural condition that does not enhance your chances of top scores on examinations nor your chances of being selected for a firefighter position. In fact, drug testing is part of the screening process for firefighters. People who test positive for drug use cannot be hired. Medical authorities warn that the intense activity of the physical performance test can cause permanent damage to the body of a drug user.

Excess alcohol and excess caffeine also are body pollutants. Moderate amounts of these stimulants probably are not harmful, though experts disagree as to the definition of "moderate." Let common sense be your guide. Do not overdo it.

**Step Three:** Control your weight. Excess weight requires your muscles, heart, and lungs to work harder. An overweight person tends to have less speed and less stamina than a well-proportioned person and tends to have a shorter productive life span as well. Rapid weight reduction, however, constitutes a shock to the system. Quick weight loss is unhealthy. Weight must be lost gradually through a combination of calorie cutback and sensible exercise.

Very low body weight also can be a problem for the prospective firefighter. A person who is severely underweight is unlikely to be able to carry heavy loads for any length of time and may tire easily. If being underweight does not stem from disease, an underweight person should be able to gain slowly by eating more food and by adding muscle weight through exercise and weight lifting.

Each department sets its own standards, but the following tables can serve as a general guide to weight requirements.

## ACCEPTABLE WEIGHT IN POUNDS ACCORDING TO FRAME

### MALE

| A Height Feet | Inches | B Small Frame | C Medium Frame | D Large Frame |
|---|---|---|---|---|
| 5 | 2 | 128–134 | 131–141 | 138–150 |
| 5 | 3 | 130–136 | 133–143 | 140–153 |
| 5 | 4 | 132–148 | 135–145 | 142–153 |
| 5 | 5 | 134–140 | 137–148 | 144–160 |
| 5 | 6 | 136–142 | 139–151 | 146–164 |
| 5 | 7 | 138–145 | 142–154 | 149–168 |
| 5 | 8 | 140–148 | 145–157 | 152–172 |
| 5 | 9 | 142–151 | 148–160 | 155–176 |
| 5 | 10 | 144–154 | 151–163 | 158–180 |
| 5 | 11 | 146–157 | 154–166 | 161–184 |
| 6 | 0 | 149–160 | 157–170 | 164–188 |
| 6 | 1 | 152–164 | 160–174 | 168–192 |
| 6 | 2 | 155–168 | 164–178 | 172–197 |
| 6 | 3 | 158–172 | 167–182 | 176–202 |
| 6 | 4 | 162–176 | 171–187 | 181–207 |

## ACCEPTABLE WEIGHT IN POUNDS ACCORDING TO FRAME

### FEMALE

| A Height Feet | Inches | B Small Frame | C Medium Frame | D Large Frame |
|---|---|---|---|---|
| 4 | 10 | 102–111 | 109–121 | 118–131 |
| 4 | 11 | 103–113 | 111–123 | 120–134 |
| 5 | 0 | 104–115 | 113–126 | 122–137 |
| 5 | 1 | 106–118 | 115–129 | 125–140 |
| 5 | 2 | 108–121 | 116–132 | 128–143 |
| 5 | 3 | 111–124 | 121–135 | 131–147 |
| 5 | 4 | 114–127 | 124–138 | 134–151 |
| 5 | 5 | 117–130 | 127–141 | 137–155 |
| 5 | 6 | 120–133 | 130–144 | 140–149 |
| 5 | 7 | 123–136 | 133–147 | 143–163 |
| 5 | 8 | 126–139 | 136–150 | 146–167 |
| 5 | 9 | 129–142 | 139–153 | 149–170 |
| 5 | 10 | 132–145 | 142–156 | 152–173 |
| 5 | 11 | 135–148 | 145–159 | 155–176 |
| 6 | 0 | 138–151 | 148–162 | 158–179 |

**NOTE**

Although the weight tables commence at specified heights, no minimum height requirement has been prescribed.

The civil service examining physician may determine that weight in excess of that shown in the tables (up to a maximum of 20 pounds) is lean body mass and not fat. Decisions as to frame size of a candidate shall be made by the examining physician.

**Step Four:** Embark on a positive program to build up your strength, agility, speed, and stamina. The following section will give you some suggestions.

## FITNESS AND PHYSICAL ABILITY TESTS

Almost all fire departments require successful candidates to pass some type of physical test. This physical test is called either a fitness test or a physical ability (or agility) test. These two types of tests are very different, yet they are both designed to ensure that individuals are able to perform the physically demanding tasks of the firefighter position.

Fitness tests are designed to measure your level of general "fitness" through sit-ups, mile-and-a-half runs, bench press repetitions, and so on. The fitness test uses different cutoff scores in these components for men and women and for various age groups; however, these cutoff scores are set at the same percentile for everyone. In other words, even if men and women are not required to bench press the same amount of weight, the amount required of each can probably be pressed by about three fourths (75 percent) of the men and three fourths (75 percent) of the women.

Physical ability (or agility) tests require candidates to perform a series of linked exercises that simulate a firefighter's job, such as swinging an ax, climbing stairs with equipment, and dragging a dummy. This type of physical test is not measuring general fitness. Rather, it is measuring your ability to perform essential job tasks. Because the tasks are essential to the job, the same score (the ability to perform these tasks in a reasonable period of time) is required of all candidates to pass regardless of sex or age. Physical ability (or agility) tests also require a high level of physical fitness to perform well.

## Defining Fitness

Physical fitness is to the human body what fine-tuning is to an engine. It enables us to perform up to our potential. Fitness can be described as a condition that helps us look, feel, and do our best. More specifically, it is:

The ability to perform daily tasks vigorously and alertly, with energy left over for enjoying leisure-time activities and meeting emergency demands. It is the ability to endure, to bear up, to withstand stress, and to carry on in circumstances where an unfit person could not continue and is a major basis for good health and well-being.

Physical fitness involves the performance of the heart and lungs and the muscles of the body. Because what we do with our bodies also affects what we can do with our minds, fitness influences to some degree qualities such as mental alertness and emotional stability.

As you undertake your fitness program, it's important to remember that fitness is an individual quality that varies from person to person. It is influenced by age, sex, heredity, personal habits, exercise, and eating practices. You can't do anything about the first three factors, but it is within your power to change and improve the others as necessary.

## Knowing the Basics

Physical fitness is most easily understood by examining its components or parts. There is widespread agreement that these five components are basic:

**Cardiorespiratory endurance:** The ability to deliver oxygen and nutrients to tissues and to remove wastes over sustained periods of time. Long runs and swims are among the methods employed in measuring this component.

**Muscular strength:** The ability of a muscle to exert force for a brief period of time. Upper-body strength, for example, can be measured by various weight-lifting exercises.

**Muscular endurance:** The ability of a muscle, or a group of muscles, to sustain repeated contractions or to continue applying force against a fixed object. Push-ups often are used to test endurance of arm and shoulder muscles.

**Flexibility:** The ability to move joints and use muscles through their full range of motion. The sit-and-reach test is a good measure of flexibility of the lower back and the backs of the upper legs.

**Body composition:** This refers to the makeup of the body in terms of lean mass (muscle, bone, vital tissue, and organs) and fat mass. An optimum ratio of fat to lean mass is an indication of fitness, and the right type of exercises can help you decrease body fat and increase or maintain muscle mass.

## A Workout Schedule

How often, how long, and how hard you exercise and what types of exercises you do should be determined by what you are trying to accomplish. Right now, your goal is to prepare your body so it can withstand the rigors of the physical performance test and make a top showing. Your goal also should be solid preparation for the physical demands of firefighting.

Your exercise program should include something from each of the five basic fitness components previously described. Each workout should begin with a warm-up and should end with a cool-down. As a general rule, space your workouts throughout the week and avoid consecutive days of hard exercise.

The following is the amount of activity necessary for the average, healthy person to maintain a minimum level of overall fitness. Included are some of the popular exercises for each category.

**Warm up:** 5–10 minutes of exercises such as walking, slow jogging, knee lifts, arm circles, or trunk rotations. Low-intensity movements that simulate movements to be used in the activity also can be included in the warm-up.

**Muscular strength:** a minimum of two 20-minute sessions per week that include exercises for all the major muscle groups. Lifting weights is the most effective way to increase strength.

**Muscular endurance:** at least three 30-minute sessions each week that include exercises such as calisthenics, push-ups, sit-ups, pull-ups, and weight training for all the major muscle groups.

**Cardiorespiratory endurance:** at least three 20-minute bouts of continuous aerobic (activity requiring oxygen), rhythmic exercise each week. Popular aerobic conditioning activities include brisk walking, jogging, swimming, cycling, rope-jumping, rowing, cross-country skiing, and some continuous-action games such as racquetball and handball.

**Flexibility:** 10–12 minutes of daily stretching exercises performed slowly and without a bouncing motion. This can be included after a warm-up or during a cool-down.

**Cool-down:** a minimum of 5–10 minutes of slow walking or low-level exercise combined with stretching.

## A Matter of Principle

The keys to selecting the right types of exercises for developing and maintaining each of the basic components of fitness are found in these principles:

**Specificity:** Pick the right type of activities to affect each component. Strength training results in specific strength changes. In addition, train for the specific activity in which you're interested. For example, optimum swimming performance is best achieved when the muscles involved in swimming are trained for the movements required. It does not follow that a good runner is a good swimmer.

**Overload:** Work hard enough, and at levels that are vigorous and long enough to overload your body above its resting level, to bring about improvement.

**Regularity:** You can't hoard physical fitness. At least three balanced workouts per week are necessary to maintain a desirable level of fitness.

**Progression:** Increase the intensity, frequency, and/or duration of activity over periods of time to improve.

Some activities can be used to fulfill more than one of your basic exercise requirements. In addition to increasing cardiorespiratory endurance, for example, running builds muscular endurance in the legs, and swimming develops the arm, shoulder, and chest muscles. If you select the proper activities, it is possible to fit parts of your muscular endurance workout into your cardiorespiratory workout and save time.

## Measuring Your Heart Rate

Heart rate is widely accepted as a good method for measuring intensity during running, swimming, cycling, and other aerobic activities. Exercise that doesn't raise your heart rate to a certain level and keep it there for 20 minutes won't contribute significantly to cardiovascular fitness.

The heart rate you should maintain is called your target heart rate. There are several ways to arrive at this figure. One of the simplest is as follows:

maximum heart rate = (220 − age)

target heart rate = maximum heart rate × 70%

Thus, the target heart rate for a 40-year-old individual would be 126.

Some methods for figuring the target rate take individual differences into consideration. Here is one such method:

❶ Subtract your age from 220 to find your maximum heart rate.

❷ Subtract your resting heart rate (see below) from your maximum heart rate to determine your heart rate reserve.

❸ Take 70 percent of your heart rate reserve to determine your heart rate raise.

❹ Add your heart rate raise to your resting heart rate to find your target rate.

Your resting heart rate should be determined by taking your pulse after sitting quietly for 5 minutes. When checking heart rate during a workout, take your pulse within 5 seconds after interrupting exercise because it starts to go down once you stop moving. Count your pulse for 10 seconds and then multiply by six to get the per-minute rate.

## Measure Your Progress

You will be able to observe the increase in your strength and stamina from week to week in many ways.

There also is a 2-minute step test you can use to measure and keep a running record of the improvement in your circulatory efficiency, one of the most important of all aspects of fitness.

The immediate response of the cardiovascular system to exercise differs markedly between well-conditioned individuals and others. The test measures the response in terms of pulse rate taken shortly after a series of steps up and down onto a bench or chair.

Although it does not take long, it is necessarily vigorous. Stop if you become overly fatigued while taking the step test. You should not try it until you have completed a number of weeks of conditioning exercises.

## The Step Test

Use any sturdy bench or chair 15–17 inches in height.

Count 1—Place right foot on bench.

Count 2—Bring left foot alongside of right and stand erect.

Count 3—Lower right foot to floor.

Count 4—Lower left foot to floor.

REPEAT the four-count movement 30 times a minute for 2 minutes.

THEN sit down on a bench or chair for 2 minutes.

FOLLOWING the 2-minute rest, take your pulse for 30 seconds.

Double the count to get the per-minute rate. (You can find your pulse by applying the middle and index fingers of one hand firmly to the inside of the wrist of the other hand, on the thumb side.)

Record your score for future comparisons. In succeeding tests—about once every two weeks—you probably will find your pulse rate becoming lower as your physical condition improves.

Three important points to remember:

❶ For best results, do not engage in physical activity for at least 10 minutes before taking the test. Take it at about the same time of day and always use the same bench or chair.

❷ Remember that pulse rates vary among individuals. This is an individual test. What is important is not a comparison of your pulse rate with that of anybody else but rather a record of how your own rate is reduced as your fitness increases.

❸ As you progress, the rate at which your pulse is lowered should gradually level off. This is an indication that you are approaching peak fitness.

## STAYING FIT

When you have reached the level of conditioning you have chosen for yourself, you will want to maintain your fitness.

Although it has been found possible to maintain fitness with three workouts a week, exercise ideally should be a daily habit.

## Broadening Your Program

There are many other activities and forms of exercise that you can use to supplement a basic program.

These activities include a variety of sports, water exercises you can use if you have access to a pool, and isometrics—sometimes called exercises without movement—which take little time (6–8 seconds each).

## Isometrics

Isometric contraction exercises take very little time and require no special equipment. They're excellent muscle strengtheners and, as such, are valuable supplements. The idea of isometrics is to work out a muscle by pushing or pulling against

**TIP**

If you can, by all means continue your workouts on a five-days-a-week basis.

an immovable object such as a wall or by pitting it against the opposition of another muscle.

The basis is the "overload" principle of exercise physiology, which holds that a muscle required to perform work beyond the usual intensity will grow in strength. Research has indicated that one hard, 6- to 8-second isometric contraction per workout can, over a period of six months, produce a significant strength increase in a muscle.

The isometric exercises described here cover the major large muscle groups of the body. They can be performed almost anywhere and at almost any time. Note that there is no set order for doing them nor do they all have to be completed at one time. You can, if you like, do one or two in the morning and others at various times during the day whenever you have half a minute or even less to spare.

For each contraction, maintain tension for no more than 8 seconds. Do little breathing during a contraction; breathe deeply between contractions. Start easily. Do not apply maximum effort in the beginning.

For the first three or four weeks, you should exert only about one half of what you think is your maximum force. Use the first three or four seconds to build up to this degree of force and use the remaining four or five seconds to hold it.

For the next two weeks, gradually increase force to more nearly approach maximum. After about six weeks, it will be safe to exert maximum effort.

Pain indicates that you're applying too much force; reduce the amount immediately. If pain continues to accompany any exercise, discontinue using that exercise for a week or two. Then try it again with about 50 percent of maximum effort; if no pain occurs, you can go on to gradually build up toward maximum.

### Neck

*Starting position:* Sit or stand with interlaced fingers of your hands on your forehead.

*Action:* Forcibly exert a forward push of your head while resisting equally hard with your hands.

*Starting position:* Sit or stand with interlaced fingers of your hands behind your head.

*Action:* Push your head backward while exerting a forward pull with your hands.

*Starting position:* Sit or stand with the palm of your left hand on the left side of your head.

*Action:* Push with your left hand while resisting with your head and neck. Reverse using your right hand on the right side of your head.

### Upper Body

*Starting position:* Stand with your back to the wall, your hands at your sides, your palms toward the wall.

*Action:* Press your hands backward against the wall, keeping your arms straight.

*Starting position:* Stand facing the wall with your hands at your sides and your palms toward the wall.

*Action:* Press your hands forward against the wall, keeping your arms straight.

*Starting position:* Stand in a doorway or with your side against a wall, with your arms at your sides and your palms toward your legs.

*Action:* Press your hand(s) outward against the wall or door frame, keeping your arms straight.

### Arms

*Starting position:* Stand with your feet slightly apart. Flex your right elbow close to your body, palm up. Place your left hand over your right.

*Action:* Forcibly attempt to curl your right arm upward, giving equally strong resistance with your left hand. Repeat with your left arm.

### Arms and Chest

*Starting position:* Stand with your feet comfortably spaced and your knees slightly bent. Clasp your hands, palms together and close to your chest.

*Action:* Press your hands together and hold.

*Starting position:* Stand with your feet slightly apart and your knees slightly bent. Grip your fingers with your arms close to your chest.

*Action:* Pull hard and hold.

### Abdominal

*Starting position:* Stand with your knees slightly flexed and your hands resting on your knees.

*Action:* Contract your abdominal muscles.

### Lower Back, Buttocks, and Back of Thighs

*Starting position:* Lie face down with your arms at your sides, your palms up, and your legs placed under a bed or another heavy object.

*Action:* With both hips flat on the floor, raise one leg, keeping your knee straight so that the heel pushes hard against the resistance. Repeat with opposite leg.

### Legs

*Starting position:* Sit in chair with your left ankle crossed over your right, with your feet resting on floor and your legs bent at a 90-degree angle.

*Action:* Forcibly attempt to straighten your right leg while resisting with the left. Repeat with opposite leg.

### Inner and Outer Thighs

*Starting position:* Sit with your legs extended, with each ankle pressed against the outside of sturdy chair legs.

*Action:* Keep your legs straight and pull toward one another firmly. For outer thigh muscles, place your ankles inside chair legs and exert pressure outward.

## Water Activities

Swimming is one of the best physical activities for people of all ages—and for many people with disabilities.

With the body submerged in water, blood circulation automatically increases to some extent. The pressure of water on the body also helps promote deeper ventilation of the lungs. With well-planned activity, both circulation and ventilation increase still more.

## Weight Training

Weight training also is an excellent method of developing muscular strength and muscular endurance. Both barbells and weighted dumbbells—complete with instructions—are available at most sporting goods stores. A good rule to follow in deciding the maximum weight you should lift is to select a weight you can lift six times without strain. If you have access to a gym with sophisticated equipment, take advantage of the advice of professional trainers in establishing weight-training programs and goals.

## Sports

Soccer, basketball, handball, squash, ice hockey, and other sports that require sustained effort can be valuable aids in building circulatory endurance.

If you have been sedentary, however, it's important to pace yourself carefully in such sports. It might even be advisable to avoid them until you are well along in your physical conditioning program. This doesn't mean you should avoid all sports.

There are many excellent conditioning and circulatory activities in which the amount of exertion can be easily controlled and in which you can progress at your own rate. Bicycling is one example. Others include hiking, skating, tennis, running, cross-country skiing, rowing, canoeing, water skiing, and skin diving.

Games should be played with full speed and vigor only when your conditioning permits doing so without undue fatigue.

On days when you get a good workout in sports, you can skip part or all of your exercise program. Use your own judgment.

If you have engaged in a sport that exercises the legs and that stimulates the heart and lungs—such as skating—you could skip the circulatory activity for that day, but you still should do some of the conditioning and stretching exercises for the upper body. On the other hand, weight lifting is an excellent conditioning activity, but it should be supplemented with running or one of the other circulatory exercises.

Whatever your favorite sport, you will find your enjoyment enhanced by improved fitness. Every weekend athlete should invest in frequent workouts.

## District of Columbia Fire and Emergency Medical Services Department Firefighter Physical Ability Test

This test consists of a series of tasks designed to assess important physical abilities necessary for effective job performance as an entry-level firefighter. These tasks were developed to mirror real situations that firefighters might encounter on the job.

In developing the Physical Ability Test, a group of firefighter experts from the D.C. FEMSD identified many of the tasks essential to the performance of the job of a firefighter. Job analysis questionnaire data also was collected from incumbent firefighters; this provided the background information necessary to develop this job-related physical ability test.

From this information, nine job-related Physical Ability Test components were developed. These nine components are:

- Aerial ladder climb
- Hydrant opening
- Ladder carry
- Ladder extension
- Charged line advance
- Stair climb with equipment
- Ceiling pole
- Victim rescue in a confined space
- Victim rescue dummy drag

### Important Information about the Test

The physical ability test consists of a series of events designed to simulate such firefighter-related activities as carrying fire equipment up flights of stairs, opening fire hydrants, rescuing a victim while in a confined space, and carrying and extending a ladder. The following points should help applicants familiarize themselves with what will take place on the day of the physical ability test.

- Applicants must wear a turnout coat, a hard hat, and a completely filled self-contained breathing apparatus (SCBA and composite bottle, excluding the facepiece and low-pressure hose) for all but the first test component (the aerial ladder climb). This equipment will be supplied to the applicants at the test site. Applicants should wear athletic shoes, and it is recommended that they wear long pants for safety.

- Timing of the testing components is to begin after the first event (the aerial ladder climb) has been successfully completed. Applicants unable to successfully complete the aerial ladder climb will not be allowed to continue the testing process.

- All remaining testing components must be completed in a series. There are no breaks between timed events.

- During the sequence of timed events, applicants will be permitted to run, walk, or rest between testing stations if they choose to do so. Given the extremely demanding nature of the test, it is recommended that applicants pace themselves so they can complete the entire series of events.

- Prior to the actual examination, a test administrator will give an orientation and thorough walk-through to all candidates. During the walk-through, applicants will not be permitted to touch any of the equipment. Applicants will be given a demonstration of each test component.

- One test monitor will time candidates as they proceed through the test stations. Additional monitors will be located at each test station to provide direction to candidates and to observe their health and safety.

- The Physical Ability Test is timed from beginning to end and must be completed within a specific period of time to pass.

## Description of the Physical Ability Test

The following nine physical tasks compose the entire Physical Ability Test. Each task reflects the types of situations that an entry-level firefighter might encounter while on the job.

### *Untimed Event*

❶ *Aerial Ladder Climb:* For this first event, applicants must climb to the top of an aerial ladder set at 60 degrees to a height of 50 feet and then return to the bottom. Applicants will be attached to a safety line and will, upon instruction, ascend and descend the ladder without stopping. This is the first event of the Physical Ability Test, and it is not timed. Applicants unable to successfully complete this event will not be allowed to continue the testing process. For this event only, applicants will not be required to wear the turnout coat or the SCBA; however, they will be required to wear the hard hat for safety.

### *Timed Events*

❷ *Hydrant Opening:* Applicants must use a hydrant wrench to open a functional hydrant completely (17 turns) and to close it completely (17 turns).

③ *Ladder Operations—Carry:* Applicants must remove a 12-foot plain ladder from a truck and carry it a distance of 50 feet, placing it against the drill tower. The ladder must then be picked up, carried back, and replaced onto the truck in its original position.

④ *Ladder Operations—Extension:* Applicants must extend a 30-foot ground ladder that is affixed to the side of the drill tower and bring it back down. The ladder cannot be dropped.

⑤ *Charged Line Advance:* Applicants must pick up and advance a charged 1½-inch hoseline with playpipe attached for a distance of 100 feet. Pump pressure is set at 130 psi.

⑥ *Stair Climb with Equipment:* Applicants must pick up a 1½-inch standpipe rack (a wrapped hose) and carry it up six flights of stairs. At the top of the stairs, applicants must place the standpipe rack on the floor. Applicants must then pick up the standpipe rack again and carry it back down to the bottom of the stairs. Note that this is an extremely difficult test component given the weight of the equipment worn and carried.

⑦ *Ceiling Pole:* Simulators have been constructed to replicate the action involved when using a ceiling pole to break apart ceilings in structures. Applicants must complete two sets of repetitions on both ceiling pole simulators (up and down). There are four total sets. Each set consists of six repetitions.

⑧ *Victim Rescue—Confined Space:* Applicants must wear a blackened facepiece to block their vision and will be directed to enter a maze. Applicants must navigate the maze on their hands and knees and must exit at the opposite end.

⑨ *Victim Rescue—Dummy Drag:* Applicants must drag a 165-pound human-form dummy for a distance of 50 feet.

The amount of time needed to successfully pass the test is based on the times of incumbent firefighters who proceeded through the course at a pace they used while performing these tasks on the job.

## The Layout of the Testing Site

The following diagram illustrates the layout of the physical ability testing sequence. Each testing component is labeled with a number that corresponds to the components described in this section.

## Clothing Worn and Equipment Used

The following is a list of clothing to be worn by candidates during the Physical Ability Test and equipment that will be used. Note the weight of the various pieces of equipment:

| Equipment | Weight (in pounds) |
| --- | --- |
| Turnout coat | 5.5 |
| SCBA (filled) | 34 |
| 12-foot ladder | 27 |
| 1½-inch standpipe rack | 57 |
| Ceiling pole simulator (up) | 75 |
| Ceiling pole simulator (down) | 75 |
| Human-form dummy | 165 |

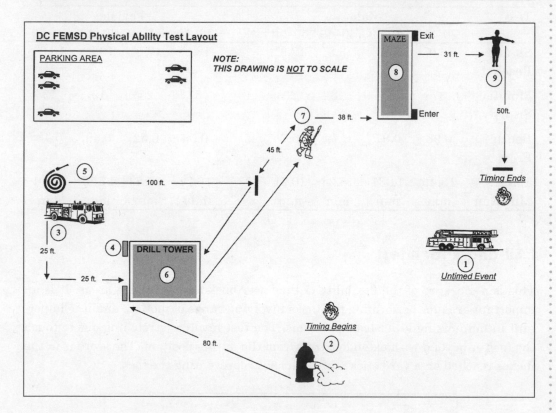

DC FEMSD Physical Ability Test Layout

## Illinois Fire and Police Recruitment Administration Personnel Aptitude Screening Test

The firefighter physical ability test consists of four pass-required events and four additional events that are assessments used for evaluation only. Each event is a scientific and valid test. The test will be given in sequence with a rest period between each event.

The required performance to pass each event is based on sex and age (decade). Although the absolute performance is different for each category, the relative level of effort is identical for each age and sex group. All candidates are required to meet the same percentile rank in terms of their respective age and sex groups. The performance requirement is the level of physical performance that approximates the 40th percentile for each age and sex group.

| Test | Male | | | | Female | | | |
|------|------|------|------|------|------|------|------|------|
| | 20–29 | 30–39 | 40–49 | 50–59 | 20–29 | 30–39 | 40–49 | 50–59 |
| Sit and Reach | 16.0" | 15.0" | 13.8" | 12.8" | 18.8" | 17.8" | 16.8" | 16.3" |
| Minute Sit-Up | 37 | 34 | 28 | 23 | 31 | 24 | 19 | 13 |
| Bench Press | 0.98 | 0.87 | 0.79 | 0.7 | 0.58 | 0.52 | 0.49 | 0.43 |
| 1.5-Mile Run | 13.46 min | 14.31 min | 15.24 min | 16.21 min | 16.21 min | 16.52 min | 17.53 min | 18.44 min |

## I. Sit-and-Reach Test

This is a measure of the flexibility of the lower-back and upper-leg areas. It is an important area for performing fire tasks involving range of motion, and it is important in minimizing lower-back problems. The test involves stretching out to touch the toes or beyond with extended arms from the sitting position. The score is in the inches reached on a yard stick with 15 inches representing the toes.

## I. 1-Minute Sit-Up Test

This is a measure of the muscular endurance of the abdominal muscles. It is an important area for performing fire tasks that might involve the use of force, and it is an important area for maintaining good posture and for minimizing lower-back problems. The score is the number of bent-leg sit-ups performed in 1 minute.

## III. Maximum Bench Press (One Repetition)

This is a maximum weight pushed from the bench-press position. It measures the amount of force the upper body can generate. It is an important area for performing fire tasks requiring upper-body strength. The score is a ratio of weight pushed divided by body weight.

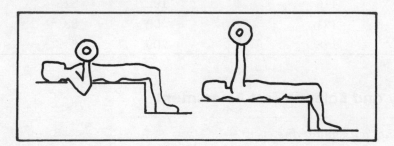

## IV. 1.5-Mile Run

This is a timed run to measure the heart's and vascular system's capability to transport oxygen. It is an important area for performing fire tasks involving stamina and endurance, minimizing the risk of cardiovascular problems. The score is in minutes and seconds. This run is measured using a treadmill within a controlled atmosphere. (The test-givers hold the option of testing on a treadmill or an indoor or outdoor track.)

## V. Threshold Weight and Skinfold Body Fat Percentage

*— This is not a disqualifying event.—*

This is the weight that has been determined as the weight necessary to perform firefighting tasks without undue effort and to minimize health problems due to excessive body fat. The weight should not be more than the recommended pounds per height in inches.

| HT/IN | Threshold weight | HT/IN | Threshold weight | HT/IN | Threshold weight |
|---|---|---|---|---|---|
| 52 | 75 | 63 | 134 | 74 | 217 |
| 53 | 80 | 64 | 141 | 75 | 226 |
| 54 | 85 | 65 | 147 | 76 | 235 |
| 55 | 89 | 66 | 154 | 77 | 245 |
| 56 | 94 | 67 | 161 | 78 | 255 |
| 57 | 99 | 68 | 168 | 79 | 265 |
| 58 | 105 | 69 | 176 | 80 | 275 |
| 59 | 110 | 70 | 184 | 81 | 285 |
| 60 | 116 | 71 | 192 | 82 | 297 |
| 61 | 121 | 72 | 200 | 83 | 307 |
| 62 | 128 | 73 | 209 | 84 | 318 |

## VI. Carry and Balance Test Assessment

*— This is not a disqualifying event.—*

This test is designed to measure how well the joint and muscle sensors react to control movement and maintain balance.

**Procedure:** The candidate, given an oblong object weighing at least 20 pounds, will walk the length of a beam (measuring more than 12 feet long by 3–4 inches wide) secured to a level floor without falling off or stepping off the beam.

## VII. Lumbar Spine Endurance Assessment

*— This is not a disqualifying event.—*

This test is designed to measure how well the lumbar spine will react under strenuous activities and movement.

**Procedure:** Holding an object weighing at least 5 pounds, standing upright, and keeping feet within a 12-inch area, the candidate will bend down and touch the floor with the weighted object at least 12 inches to the left of the left foot, return to the upright position, and then do the same on the right side. This counts as one repetition. The score is a minimum of 20 repetitions in 1 minute.

## VIII. 5-Minute Claustrophobia Assessment (Covered Mask)

*— This is not a disqualifying event.—*

This event is designed to measure how the candidate will react in a closed, tightly sealed area. It is important for performing fire tasks involving face masks in hazardous environments.

**Procedure:** Given a covered mask, the candidate will secure a mask to his or her face for a minimum of 5 minutes without interruption.

# The Candidate Physical Ability Test™

Many departments have recently chosen to adopt the Candidate Physical Ability Test (CPAT)™ to measure candidate fitness, strength, and endurance. Developed by the International Association of Fire Fighters and the International Association of Fire Chiefs, the CPAT is a job-simulation, physical ability examination. In other words, applicants complete tasks that simulate as closely as possible the most essential firefighting duties.

The CPAT consists of eight events: the stair climb; the hose drag; the equipment carry; the ladder raise and extension; the forcible entry exercise; the search event; the rescue exercise; and the ceiling breach and pull.

During the test, you will wear a 50-pound vest to simulate the weight of self-contained breathing apparatus (SCBA) and protective garments. You must also wear long pants, a hard hat with a chinstrap, work gloves, and shoes with a closed heel or toe. You are not allowed to wear a watch or jewelry that is restrictive or loose.

You will be required to walk—not run—from event to event along a marked path. Most departments require you to complete all eight events within **10 minutes and 20 seconds** (although you may find that some departments will use a slightly different cut-off time).

**NOTE**

In addition to the 50-pound vest, you also will wear two 12.5-pound weights on your shoulders, which simulate the weight of a high-rise hose bundle.

Following is a description of the eight CPAT events:

❶ **The Stair Climb.** This event is designed to replicate the task of climbing up stairs while wearing protective gear and carrying firefighting equipment.

During this event, you will be stepping on a StairMaster StepMill. Before the test begins, you are allowed to warm up for 20 seconds at a set stepping rate of 50 steps per minute. During the warm-up, you are allowed to step off of the StepMill, touch the wall, or hold onto the rail as you establish your walking rhythm. If you fall or step off of the machine, you are able to restart your 20-second warm-up period. You are allowed to restart the warm-up period twice, if need be. However, if you fall or dismount three times during the warm-up period, you fail the test. Note that you will not have a break between the warm-up period and the actual test.

When the proctor says "Start," begin the timed test. Walk on the StepMill at a set stepping rate of 60 steps per minute for 3 minutes. If at any time during the timed test you fall, step off of the StepMill, or grasp any testing equipment, you fail the test. You may momentarily touch the wall or handrail to keep your balance; however, you will be warned if you grasp the wall or rail for an extended period of time or use them to bear your weight. You will be given only two warnings. You fail the test if you violate this rule a third time.

❷ **The Hose Drag.** During this event, you perform tasks that simulate dragging an uncharged hoseline from the fire apparatus to the fire occupancy and pulling the hoseline around obstacles while remaining in a stationary kneeling position.

You will grasp an automatic nozzle attached to 200 feet of 1-inch hose. Place the uncharged hoseline over your shoulder or across your chest; note that you are not allowed to grab the hose past the 8-foot mark. You may run while dragging the hose 75 feet to a prepositioned drum. Make a 90-degree turn around the drum and continue dragging the hose for an additional 25 feet. If you fail to go around the drum or stray outside of the marked path, you fail the test.

Stop within a 5-foot × 7-foot box, drop to at least one knee, and pull the hoseline until the 50-foot mark is across the finish line. You must keep at least one knee on the ground during the hose pull, and you must stay within the marked boundary lines. You will receive one warning if you do not keep at least one knee on the ground; you fail the test if you violate this rule a second time. Furthermore, you will also receive one warning if your knees go outside of the marked boundary lines; a second violation constitutes a failure of the CPAT.

❸ **The Equipment Carry.** While on the job, firefighters frequently remove power tools from the fire apparatus, carry them to the fire occupancy, and then later return them to the apparatus. This event simulates that task. You have to remove two rescue circular saws (approximately 32 pounds each) from a cabinet one at a time and place them on the ground. You will then pick up both saws at once and carry them both (one in each hand) while walking 75 feet, going around a prepositioned drum and then back to the starting point. At any point during this equipment carry, you are able to put the saws back on the ground in order to adjust your grip on them. However, if you drop either saw during the carry, you fail the test. You will not be allowed to run during this event. You will receive one warning only, and if you violate the rule a second time, you fail the test. When you return to the tool cabinet, place both saws on the ground, one at a time, and replace them in the proper place in the cabinet.

❹ **Ladder Raise and Extension.** During this event, you replicate the task of placing a ground ladder near a building and raising the ladder to the building's roof or window. Walk to the top rung of a 24-foot aluminum extension ladder, lift the unhinged end from the ground, and walk the ladder up in a hand-over-hand fashion until it is stationary against the wall. You may not use the ladder rails to raise the ladder. If you miss any rung during the raise, you will be given a single warning. You fail if you violate this rule twice. Next, place both feet inside a 36-inch × 36-inch box that is marked on the ground and extend the fly section of the prepositioned and secured 24-foot aluminum extension ladder in a hand-over-hand fashion until the fly section hits the stop. Lower the fly section hand over hand to the starting position. If at any time your feet do not remain within the marked boundary lines, you will be given a warning. Straying outside of the boundary lines twice will constitute a failure of the CPAT. Also, if you fail to raise or lower the ladder in a hand-over-hand manner or drop the ladder at any time during this event, you fail the test.

❺ **The Forcible Entry Exercise.** When entering a fire occupancy, firefighters often have to open a locked door forcibly or breach a wall. This exercise is designed to simulate that task.

You will use a 10-pound sledgehammer to strike a measuring device in the target area until the buzzer sounds. While striking the device, your feet must remain outside of the toe-box. After the buzzer is activated, place the sledgehammer on the ground. You fail the test if you lose control of the sledgehammer. You will receive one warning if you step inside the toe-box; a repeat of this violation will cause you to fail the test.

❻ **The Search Event.** This task simulates searching for a victim in an unfamiliar area with low visibility. During this exercise, you will crawl on your hands and knees through a tunnel maze that is 3 feet high, 4 feet wide, and 64 feet in length with two 90-degree turns. You will face some obstacles at several points in the maze and will have to navigate your way over and under these obstacles. In two locations, the dimensions of the tunnel are reduced. Proctors will observe your movements. If at any time you need to exit the tunnel before the event is completed, you may call out or knock on the wall or ceiling of the maze, and the proctors will assist you. If you request such assistance, however, you fail the test.

## NOTE

To prevent candidate injury, a safety lanyard is attached to the 24-foot aluminum ladder and activated in case the candidate loses control of the ladder.

❼ **The Rescue Exercise.** The purpose of this event is to replicate the experience of removing a victim or injured partner from the scene of a fire. You will grasp a 165-pound human-form dummy by the handle that is on the harness on the dummy's shoulders. You may grasp either one or both of the handles. Drag the dummy 35 feet to a pre-positioned drum, turn 180 degrees around the drum, and drag the dummy for an additional 35 feet to the finish line. The dummy may touch the drum as you perform this task. However, you may not grasp or rest on the drum; a single warning will be given if you violate this rule, and you fail the test if you do it twice. You are allowed to drop the dummy and adjust your grip. Keep in mind that the **entire** dummy must be dragged across the finish line in order for you to successfully complete the event.

❽ **The Ceiling Breach and Pull.** In order to see if a fire has extended to other parts of a structure, firefighters often breach and pull down ceilings. This event is designed to measure your ability to perform this critical task.

You will remove the pike pole from the bracket and, while standing within the boundary established by the frame of the equipment, you will place the tip of the pole on the painted area of the hinged door in the ceiling. Completely push up the hinged door in the ceiling with the pike pole three times. Then hook the pike pole to the ceiling device and pull the pole down five times. Three pushes and five pulls constitute a "set." You must repeat this set four times. When you finish all of your repetitions, the proctor will call out "Time!"

During this event, you will be allowed to stop and adjust your grip on the pike pole, if needed. If you let your grip on the pike pole handle slip, you may readjust your grip without a warning or failure, **but only if the pike pole does not fall to the ground.** If you drop the pike pole, you will receive only one warning. You must pick up the pike pole without help from the proctors and resume the event. If you violate this rule a second time, you fail the test. If you do not successfully complete a repetition, the proctor will call out "Miss!" and you will have to push or pull the apparatus again to complete the repetition.

## How Do I Prepare for the CPAT?

Many candidates have been overheard saying, "I haven't run in a long time" or "I've never lifted this much weight before." Test administrators are not surprised when these candidates quickly fail. Make sure that you are physically fit and can meet all test requirements **before** the examination date. Spend the rest of the time before the test maintaining that level of fitness.

Some of the CPAT events may be recreated at home. You can practice climbing on a StepMill at a local health club, for example. Or you can practice carrying weights for the equipment carry. But please remember—**safety first!** Some of the testing equipment, such as the 24-foot aluminum extension ladder, is outfitted with special safety gear in the event that the candidate drops or loses control of the device. If

**TIP**

The most important thing to keep in mind when preparing for any physical ability test is to practice and train well in advance of the examination day.

you are going to try to practice CPAT events at home, please be sure to **take all necessary safety precautions.** You will not be able to pass the test if you injure yourself while practicing!

Better still, many departments offer practice sessions on the CPAT course for candidates. Contact the department where you are applying for details.

Also, be sure you know the specifics about the CPAT you will be taking. Make sure you know your department's cut-off time in case it differs from the 10 minutes and 20 seconds passing time. Be sure to attend all orientation meetings and listen carefully to any instructions that the department provides for you. The more you know about the test, the better prepared you will be.

## PERSONALITY TESTING AND PSYCHOLOGICAL EVALUATIONS

Many fire departments include personality testing and/or psychological evaluations as part of their screening process. Personality testing can help departments identify candidates who are really motivated to be firefighters and who have a good attitude about their jobs. Psychological evaluations are much more in-depth and often include both personality testing and an interview with a psychologist. These psychological evaluations are designed to determine a candidate's ability to deal with the extreme stress and other emotionally challenging aspects of being a firefighter. The following are some of the personality characteristics and psychological traits required of a firefighter. A firefighter must:

- Stay calm and handle stress.

- Maintain a positive attitude.

- Maintain emotional control.

- Be flexible.

- Be decisive.

- Be resourceful.

- Determine priorities.

- Deal with critically injured or ill people.

- Handle critical decision-making under life-threatening conditions.

- Perform complex tasks under life-threatening conditions.

- Work with little or no supervision.

- Accept constructive criticism from others.

- Take charge when needed.

- Counsel, support, and be empathetic toward others.
- Perform tasks requiring long periods of intense concentration.
- Work under tight time frames.

Because there are no right or wrong answers to questions on personality tests, it is very difficult to prepare for them. There are a few guidelines, however, that can certainly improve your chances of doing well. Keep the following in mind when answering questions on a personality test or when speaking with a psychologist.

## Always Tell the Truth

Personality tests are designed to identify people who are not providing candid information. One of the main reasons people do poorly on personality tests is that they answer questions in a way that they think makes them look perfect. This is a big mistake because very few, if any, of us are perfect.

## Go with Your First Thought

Because there are no right or wrong answers for personality test questions, it's easy to read too much into them. The questions usually are pretty straightforward and should be answered as such. Don't try to second-guess the questions or look for hidden meanings; doing so will certainly lower your performance.

## Don't Be Afraid to Say How You Feel

Many questions on personality tests are answered in terms of how much you agree or disagree with a particular statement. Your choices to such a question might be something like: (1) Strongly Agree; (2) Agree; (3) Not Sure; (4) Disagree; (5) Strongly Disagree. If you feel strongly about a particular statement, don't be afraid to answer with a 1 or a 5. It's also not a good idea to answer with too many 3's (Not Sure) unless you're really not sure about how you feel.

## THE INTERVIEW

An interview can take place at any time during the screening process, but when there are many applicants, the employment interview tends to be the last step. An interview takes a great deal of the interviewer's time, so it usually is offered only to candidates who appear to be fully qualified on the basis of all other measures—written test, background investigation, medical examination, and physical fitness performance test.

If you have passed all the steps up to the interview, you are very close, but you do not yet have a place on the fire force. The interview is a very important step in the firefighter screening process.

The purpose of the interview is to gather information along a number of dimensions and to fill you out as a whole person. Some of the aims of the interview are to:

① **Supplement the application form.** The interviewer will ask about your childhood, education, and prior employment. He, she, or they will give you an opportunity to explain employment gaps and abrupt terminations. If your record is anything less than perfect, this is your chance to indicate that you have learned from your prior experience. Now is the time to let the interviewer know how you have matured. Do not blame others for your mistakes. Take responsibility for impetuous behavior, personality clashes, and brushes with the law. Make clear that you have developed the self-control to keep from such misbehaviors in the future.

You also might have an opportunity to expand on the kinds of work you did at prior jobs. If the application form gave you limited space in which to describe duties and responsibilities, you can now fill in details. You can convey enthusiasm for tasks you especially liked or at which you were notably skilled.

② **Learn of your motivation.** Firefighting is not easy work. The interviewer wants to know why you want to be a firefighter. Is your motivation strictly financial? Do you have an unhappy home life that you want to escape for long stretches? Do you care about people or are you just seeking excitement? How do your interests coincide with those of current firefighters? Do your patterns of interest and motivation match those of successful, satisfied firefighters?

③ **Assess your stability and personality.** This is a tall order, and your interview might or might not do it successfully. The firefighter lives and works under stress. In the firehouse, the firefighter lives in close quarters with other men and women, sharing household and maintenance chores and sacrificing privacy. The firefighter must be able to cooperate with all members of the shift without showing undue irritation at a person who might prove annoying. Likewise, the firefighter must not display habits or mannerisms that might prove irritating to others or create discord in the firehouse. In short, the firefighter must "get along" and must "fit in." The stress that firefighters face is even greater when they fight fires. Interview questions will try to gauge how well you can follow orders and how you will react under conditions of real physical pressure. The interviewer will look for signs as to how you can juggle order-taking with initiative. Some of the questions might be hard to answer. The interview itself is a stressful situation. Consider alternatives and give a decisive answer. A firefighter must think before acting but must not spend too much time thinking about next steps. Try to convey this same balance between deliberation and quick thinking as you answer tough questions. Give the impression that you have the self-confidence both to sustain difficult activity under pressure and to size up situations and deviate from prescribed routine when warranted by extreme emergency. Some fire departments supplement the interview with psychological tests to further screen out candidates who give signs of a tendency to "crack" when in real danger or when faced with frustrating decisions.

Because this interview usually is the final step—the moment of decision as to whether you will be accepted on the force—you want to come across at your very best. There are bound to be some surprises, but to a large extent, you can prepare yourself for an interview.

First, write down the date, time, and place on a wall calendar at home and in your pocket calendar. Fold the notice of interview into your wallet so you cannot make a mistake. If you are not sure how to get to the interview site, look into public transportation or automobile routes ahead of time. Do a dry run if necessary. On interview day, leave enough time for the unexpected. A 3-hour pile up on the highway will be reported in the newspaper and is an acceptable reason for missing an appointment, but a 20-minute bus breakdown is no excuse. If you arrive much too early, go for a cup of coffee. Err on the side of time to spare. You should go to the floor of your interview about 10 minutes ahead of time so you can go to the restroom to wash your hands and comb your hair.

Next, make sure you dress for the occasion. You are not being interviewed for an executive position. You need not wear a three-piece suit. Your clothing should be businesslike, however, and clean and neat. You should be well groomed. You want to impress the interviewer as someone he or she would find pleasant to spend a lot of time with. You will add to this impression with a firm handshake, an easy smile, and frequent eye contact with the interviewer or interviewers. Do not chew gum and do not smoke even if you are offered a cigarette.

Prepare yourself with answers to various questions. You might wonder how you can prepare the answers to questions you have not seen. Actually, this is easy. The questions are quite predictable. Begin your preparation by looking over the application forms you filled out and any other papers you were required to file. You should be able to pick out points that an interviewer will want you to amplify or explain.

Questions that might arise from the information you already have given include:

- Why did you choose your area of concentration in school?
- What particularly interests you about _____ (particular subject)?
- Why did you transfer from X school to Y school?
- How did you get the job with _____?
- Which of the duties described in your second job did you like best? Least? Why?
- What did you do during the nine months between your second and third jobs?
- Please explain your attendance pattern at your first job.
- Explain the circumstances surrounding your departure from a particular job.
- Please clarify your armed forces service, arrest record, hospitalization record, and so on, as applicable.

Other questions are almost routine to all interviews. They can be anticipated and prepared for as well.

- Why do you want to leave the kind of work you are doing now?
- Why do you want to be a firefighter?
- How does your family feel about you becoming a firefighter?
- What do you do in your leisure time?
- Do you have hobbies? What are they? What do you particularly like about _____?
- What is your favorite sport? Would you rather play or watch?
- How do you react to criticism? If you think the criticism is reasonable? If you consider the criticism unwarranted?
- What is your pet peeve?
- Name your greatest strengths and weaknesses.
- What could make you lose your temper?
- Of what accomplishment in your life are you most proud?
- What act do you most regret?
- If you could start over, what would you do differently?
- What traits do you value most in a coworker? In a friend?
- What makes you think you would make a good firefighter?

Still other questions might be more specific to a firefighter interview. You should have prepared answers to the following questions:

- How much sleep do you need?

- Are you afraid of heights?

- What is your attitude toward irregular hours?

- Do you prefer working alone or on a team?

- Are you afraid of dying?

- What would you do with the rest of your life if your legs were crippled in an accident?

- How do you deal with panic? Your own? That of others?

- What is your attitude toward smoking? Drinking? Drugs? *Playboy* magazine? Gambling?

- What is your favorite TV program? How do you feel about watching the news? Sports? Classical drama? Rock music? Opera? Game shows?

Now make a list of your own. The variety of interview questions is endless, but most can be answered with ease. Preparation makes the whole process much more pleasant and less frightening.

There is one question that strikes terror into the heart of nearly every job candidate. This question is likely to be the first and, unless you are prepared for it, might well throw you off guard. The question is "Tell me about yourself." For this question, you should have a prepared script (in your head, not in your hand). Think well ahead of time about what you want to tell. What might the interviewer be interested in? This question is not seeking information about your birth weight or your food preferences. The interviewer wants you to tell about yourself with relation to your interest in and qualifications for firefighting. Think about how to describe yourself with this goal in mind. What information puts you in a good light with reference to the work for which you are applying? Organize your presentation. Analyze what you plan to say. What is an interviewer likely to pick up on? To what questions will your little speech lead? You must prepare to answer these questions to which you have opened yourself.

The temptation to talk too much is greatest with this open-ended question, but it exists with many other questions as well. Give full, complete answers, but stop when you have answered the questions. Do not tell the interviewer more than he or she needs to know. Do not volunteer superfluous information. Do not get anecdotal, chatty, or familiar. Resist the urge to ramble on and on. Remember that the interview is a business situation, not a social one.

Toward the end of the interview, the interviewer most likely will ask if you have any questions. You undoubtedly will have had some before the interview and should come prepared to ask. If all your questions have been answered during the course of the interview, you might tell this to the interviewer. If not, or if the interview has raised new questions in your mind, by all means ask them. The interview should serve for your benefit as well; it is not just to serve the purposes of the Personnel Division or the Fire Department.

The invitation of your questions tends to be the signal that the interview is nearly over. The interviewer is satisfied that he or she has gathered enough information. The time allotted to you is up. Be alert to the cues. Do not spoil the good impression you have made by trying to prolong the interview.

At the end of the interview, smile, shake hands, thank the person for the opportunity to have the interview, and leave. A brief thank-you note to the interviewer or to the chairperson of the interviewing panel would not be out of order. Such a note brands you as a courteous, thoughtful person and indicates your continued interest in appointment.

# PART II
## TEST PREPARATION

# Reading Comprehension

## OVERVIEW

- Sample reading program to improve your skills
- Common forms of reading comprehension questions

A recent survey of firefighter examinations given nationwide indicates that there is a wide variation in the subject matter of these exams. The single topic that is common to all exams, however, is reading. Some exams include classic reading comprehension questions that present a passage and then ask questions about the details of the passage and, perhaps, its meaning. Other exams require candidates to indicate proper behavior based on their reading of printed procedures and regulations. Another type of reading-based question requires candidates to reason and predict next steps on the basis of information presented in a reading passage. Of course, questions of judgment in emergency and nonemergency situations rely heavily on reading as well. Actually, there are nearly as many variations of the reading-based question as there are test-makers.

Before you devote attention to strategies for dealing with reading-based questions, give some thought to your reading habits and skills. Of course, you already know how to read. But how well do you read? Do you concentrate? Do you get the point during your first reading? Do you notice details?

Between now and the test day, work to improve your reading concentration and comprehension. Your daily newspaper provides excellent material to improve your reading. Make a point of reading all the way through any article you begin. Do not be satisfied with the first paragraph or two. Read with a pencil in hand. Underscore details and ideas that seem to be crucial to the meaning of the article. Notice points of view, arguments, and supporting information. When you have finished the article, summarize it for yourself. Do you know the purpose of the article? The main idea presented? The attitude of the writer? The points over which there is controversy? Did you find certain information lacking? As you answer these questions, skim back over what you underlined. Did you focus on important words and ideas? Did you read with comprehension?

chapter 4

As you repeat this process day after day, you will find that your reading becomes more efficient. You will read with greater understanding and will "get more" from your newspaper.

You can't sit down the night before a test involving reading comprehension and cram for it. The only way you can build up your reading skill is to practice systematically. The gains you make will show up not only in an increased score on the test, but also in your reading for study and pleasure.

Trying to change reading habits you have had for a long time can be difficult and discouraging. Do not attempt to apply all the suggestions we have given to all your reading all at once.

## SAMPLE READING PROGRAM TO IMPROVE YOUR SKILLS

**❶** Set aside 15 minutes each day to practice new reading techniques.

**❷** Start with a short, easy-to-read article from a magazine or a newspaper. Time yourself. At the end of your practice session, time yourself on another short article and keep a record of both times.

**❸** Select a news story. Read it first and then practice an eye-scan exercise. Work toward reducing your eye fixations to no more than two per line the width of a newspaper column.

**❹** Read an editorial, a book review, or a movie or drama review in a literary magazine or newspaper. This type of article always expresses the author's (or the paper's) point of view and therefore is good practice for searching out the main idea. After you read, see whether you can write a good title for the article and jot down in one sentence the author's main idea. You also can try to make up a question based on the article with four choices. This is excellent practice for determining main ideas, and you can use the questions to test your friends.

**❺** Find one new word and write the sentence in which it appears. Guess at its meaning from the context and then look up the definition in a dictionary. Try to make up a sentence of your own using the word. Then try to use the word in your conversation at least twice the following day.

A major aspect of your daily reading that deserves special attention is vocabulary building. The most effective reader has a rich, extensive vocabulary. As you read, make a list of unfamiliar words. Include words that you understand within the context of the article but that you cannot define. In addition, mark words you do not understand at all. When you put aside your newspaper, go to the dictionary and look up every new word. Write the word and its definition in a special notebook. Writ-

---

**NOTE**

If you follow this program daily, your test score will show the great gains you have made in reading comprehension.

**TIP**

Vocabulary building is a good lifetime habit to develop.

ing down the words and their definitions helps seal them in your memory far better than just reading them, and the notebook serves as a handy reference for your own use. A sensitivity to the meaning of words and an understanding of more words will make reading easier and more enjoyable even if none of the words you learn in this way crops up on your exam. Mastering reading-based questions depends on more than just reading comprehension. You also must know how to draw the answers from the reading selection and be able to distinguish the best answer from a number of answers that all seem to be good ones or that all seem to be wrong.

Strange as it may seem, it's a good idea to approach reading comprehension questions by reading the questions—not the answer choices, just the questions themselves—before you read the selection. The questions will alert you to look for certain details, ideas, and points of view. Use your pencil and underscore key words in the questions. These will help you direct your attention as you read.

Next, skim the selection very rapidly to get an idea of its subject matter and its organization. If key words or ideas pop out at you, underline them but do not consciously search out details in the preliminary skimming.

Now read the selection carefully with comprehension as your main goal and underscore the important words as you have been doing in your newspaper reading.

Finally, return to the questions. Read each question carefully. Be sure you know what it asks. Misreading of questions is a major cause of error on reading comprehension tests. Read all the answer choices. Eliminate the obviously incorrect answers. You might be left with only one possible answer. If you find yourself with more than one possible answer, reread the question. Focus on catch words that might destroy the validity of a seemingly acceptable answer. These include expressions such as under *all circumstances, at all times, never, always, under no condition, absolutely, entirely,* and *except when.* Skim the passage once more, focusing on the underlined segments. By now, you should be able to conclude which answer is best.

## COMMON FORMS OF READING COMPREHENSION QUESTIONS

❶ **Question of fact or detail.** You might have to mentally rephrase or rearrange, but you should find the answer stated in the body of the selection.

❷ **Best title or main idea.** The answer might be obvious, but the incorrect choices to the "main idea" question often are half-truths that are easily confused with the main idea. They might misstate the idea, omit part of the idea, or even offer a supporting idea quoted directly from the text. The correct answer is the one that covers the largest part of the selection.

❸ **Interpretation.** This type of question asks you what the selection means, not just what it says. On firefighter exams, the questions based on categories of building styles, for example, might fall into the realm of interpretation.

❹ **Inference.** This is the most difficult type of reading-based question. It asks you to go beyond what the selection says and to predict what might happen next. You might have to choose the best course of action to take based on given procedures and a factual situation, or you might have to judge the actions of others. Your answer must be based on the information in the selection and your own common sense but not on any other information you might have about the subject. A variation of the inference question might be stated as, "The author would expect that…." To answer this question, you must understand the author's point of view and then make an inference from that viewpoint based on the information in the selection.

❺ **Vocabulary.** Some firefighter reading sections, directly or indirectly, ask the meanings of certain words as used in the selection.

Let's now work together on some typical reading comprehension selections and questions.

> **Directions:** Each passage below is followed by questions based upon its content. After reading the passage, choose the best answer to each question. Answer all of the questions on the basis of what is *stated* or *implied* in the passage.

The best kind of limestone for printing is Bavarian. Light-colored and perfectly smooth, it is porous and absorbs both water and greasy substances equally well. The stone used is about 6 inches thick and is fairly big, up to 90 x 65 cm (35 × 25 inches), and can weigh up to 150 or 175 pounds. Grinding the stone smooth creates a clean printing surface. This allows a drawing to be made on the surface with a greasy lithographic pencil or crayon. The drawing is then fixed by rinsing the stone with a very weak solution of nitric acid and gum arabic. The stone is wiped with water before each impression is taken and, for each print, it is inked with a leather-covered roller. During this process, the porous limestone retains the grease of the crayon where the drawing has been made, and the parts that are not drawn upon become impregnated with water. The ink, which is greasy, is repelled by the water-wet areas and adheres only to the areas marked by the crayon.

1. The best title for the preceding paragraph is

   **(A)** "Where Good Limestone Is Found."

   **(B)** "The Process of Lithographic Printing."

   **(C)** "Water Color Drawings on Limestone."

   **(D)** "How to Make a Printing Stone."

2. According to the preceding paragraph, Bavarian limestone is used because it

   **(A)** can be cut into very large blocks.

   **(B)** has a sponge-like surface.

   **(C)** contains nitric acid and gum arabic.

   **(D)** will repel grease.

3. The aspect of this process NOT described by the preceding paragraph is

   **(A)** the finished product.

   **(B)** how to ink the stone.

   **(C)** creating a printing surface.

   **(D)** what kind of limestone to use.

4. According to the preceding paragraph, the purpose of the nitric acid and gum arabic is to

   **(A)** make the limestone light-colored.

   **(B)** erase previous drawings.

   **(C)** make the drawing permanent.

   **(D)** repel the printing ink.

Begin by skimming the questions and underscoring key words. Your underscored questions should look more or less like this:

❶ The <u>best title</u> for the preceding paragraph is

❷ According to the preceding paragraph, <u>Bavarian limestone</u> is used because it

❸ The <u>aspect</u> of this process <u>NOT described</u> by the preceding paragraph is

❹ According to the preceding paragraph, the purpose of the <u>nitric acid</u> and <u>gum arabic</u> is to

Now skim the selection. This quick reading should give you an idea of the structure of the selection and its overall meaning.

Next, read the selection carefully and underscore words that seem important or that you think hold clues to the question's answers. Your underscored selection should look something like this:

The best kind of <u>limestone for printing is Bavarian</u>. Light-colored and perfectly smooth, it is <u>porous and absorbs both water and greasy substances</u> equally well. The stone used is about 6 inches thick and is fairly big, up to 90 × 65 cm (35 × 25 inches), and can weigh up to 150 or 175 pounds. Grinding the stone smooth <u>creates a clean printing surface</u>. This allows a drawing to be made on the surface with a greasy lithographic pencil or crayon. The drawing is then fixed by rinsing the stone with a very weak solution of nitric acid and gum arabic. The stone is wiped with water before each impression is taken and, for each print, it is inked with a leather-covered roller. During this process, the porous limestone retains the grease of the crayon where the drawing has been made, and the parts that are not drawn upon become impregnated with water. The ink, which is greasy, is repelled by the water-wet areas and adheres only to the areas marked by the crayon.

## Answers and Explanations

1. **The correct answer is (B).** The best title for any selection is the one that takes in all the ideas presented without being too broad or too narrow. Choice (B) provides the most inclusive title for this passage. A look at the other choices shows you why. The first sentence mentions where to find good limestone, but the rest of the paragraph clearly indicates that this is not the main idea. Therefore, choice (A) can be eliminated. Watercolor drawing was not even mentioned, so choice (C) can be eliminated. Although the paragraph emphasizes the importance of the stone, it does not specifically discuss how to create a printing stone. The passage is clearly devoted to the process of creating a print, not to creating a printing stone. Therefore, choice (D) also can be eliminated.

2. **The correct answer is (B).** The capability of Bavarian limestone to absorb water and greasy substances indicates that its sponge-like surface is important for the printing process. Therefore, choice (B) is the best answer. Although the paragraph indicates that the limestone often is cut into very large blocks, nothing in the passage indicates that this is important to the printing process. Therefore, choice (A) can be eliminated. The passage states that the stone is rinsed with a solution of nitric acid and gum arabic. Nothing indicates that the nitric acid and gum arabic come from the stone itself, so choice (C) can be eliminated. The beginning of the passage clearly states that Bavarian limestone absorbs greasy substances. Because the meaning of "repel" is the opposite of "absorb," choice (D) can be eliminated.

3. **The correct answer is (A).** The ideas you underlined should help you with this question of fact. The finished product is never mentioned in the passage. How to ink the stone, how to create a drawing surface, and what kind of limestone to use are all mentioned, so the other three options can be eliminated.

4. **The correct answer is (C).** The answer to this question is provided by the sentence "The drawing is then fixed by rinsing the stone with a very weak solution of nitric acid and gum arabic." In this sentence, the word "fixed" means "to make permanent." The best way to figure out the meaning of a word used in an unfamiliar way is to decide what meaning would make the most sense based on the rest of the passage. The sentence makes the most sense when we replace the word "fixed" with choice (C), or "made permanent."

> **Directions:** The passage below is followed by questions based upon its content. After reading the passage, choose the best answer to each question. Answer all of the questions on the basis of what is *stated* or *implied* in the passage.

Weatherproofing wooden deck surfaces with the XYZ Process results in a surface that is far superior to those treated with other, less effective methods. XYZ Process–treated surfaces are durable to heavy walking traffic, are resistant to wear from heavy rains and runoff from overhanging roofs, are resistant to bleaching from excessive exposure to the sun, and maintain a "fresh wood" look for several years. In addition to producing a wooden surface that is attractive and durable, the XYZ Process also has several safety and environmental benefits. Surfaces treated with the XYZ Process are noncombustible, odorless, safe for children and pets, and free from splinters. Any wooden surface can be treated with the XYZ Process and can be treated by anyone with access to a garden hose, a scrub brush, and a paintbrush. The application process is safe for surrounding vegetation, animal life, and anyone without major respiratory problems.

5. It is most accurate to state that the author in the preceding selection presents

    **(A)** facts but reaches no conclusion concerning the value of the process.

    **(B)** a conclusion concerning the value of the process unsupported by facts.

    **(C)** neither facts nor conclusions but merely describes the process.

    **(D)** a conclusion concerning the value of the process and facts to support that conclusion.

6. If the XYZ Process were used for a surface other than a deck, it would be most useful for a

    **(A)** cement sidewalk.

    **(B)** plastic mailbox.

    **(C)** wooden door.

    **(D)** metal handrail.

7. The aspect of the XYZ Process not discussed in the preceding paragraph is

    **(A)** resistance to wear from snow and ice.

    **(B)** safety concerns for animals.

    **(C)** effects on surrounding plants.

    **(D)** how it compares to other methods.

8. The main reason for treating wooden surfaces with the XYZ Process is to protect

   **(A)** children from harmful odors.

   **(B)** surrounding vegetation.

   **(C)** wooden surfaces from weathering.

   **(D)** animals from splinters.

9. Of the following, the one that might be negatively affected by the XYZ Process is

   **(A)** plants surrounding the treated area.

   **(B)** someone with respiratory problems.

   **(C)** children walking on a treated surface.

   **(D)** animals walking on a treated surface.

Skim the questions and underscore the words you consider to be important. The questions should look something like this:

❶ It is most accurate to state that the author in the preceding selection <u>presents</u>

❷ If the <u>XYZ Process</u> were used for a <u>surface other than a deck</u>, it would be most useful for a

❸ The <u>aspect</u> of the XYZ Process not <u>discussed</u> in the preceding paragraph is

❹ The <u>main reason for treating</u> wooden surfaces with the XYZ Process is to protect

❺ Of the following, the one that might be <u>negatively affected</u> by the XYZ Process is

Skim the reading selection. Get an idea of the subject matter of the selection and how it is organized.

Now read the selection carefully and underscore the words that you think are especially important. This fact-filled selection might be underlined like this:

<u>Weatherproofing</u> wooden deck surfaces <u>with the XYZ</u> Process results in a surface that is <u>far superior</u> to those treated with other, <u>less effective methods</u>. XYZ Process–treated surfaces are <u>durable to heavy walking traffic</u>, are <u>resistant to wear</u> from heavy rains and runoff from overhanging roofs, are <u>resistant to bleaching</u> from excessive exposure to the sun, and <u>maintain a "fresh wood" look</u> for several years. In addition to producing a wooden surface that is <u>attractive and durable</u>, the XYZ Process also has several <u>safety and environmental benefits</u>. Surfaces treated with the XYZ Process are <u>noncombustible</u>, <u>odorless</u>, <u>safe</u> for <u>children and pets</u>, and free from <u>splinters</u>. <u>Any wooden surface</u> can be treated with the XYZ Process and can be treated by anyone with access to a garden hose, a scrub brush, and a paintbrush. The application process is <u>safe</u> for <u>surrounding vegetation</u>, <u>animal life</u>, and anyone without <u>major respiratory problems</u>.

## Answers and Explanations

5.  **The correct answer is (D).** This is a combination main idea and interpretation question. If you cannot easily answer this question, reread the selection. The author clearly states his or her opinion about the XYZ Process in the first sentence. The rest of the selection provides an abundance of facts to support his conclusion.

6.  **The correct answer is (C).** Toward the end of the selection, the author states, "Any wooden surface can be treated with the XYZ Process." Choice (C) is the only one that mentions a wooden surface.

7.  **The correct answer is (A).** The ideas you underlined should help you with this factual statement. Resistance to wear from snow and ice is the only aspect not mentioned. Safety concerns for animals, effects on surrounding plants, and how it compares to other methods are all mentioned.

8.  **The correct answer is (C).** Protecting wooden surfaces from weathering is the main reason for treating wooden surfaces with the XYZ Process. The fact that the XYZ Process is safe for children, plants, and animals is just another benefit of using this product.

9.  **The correct answer is (B).** The only negative effect mentioned is that the XYZ Process might not be safe for people with major respiratory problems.

**Directions:** The passage below is followed by questions based upon its content. After reading the passage, choose the best answer to each question. Answer all of the questions on the basis of what is *stated* or *implied* in the passage.

Most people first encounter grammar in connection with the study of their own language in school. This kind of grammar is called prescriptive grammar because it defines the role of the various parts of speech (such as nouns, verbs, adjectives, and so on). The purpose of prescriptive grammar is to define the norms, or rules, of correct usage. This kind of grammar states how words and sentences are to be put together in a language so the speaker will be perceived as having good grammar. When people are said to have good grammar, the inference is that they obey the rules of accepted usage associated with the language they speak.

**10.** According to the preceding paragraph, what is the purpose of grammar?

   **(A)** It is actually not important in our language.

   **(B)** It is the "glue" that holds sentences together.

   **(C)** It indicates how to use language properly.

   **(D)** It obeys the rules of proper language.

**11.** In the preceding paragraph, the author is primarily concerned with

   **(A)** the different parts of speech.

   **(B)** the purpose of prescriptive grammar.

   **(C)** who has good grammar.

   **(D)** the study of grammar in school.

**12.** According to the preceding paragraph, someone who is considered to use the rules of prescriptive grammar properly

   **(A)** has studied his or her own language in school.

   **(B)** defines the various parts of speech.

   **(C)** did well in language class.

   **(D)** obeys the rules of accepted usage.

## Answers and Explanations

**10.**  **The correct answer is (C).** The paragraph states that the purpose of prescriptive grammar is to define the rules of correct usage. This is just another way of saying that grammar indicates how to use language properly.

**11.**  **The correct answer is (B).** The purpose of this selection should be clear as you read through the paragraph. If you were unsure of the answer, you can easily work backward by eliminating the wrong answers.

**12.**  **The correct answer is (D).** The answer to this question can be found in the last sentence. The wording of the question is slightly different from the last sentence, but the meaning is the same. If you had difficulty with the meaning of this question, you should have no difficulty answering the question by eliminating the wrong answers.

**Directions:** The passage below is followed by questions based upon its content. After reading the passage, choose the best answer to each question. Answer all of the questions on the basis of what is *stated* or *implied* in the passage.

Information theory deals with the numerical measurement of information, the representation of information (such as encoding), and the capacity of communication systems to transmit, receive, and process information. We use, and may even be a part of, many communication systems on a daily basis. A typical communication system consists of several components. First, there must be a source that produces the information, or message, to be transmitted. A source with which we are familiar is human speech. Second, a transmitter, such as a telephone and an amplifier or a television camera and a transmitter, converts the message into electronic or electromagnetic signals. Third, these signals are transmitted through a channel, or medium, such as a wire or the atmosphere. The channel is the component of the system that is most susceptible to interference. Distortion and degradation of the signal can originate from many sources. (Some examples of interference, known as noise, include the static in radio and telephone reception, the fading of a cellular phone signal, and the "snow" experienced in poor TV reception.) The fourth component of a communication system is the receiver, such as a television or a radio. A TV or radio can act as a receiver by reconstructing the signal into the original message. The final component is the destination, such as a person watching or listening to the message that the television or radio receives.

13. Which of the following is the best title for the preceding selection?

    **(A)** "The Basics of Information Theory"

    **(B)** "Humans and Data Transmission"

    **(C)** "Communication System Components"

    **(D)** "Transmission and Signal Interference"

14. According to the preceding selection, the signal is converted by the

    **(A)** source.

    **(B)** transmitter.

    **(C)** receiver.

    **(D)** channel.

**15.** According to the preceding selection, "degradation of the signal" means the signal

(A) originates from many sources.

(B) is successfully transmitted.

(C) is graded as a clear signal.

(D) is experiencing interference.

**16.** According to the preceding selection, a human being can act as a

(A) source only.

(B) transmitter and a receiver.

(C) source and a destination.

(D) destination only.

**17.** According to the preceding selection, what is likely to happen to a telephone signal transmitted through a faulty wire?

(A) It will result in static.

(B) It will sound the same.

(C) It will be amplified.

(D) It will result in "snow."

**18.** According to the preceding selection, an example of the representation of information is

(A) distortion.

(B) encoding.

(C) "snow."

(D) transmitting.

## Answers and Explanations

**13.** **The correct answer is (C).** Although this selection starts out with a comment about information theory, this is not the main focus of the selection. Therefore, you can eliminate choice (A). The topics of transmission and interference are discussed, but they are not the focus of the selection. In the same way, the relationship between humans and this topic is discussed but is not the focus. Therefore, choices (B) and (D) can be eliminated.

14. **The correct answer is (B).** The selection clearly states that the second component of a communication system, the transmitter, converts the message into a signal.

15. **The correct answer is (D).** If you were unsure of the meaning of the word "degradation," you could try to guess its meaning by looking at the context in which the word appears. The sentence containing the word "degradation" appears between two sentences that both discuss interference. In addition, because the word "degradation" is used to describe interference, you can infer that it is just another word for it.

16. **The correct answer is (C).** Human beings are mentioned as possible sources and destinations of a communication system.

17. **The correct answer is (A).** This question requires you to go beyond the available information and guess which answer is most likely to be true given the information provided. A damaged wire is likely to have some negative effects, so choice (B) can be eliminated. Wire damage also is not likely to be beneficial, so choice (C) can be eliminated. The experience of "snow" is specifically described as an effect of interference on TV reception, so choice (D) can be eliminated.

18. **The correct answer is (B).** The first sentence of the selection clearly indicates that encoding is an example of the representation of information.

On firefighter exams, many reading passages will relate to standard designations or to specified rules of procedure. When reading these passages, you must pay special attention to details relating to exceptions, special preconditions, combinations of activities, choices of actions, and prescribed time sequences. Sometimes the printed procedure specifies that certain actions are to be taken only when there is a combination of factors (such as water pressure has dropped AND staircase collapse is imminent). At other times, the procedures give choices of action under certain circumstances. You must read carefully to determine whether the passage requires a combination of factors or gives a choice and then make the appropriate judgment. When a time sequence is specified, be certain to follow that sequence in the prescribed order.

The remaining reading selections in this chapter are of a style often found on firefighter exams. Beyond requiring routine reading comprehension skills, they require the special attention to sequence, combinations, and choices we just discussed.

**Directions:** The passage below is followed by questions based upon its content. After reading the passage, choose the best answer to each question. Answer all of the questions on the basis of what is *stated* or *implied* in the passage.

### PREPARING FOODS FOR STORAGE: MEAT, POULTRY, AND FISH

Always keep your refrigerator at about 40°F and your freezer at 0°F or colder. You should check the temperature periodically with an appliance thermometer. This is very important because, if you do not store these items properly, your customers could become ill from contaminated food. Refrigerate meat and poultry in the original packaging. For long-term freezer storage, it is best to remove the original packaging and wrap it tightly in moisture- and vapor-proof material. You can use heavy foil, freezer bags, freezer plastic wrap, or freezer containers. Tightly wrap fresh fish in moisture- and vapor-proof material before refrigerating or freezing.

19. Assume that you, as the head chef of a major restaurant chain, are responsible for the proper storage of large amounts of meat, poultry, and fish. One of your duties as head chef is to teach your cooks how to store these items. Today, one of your cooks asks you what she should do with a new shipment of fresh fish. You should tell her to

    (A) refrigerate the fish in its original packaging.

    (B) store it in the freezer at about 40°F.

    (C) freeze the fish in its original packaging.

    (D) wrap it in heavy foil and put it in the freezer.

20. What the storage instructions mean when they talk about using vapor-proof and moisture-proof materials is using materials that

    (A) do NOT allow liquids or odors to leak out.

    (B) allow liquids but NOT odors to leak out.

    (C) allow odors but NOT liquids to leak out.

    (D) allow odors and liquids to leak out.

21. According to the instructions for long-term storage, removing meat or poultry from the original package and wrapping it tightly in freezer plastic wrap is

    **(A)** inadvisable because you should freeze meat or poultry in the original packaging.

    **(B)** advisable because freezer plastic wrap is moisture- and vapor-proof.

    **(C)** inadvisable because you should not store meat or poultry in the freezer.

    **(D)** advisable because you should not refrigerate meat or poultry in the original packaging.

22. According to the storage instructions, the one that would NOT be considered proper food storage is

    **(A)** storing food in a freezer with a temperature that is below 0°F.

    **(B)** wrapping meat or poultry in heavy foil before freezing.

    **(C)** removing fish from its packaging and storing it in a freezer container.

    **(D)** freezing meat or poultry in the original packaging.

## Answers and Explanations

19. **The correct answer is (D).** The last sentence says to wrap fresh fish in moisture- and vapor-proof material. The previous sentence mentions that heavy foil is a vapor-proof material.

20. **The correct answer is (A).** This is somewhat of a vocabulary question. Storage materials that do NOT let odors leak out are called vapor-proof. Storage materials that do NOT let liquids leak out are called moisture-proof.

21. **The correct answer is (B).** The instructions clearly state that meat and poultry should be removed from their original packaging and wrapped in moisture- and vapor-proof material before long-term storage. Freezer plastic wrap is moisture- and vapor-proof and is therefore suitable for long-term storage.

22. **The correct answer is (D).** The instructions clearly state that meat and poultry should be removed from their original packaging and wrapped in moisture- and vapor-proof material before freezing. Freezing meat or poultry in the original packaging is not considered proper food storage.

**Directions:** The passage below is followed by questions based upon its content. After reading the passage, choose the best answer to each question. Answer all of the questions on the basis of what is *stated* or *implied* in the passage.

Citizens interested in filing a property damage claim that totals less than $5,000 are required to fill out Minor Claim Form LG-203 in duplicate and return it to the Secretary of Minor Claims in the District Court Building. If the citizen seeks compensation for emotional damage in addition to the property damage, he or she also is required to fill out Intermediate Claim Form K-46. This form is to be filled out in duplicate and sent to the Secretary of Minor Claims. Minor Claim Form LG-203 will then be compared to Intermediate Claim Form K-46. The Secretary of Minor Claims will then forward both copies of Form K-46 to the Division Office, which will send one copy to the State Office of Minor Claims. When the information on form LG-203 indicates that the claim is of a minor nature, the case is thoroughly reviewed by the Secretary of Minor Claims. If this review indicates that the complaining party is entitled to an emotional-damage hearing, Witness Form R-11 must be completed by a witness to the incident. Once this form is returned to the Secretary of Minor Claims, the defending party is sent a copy of the Summons Form S-4 and is asked to return it to the State Office of Minor Claims. When the State Office of Minor Claims receives the Summons Form, they will compare it to Form K-46 and decide whether the case may be heard in the District Court.

23. According to the preceding paragraph, in order for the State Office of Minor Claims to decide whether a case may be heard in the District Court, they must have received which documents?

    **(A)** Form K-46

    **(B)** Forms K-46 and LG-203

    **(C)** Forms K-46 and S-4

    **(D)** Forms K-46, LG-203, R-11, and S-4

24. According to the preceding paragraph, the requirement of Witness Form R-11 depends on

    **(A)** a review by the Secretary of Minor Claims.

    **(B)** whether the citizen does NOT claim emotional damage.

    **(C)** whether the citizen wants to have a witness in court.

    **(D)** a review by the State Office of Minor Claims.

**25.** Of the forms mentioned in the preceding paragraph, the defending party is responsible for preparing the

   **(A)** Intermediate Claim Form.

   **(B)** Summons Form.

   **(C)** Witness Form.

   **(D)** Minor Claim Form.

**26.** According to the preceding paragraph, the Division Office

   **(A)** keeps one copy of Form LG-203.

   **(B)** sends out both copies of Form K-46.

   **(C)** keeps one copy of Form K-46.

   **(D)** sends out both copies of Form LG-203.

## Answers and Explanations

**23.** **The correct answer is (C).** The State Office of Minor Claims must receive Form K-46 from the Division Office and Form S-4 from the defending party.

**24.** **The correct answer is (A).** If a review by the Secretary of Minor Claims indicates that the complaining party is entitled to an emotional-damage hearing, Witness Form R-11 must be completed.

**25.** **The correct answer is (B).** The defending party is only responsible for the Summons Form.

**26.** **The correct answer is (C).** The Division Office receives both copies of Form K-46 and sends one copy to the State Office of Minor Claims. This implies that the Division Office keeps a copy of Form K-46.

# EXERCISE: READING COMPREHENSION

> **Directions:** Each passage is followed by questions based upon its content. After reading the passage, choose the best answer to each question. Answer all of the questions on the basis of what is *stated* or *implied* in the passage.

The unadjusted loss per $1,000 valuation has only a very limited usefulness in evaluating the efficiency of a fire department, for it depends on the assumption that other factors will remain constant from time to time and city to city. It might be expected that high fire department operation expenditures would tend to be associated with a low fire loss. A statistical study of the loss and cost data in more than 100 cities failed to reveal any such correlation. The lack of relationship, although to some extent due to failure to make the most efficacious expenditure of fire protection funds, must be attributed at least in part to the obscuring effect of variations in the natural, physical, and moral factors that affect fire risk.

1. One reason for the failure to obtain the expected relationship between fire department expenditures and fire loss data is the

   **(A)** changing dollar valuation of property.

   **(B)** unsettling effects of rapid technological innovations.

   **(C)** inefficiency of some fire department activities.

   **(D)** statistical errors made by investigators.

2. We can conclude that the "unadjusted loss per $1,000" figure is useful in comparing the fire departments of two cities

   **(A)** only if the cities are of comparable size.

   **(B)** only if adjustments are made for other factors that affect fire loss.

   **(C)** under no circumstances.

   **(D)** only if properly controlled experimental conditions can be obtained.

3. Of the following factors that affect fire risk, the one that is most adequately reflected in the "unadjusted loss per $1,000 valuation" index is

   **(A)** fire department operation expenditures.

   **(B)** physical characteristics of the city.

   **(C)** type of structures most prevalent in the city.

   **(D)** total worth of property in the city.

4. According to the paragraph, cities that spend larger sums on their fire departments

   **(A)** tend to have lower fire losses than cities that spend smaller sums on their fire departments.

   **(B)** do not tend to have lower fire losses than cities that spend smaller sums on their fire departments.

   **(C)** tend to have higher fire losses than cities that spend smaller sums on their fire departments.

   **(D)** do not tend to have the same total property valuation as cities that spend smaller sums on their fire departments.

### QUESTIONS 5–7 ARE BASED ON THE FOLLOWING PASSAGE:

Shafts extending into the top story, except those stair shafts in which the stairs do not continue to the roof, shall be carried through and at least 2 feet above the roof. Every shaft extending above the roof, except open shafts and elevator shafts, shall be enclosed at the top with a roof of materials having a fire-resistant rating of 1 hour and a metal skylight covering at least three quarters of the area of the shaft in the top story. Skylights over stair shafts shall have an area not less than one tenth the area of the shaft in the top story but shall have an area not less than 15 square feet in area. Any shaft terminating below the top story of a structure and those stair shafts not required to extend through the roof shall have the top enclosed with materials having the same fire-resistant rating as required for the shaft enclosure.

5. The paragraph states that the elevator shafts that extend into the top story are

   **(A)** not required to have a skylight but are required to extend at least 2 feet above the roof.

   **(B)** neither required to have a skylight nor to extend above the roof.

   **(C)** required to have a skylight covering at least three quarters of the area of the shaft in the top story and to extend at least 2 feet above the roof.

   **(D)** required to have a skylight covering at least three quarters of the area of the shaft in the top story but are not required to extend above the roof.

6. Of the following skylights, the one that meets the requirements of the paragraph is a skylight measuring

   **(A)** 4 feet by 4 feet over a stair shaft that, on the top story, measures 20 feet by 9 feet.

   **(B)** $4\frac{1}{2}$ feet by $3\frac{1}{2}$ feet over a pipe shaft that, on the top story, measures 5 feet by 4 feet.

   **(C)** $2\frac{1}{2}$ feet by $1\frac{1}{2}$ feet over a dumbwaiter shaft that, on the top story, measures $2\frac{1}{2}$ feet by $2\frac{1}{2}$ feet.

   **(D)** 4 feet by 3 feet over a stair shaft that, on the top story, measures 15 feet by 6 feet.

7. Suppose a shaft that does not go to the roof is required to have a 3-hour fire-resistant rating. In regard to the material enclosing the top of this shaft, the paragraph

   **(A)** states that a 1-hour fire-resistant rating is required.

   **(B)** states that a 3-hour fire-resistant rating is required.

   **(C)** implies that no fire-resistant rating is required.

   **(D)** neither states nor implies anything about the fire-resistant rating.

### QUESTIONS 8–14 ARE BASED ON THE FOLLOWING PASSAGE:

Fire regulations require that every liquefied petroleum gas installation should be provided with the means for shutting off the supply to a building in case of an emergency. The installation of a shutoff valve immediately inside a building, which sometimes is done for the convenience of the user, does not comply with this regulation. An outside shutoff valve just outside the building seems to be the logical solution. However, the possibility of tampering illustrates the danger of such an arrangement. A shutoff valve so located might be placed in a locked box. This has no advantage, however, over a valve provided within the locked cabinet containing the cylinder or an enclosure provided over the top of the cylinder. Keys can be carried by firefighters or, in an emergency, the lock can be broken. Where no valve is visible, the firefighters should not hesitate to break the lock to the cylinder enclosure. The means for shutting off the gas varies considerably in the numerous types of equipment in use. When the cover to the enclosure has been opened, the gas can be shut off as follows:

Close the tank or cylinder valves to which the supply line is connected. Such valves always turn to the right. If the valve is not provided with a handwheel, an adjustable wrench can be used. If conditions are such that shutting off the supply at once is imperative and this cannot be accomplished as previously mentioned, the tubing that is commonly employed as the supply line can be flattened to the extent of closure by a hammer. If the emergency requires the removal of the cylinder, the supply line should be disconnected and the cylinder removed to a safe location. A tank buried in the ground is safe against fire. When conditions indicate the need to remove a cylinder or tank and this cannot be done due to the severity of exposure, pressures within the container can be kept within control of the safety valve by means of a hose stream played on the surface of the container. The melting of the fuse plug also can be prevented in this way.

8. According to the preceding paragraphs, in an emergency, a firefighter should break the lock of a cylinder enclosure whenever the shutoff valve

    **(A)** fails to operate.

    **(B)** has no handwheel.

    **(C)** has been tampered with.

    **(D)** cannot be seen.

9. According to the preceding paragraphs, shutoff valves for liquefied petroleum gas installations

    **(A)** always turn to the right.

    **(B)** always turn to the left.

    **(C)** sometimes turn to the right and sometimes turn to the left.

    **(D)** generally are pulled up.

10. According to the preceding paragraphs, if a cylinder needs to be moved but cannot be because of the severity of exposure, the pressure can be kept under control by

    **(A)** opening the shutoff valve.

    **(B)** playing a hose stream on the cylinder.

    **(C)** disconnecting the supply line into the cylinder.

    **(D)** removing the fuse plug.

11. The preceding paragraphs state that the supply line should be disconnected when the

    **(A)** fuse plugs melt.

    **(B)** cylinder is removed to another location.

    **(C)** supply line becomes defective.

    **(D)** cylinder is damaged.

12. The preceding paragraphs state that the shutoff valves for liquefied petroleum gas installations are sometimes placed inside buildings

   **(A)** so that firefighters will be able to find the valves more easily.

   **(B)** because it is more convenient for the occupants.

   **(C)** to hide the valves from public view.

   **(D)** because this makes it easier to keep the valves in good working condition.

13. It is suggested in the preceding paragraphs that, during an emergency, the supply line tubing should be flattened to the extent of closure when the

   **(A)** supply line becomes defective.

   **(B)** shutoff valve cannot be opened.

   **(C)** shutoff valve cannot be closed.

   **(D)** supply line is near a fire.

14. According to the preceding paragraphs, fire regulations require that liquefied petroleum gas installations should

   **(A)** be made in safe places.

   **(B)** be tamperproof.

   **(C)** have shutoff valves.

   **(D)** not exceed a certain size.

### QUESTIONS 15–24 ARE BASED ON THE FOLLOWING PASSAGE:

Air-conditioning systems are complex and are made up of several processes. The circulation of air is produced by fans and ducts; the heating is produced by steam, hot water coils, coal, gas, or oil fire furnaces; the cooling is done by ice or mechanical refrigeration; and the cleaning is done by air washers or filters.

Air-conditioning systems in large buildings generally should be divided into several parts with wholly separate ducts for each part or floor. The ducts are then extended through fire partitions. As a safeguard, whenever ducts pass through fire partitions, automatic fire dampers should be installed in the ducts. Furthermore, the ducts should be lined on the inside with fire-resistant materials. In addition, a manually operated fan shutoff should be installed at a location that is readily accessible under fire conditions.

Most air-conditioning systems recirculate a considerable portion of the air. When this is done, an additional safeguard has to be taken to have the fan arrange to shut down automatically in case of fire. A thermostatic device in the return air duct will operate the shut-off device whenever the temperature of the air coming to the fan becomes excessive. The air filters frequently are coated with oil to help catch dust. Such oil should be of a type that does not ignite readily. Whenever a flammable or toxic refrigerant is employed for air cooling, coils containing such a refrigerant should not be inserted in any air passage.

15. According to the preceding paragraphs, fan shutoffs in the air-conditioning system should be installed

    (A) near the air ducts.

    (B) next to fire partitions.

    (C) near the fire dampers.

    (D) where they can be reached quickly.

16. On the basis of the preceding paragraphs, whenever a fire breaks out in a building containing an air-conditioning system that recirculates a portion of the air, the

    (A) fan will shut down automatically.

    (B) air ducts will be opened.

    (C) thermostat will cease to operate.

    (D) fire partitions will open.

17. The preceding paragraphs state that, on every floor of a large building in which air-conditioning systems are used, there should be a(n)

    (A) automatic damper.

    (B) thermostatic device.

    (C) air filter.

    (D) separate duct.

18. From the preceding paragraphs, the conclusion can be drawn that, in an air-conditioning system, flammable refrigerants

    (A) can be used if certain precautions are observed.

    (B) should be used sparingly and only in air passages.

    (C) should not be used under any circumstances.

    (D) might be more effective than other refrigerants.

19. According to the preceding paragraph, the spreading of dust by means of fans in the air-conditioning system is reduced by

    (A) shutting down the fan automatically.

    (B) lining the inside of the air duct.

    (C) cleaning the circulated air with filters.

    (D) coating the air filters with oil.

20. According to the preceding paragraphs, the purpose of a thermostatic device is to

    (A) regulate the temperature of the air-conditioning system.

    (B) shut off the fan when the temperature of the air rises.

    (C) operate the fan when the temperature of the air falls.

    (D) assist in the recirculation of the air.

21. According to the preceding paragraphs, hot water coils in the air-conditioning system are limited to

    (A) cooling.

    (B) heating.

    (C) heating and cooling.

    (D) heating and cleaning.

22. The parts of an air-conditioning system that the preceding paragraphs state should be made of fire-resistant materials are the

    (A) hot water coils.

    (B) automatic fire dampers.

    (C) air duct linings.

    (D) thermostatic devices.

23. According to the preceding paragraphs, automatic fire dampers should be installed

    (A) on oil-fired furnaces.

    (B) on every floor of a large building.

    (C) in ducts passing through fire partitions.

    (D) next to the hot water coils.

24. On the basis of the preceding paragraphs, the most accurate statement is that the coils containing toxic refrigerants should be

    (A) used only when necessary.

    (B) lined with fire-resistant materials.

    (C) coated with nonflammable oil.

    (D) kept out of any air passage.

exercises

**QUESTIONS 25–27 ARE BASED ON THE FOLLOWING PASSAGE:**

It shall be unlawful to place, use, or maintain in a condition intended, arranged, or designed for use any gas-fired cooking appliance, laundry stove, heating stove, range, water heater, or combination of such appliances in any room or space used for living or sleeping in any new or existing multiple dwelling unless such room or space has a window opening to the outer air or such gas appliance is vented to the outer air. All automatically operated gas appliances shall be equipped with a device that shuts off automatically the gas supply to the main burners when the pilot light in such appliances is extinguished. A gas range or the cooking portion of a gas appliance incorporating a room heater shall not be deemed an automatically operated gas appliance. However, burners in gas ovens and broilers that can be turned on and off or ignited by nonmanual means shall be equipped with a device that shall shut off automatically the gas supply to those burners when the operation of such nonmanual means fails.

**25.** According to this paragraph, an automatic shutoff device is NOT required on a gas

   **(A)** hot water heater.

   **(B)** laundry dryer.

   **(C)** space heater.

   **(D)** range.

**26.** According to this paragraph, a gas-fired water heater is permitted

   **(A)** only in kitchens.

   **(B)** only in bathrooms.

   **(C)** only in living rooms.

   **(D)** in any type of room.

**27.** An automatic shutoff device shuts off

   **(A)** the gas range.

   **(B)** the pilot light.

   **(C)** the gas supply.

   **(D)** All of the above

**QUESTIONS 28–30 ARE BASED ON THE FOLLOWING PASSAGE:**

A utility plan is a floor plan that shows the layout of a heating, electrical, plumbing, or other utility system. Utility plans are used primarily by the persons responsible for the utilities, but they are important to the craftsman as well. Most utility installations require the leaving of openings in walls, floors, and roofs for the admission or installation of utility features. The craftsman pouring a concrete foundation wall, for example, must study the utility plans to determine the number, sizes, and locations of the openings he must leave for piping, electrical lines, and the like.

28. Of the following items of information, the one that is least likely to be provided by a utility plan is the

(A) location of the joists and frame members around stairwells.

(B) location of the hot water supply and return piping.

(C) location of light fixtures.

(D) number of openings in the floor for radiators.

29. According to the paragraph, of the following, the persons who most likely will have the greatest need for the information included in a utility plan of a building are those who

(A) maintain and repair the heating system.

(B) clean the premises.

(C) put out the fires.

(D) advertise property for sale.

30. According to the paragraph, a repair crew member should find it most helpful to consult a utility plan when information is needed about the

(A) thickness of all doors in the structure.

(B) number of electrical outlets located throughout the structure.

(C) dimensions of each window in the structure.

(D) length of a roof rafter.

## ANSWERS AND EXPLANATIONS

| | | | | | | | |
|---|---|---|---|---|---|---|---|
| 1. | C | 9. | A | 17. | D | 25. | D |
| 2. | B | 10. | B | 18. | A | 26. | D |
| 3. | D | 11. | B | 19. | D | 27. | C |
| 4. | B | 12. | B | 20. | B | 28. | A |
| 5. | A | 13. | C | 21. | B | 29. | A |
| 6. | B | 14. | C | 22. | C | 30. | B |
| 7. | B | 15. | D | 23. | C | | |
| 8. | D | 16. | A | 24. | D | | |

1.  **The correct answer is (C).** The last sentence states that part of the lack of relationship between the cost of running a fire department and dollar losses in fires is due to inefficiencies in the use of fire department funds.

2.  **The correct answer is (B).** The paragraph makes it clear that many factors affect the "unadjusted loss per $1,000" figure. Comparisons to the fire departments of different cities must allow for these other factors.

3.  **The correct answer is (D).** The "loss per $1,000" figure is based on the total value of property in the city.

4.  **The correct answer is (B).** The paragraph simply states that spending more on the fire department does not guarantee lower total property loss. It does NOT, however, state that the city that spends more has greater losses.

5.  **The correct answer is (A).** The first sentence states that all shafts except stair shafts in which the stairs do not continue to the roof must be carried at least 2 feet above the roof. Elevator shafts must extend those 2 feet. The second sentence makes the exception that elevator shafts need not have a skylight.

6.  **The correct answer is (B).** The 15.75-square-foot skylight over the 20-square-foot pipe shaft covers more than three quarters of the area of the shaft. All other skylights are inadequate.

7.  **The correct answer is (B).** According to the last sentence, a shaft terminating below the top story of a structure must have a top with the same fire-resistant rating as that required for the shaft enclosure (here 3 hours).

8.  **The correct answer is (D).** The third-from-the-last sentence of the first paragraph states that, where no valve is visible, the firefighters should break the lock to the cylinder enclosure.

9.  **The correct answer is (A).** The second sentence of the second paragraph states that these tank valves always turn to the right.

10. **The correct answer is (B).** This information is given in the next-to-last sentence.

11. **The correct answer is (B).** See the fourth sentence of the second paragraph.

12. **The correct answer is (B).** The second sentence states that the shutoff valve placed inside a building for the convenience of the user does not satisfy the emergency shutoff requirements.

13. **The correct answer is (C).** If the shutoff valve cannot be closed, the supply of gas can be shut off by flattening the supply line tubing.

14. **The correct answer is (C).** See the first sentence.

15. **The correct answer is (D).** The last sentence of the second paragraph states that manual fan shutoffs should be installed where they are readily accessible.

16. **The correct answer is (A).** The second sentence of the last paragraph describes a thermostatic device that automatically shuts off recirculation of air in case of fire.

17. **The correct answer is (D).** See the first sentence of the second paragraph.

18. **The correct answer is (A).** The last sentence states that, when flammable refrigerants are used, the coils containing the refrigerant should not be inserted in the air passages. This constitutes a precaution that makes the use of flammable refrigerants acceptable.

19. **The correct answer is (D).** See the third-from-the-last sentence.

20. **The correct answer is (B).** This is detailed in the second sentence of the last paragraph.

21. **The correct answer is (B).** See the second sentence.

22. **The correct answer is (C).** See the next-to-the-last sentence of the second paragraph.

23. **The correct answer is (C).** See the third sentence of the second paragraph.

24. **The correct answer is (D).** Again, this answer is in the last sentence.

25. **The correct answer is (D).** An automatic shutoff device is required on all automatically operated gas appliances; however, a gas range is not considered to be an automatically operated gas appliance.

26. **The correct answer is (D).** A gas-fired water heater is permitted anywhere provided that the space has a window opening to the outer air or the appliance is vented to the outer air.

27. **The correct answer is (C).** The automatic shutoff device shuts off the gas supply to the burners when the pilot light goes out. This prevents the room from filling with gas.

28. **The correct answer is (A).** The utility plan shows the layout of a utility system, not the structure of the building itself.

29. **The correct answer is (A).** The heating system is one of the utility systems, so the people concerned with its maintenance and repair would find it most useful.

30. **The correct answer is (B).** Again, consider the definition of a utility. The electrical system is a utility system. Structural information appears on a utility plan only where it is incidental to the layout of utilities.

# Reasoning and Judgment

## OVERVIEW

- What are reasoning and judgment questions?
- Tips for answering reasoning and judgment questions

## WHAT ARE REASONING AND JUDGMENT QUESTIONS?

Reasoning and judgment questions are somewhat similar to deductive reasoning questions. Deductive reasoning questions start with a rule. Reasoning and judgment questions tend to start with a description of some general practice among firefighters. Then the question asks you what might be the best reason for that common practice. For example, a question might start by telling you that firefighters wear helmets made from hard leather, not metal or plastics. The answer choices might give various possible reasons for this; for example, it is lighter, it is less costly, or it is less likely to get hot or melt. You must pick the answer which gives the best reason for the practice.

## TIPS FOR ANSWERING REASONING AND JUDGMENT QUESTIONS

A fire department is expected to protect life and property and to run efficiently. Being efficient is important in any kind of work, but it is less important than protecting life and property. Hence, if we list the goals of a fire department in the order of their importance, the list would be:

1. Protecting life
2. Protecting property
3. Efficiency

As the list of goals suggests, protecting life is the top priority. If the protection of life is a real issue in the "fact pattern" of the question, then the protection of life is the best reason to justify any practice.

## chapter 5

Sometimes the protection of life is not an issue. In that case, the best reason for any practice is to protect property. However, "property" includes fire department property. Property can be sacrificed when someone's life is at stake. But if there is no real threat to one's life, one must protect property.

If there is no real question of life or death and no real threat to property, then the best reason for doing something is that it is efficient. If several answer choices are based on efficiency, you will be judging which answer choice would be most efficient.

Any answer which suggests that something should be done because it will bring praise or benefit to the firefighter is not likely to be a correct answer to a test question. Financial benefits to other people are not usually good enough reasons either. Likewise, an answer choice which tries to justify something only on the grounds that it will make the fire department "look good" is not likely to be a correct answer.

Remember that reasons must be realistic. To justify a practice on the basis of protecting life, there must be something in the question situation to support the idea that someone's life is at stake. Similarly, a proposed answer based on efficiency should really have the appearance of being possible and efficient.

It is especially difficult to choose between some of the answer choices with these questions about the best reasons for doing things. Reasoning and judgment questions deal with more ambiguous problems than other kinds of questions. But there is an old test-taking strategy that may help you here. You should remember that, in a sense, it is ultimately the Mayor or the Fire Chief who is testing you for the firefighter job. When faced with difficult choices on a question of this type, imagine that the question is being asked personally by the Mayor or Fire Chief. Choose the answer you would give to the Mayor or Chief in face-to-face questioning.

## EXERCISE: REASONING AND JUDGMENT

1. Of the following, the chief advantage of having firefighters under competitive civil service is that

   **(A)** fewer fires tend to occur.

   **(B)** fire prevention becomes a reality instead of a distant goal.

   **(C)** the efficiency level of the personnel tends to be raised.

   **(D)** provision can then be made for training by the municipality.

2. Of the following, it is least likely that fire will be caused by

   **(A)** arson.

   **(B)** poor building construction.

   **(C)** carelessness.

   **(D)** inadequate supply of water.

### QUESTIONS 3–5 ARE TO BE ANSWERED BASED ON THE FOLLOWING TABLE:

|              | Last Year | This Year | Increase |
| ------------ | --------- | --------- | -------- |
| Firefighters | 9,326     | 9,744     | —        |
| Lieutenants  | 1,355     | 1,417     | —        |
| Captains     | —         | 433       | 107      |
| Others       | —         | —         | —        |
| TOTAL        | 11,469    | 12,099    | —        |

3. The number in the *Others* group last year was most nearly

   **(A)** 450

   **(B)** 475

   **(C)** 500

   **(D)** 525

4. The group that had the largest percentage of increase was

   **(A)** *Firefighters*.

   **(B)** *Lieutenants*.

   **(C)** *Captains*.

   **(D)** *Others*.

5. This year, the ratio between *Firefighters* and all other ranks of the uniformed force was most nearly

   **(A)** 5:1

   **(B)** 4:1

   **(C)** 2:1

   **(D)** 1:1

6. The ideal is that, in big cities, a fire alarm box can be seen from any corner. This is a desirable condition mainly because

   **(A)** several alarms can be sounded by one person running from one box to another.

   **(B)** little time is lost in sounding an alarm.

   **(C)** an alarm can be sounded from a different box if the nearest one is out of order.

   **(D)** fire apparatus can quickly reach any box.

7. Suppose hydrants with a flowing capacity of less than 500 gallons per minute were to be painted red, hydrants with a flowing capacity between 500 and 1,000 gallons per minute were to be painted yellow, and hydrants with a flowing capacity of 1,000 gallons or greater per minute were to be painted green. The principal advantage of such a scheme is that

   **(A)** fewer fires would occur.

   **(B)** more water would become available at a fire.

   **(C)** citizens would become more acutely aware of the importance of hydrants.

   **(D)** firefighters would save time.

8. "At 2 o'clock in the morning, Mrs. Smart awakened her husband and told him the house was on fire. Mr. Smart dressed hurriedly and ran 17 blocks—past five fire alarm boxes—to the fire station to tell the firefighters that his house was on fire. When Mr. Smart and the firefighters returned, the house had burned down." This is an illustration of the

   **(A)** need for a plentiful supply of fire alarm boxes.

   **(B)** need for more fire stations.

   **(C)** necessity of preventing fires.

   **(D)** desirability of educating the public.

9. "Installation of a modern fire alarm system will mean smaller fires." Of the following, the best justification for this statement is that

   **(A)** if summoned quickly, firefighters can control a fire before it has a chance to spread.

   **(B)** if the alarm system is modern, firefighters can be given a complete picture of a fire even before they respond.

   **(C)** some fires, such as fires resulting from explosions, assume large proportions in a few seconds.

   **(D)** most industrial establishments depend on more than one method of transmitting fire alarms.

10. "The firefighter must not only know how to make the proper knots, but he must also know how to make them quickly." Of the following, the chief justification for this statement is that

    **(A)** a firefighter uses many kinds of knots.

    **(B)** a slowly tied knot slips more readily than one tied quickly.

    **(C)** haste makes waste.

    **(D)** in most fire operations, speed is important.

11. A survey of dip tank fires in large cleaning establishments shows that the majority of these firms started with an obvious hazard, such as cutting and welding torches. This experience indicates most strongly a need for

    **(A)** a law prohibiting the use of welding torches.

    **(B)** fire safety programs in this industry.

    **(C)** a law requiring welders to be licensed.

    **(D)** eliminating the use of dip tanks.

12. To reduce the number of collisions between fire apparatus and private automobiles, a number of suggestions have been made regarding the training and selection of drivers. Of the following courses of action, the one that is least likely to lead to the reduction of accidents is to

    **(A)** select drivers who have the longest driving experience.

    **(B)** require all members of a company to drive fire apparatus.

    **(C)** select drivers who have experience in driving trucks.

    **(D)** select an alternate driver for every piece of fire apparatus.

13. "Automobile parking and double-parking on city streets is daily becoming a greater menace to effective firefighting." Of the following effects of automobile parking, the least serious is that

    **(A)** access to the fire building might be obstructed.

    **(B)** ladder and rescue work might be delayed.

    **(C)** traffic congestion might make it difficult for fire apparatus to get through.

    **(D)** greater lengths of hose might be needed in fighting a fire.

14. "Sometimes at a fire, a firefighter endures great punishment to reach a desired position that could have been reached more quickly and with less hardship." Of the following, the chief implication of the preceding statement is that

    **(A)** one should take time off during a fire to completely think through problems presented.

    **(B)** courage is an important asset in a firefighter.

    **(C)** training is an important factor in firefighting.

    **(D)** modern firefighting requires alert firefighters.

15. The statement "Municipal inspections should be coordinated" suggests most nearly that

    **(A)** fire hazards are dangerous.

    **(B)** the detection of fire hazards is connected with other problems found in a municipality.

    **(C)** municipal and fire inspection services are of unequal importance.

    **(D)** municipal inspections are closely tied to noninspectional services.

16. "With fireproof schools, it would appear that drills are unnecessary." The main reason for believing this statement to be false is that

    **(A)** panic sometimes occurs.

    **(B)** fire extinguishers are available in every school.

    **(C)** fire alarms are easily sounded.

    **(D)** children are accustomed to drilling.

17. "Flame is of varying heat according to the nature of the substance producing it." This means most nearly that

    **(A)** some fires are larger than others.

    **(B)** the best measure of the heat produced by a particular substance is its temperature.

    **(C)** there can be no fire without flame or flame without a fire.

    **(D)** the degree of heat evolved by the combustion of different materials is not identical.

18. "No firefighter is expected to use a measuring stick to determine the exact amount of hose needed to stretch a line from the entrance of a burning building to the seat of a fire. He should be guided generally by the following rule—one length of hose for every story." This rule assumes most directly some degree of uniformity in

    **(A)** fire hazards.

    **(B)** building construction.

    **(C)** causes of fires.

    **(D)** window areas.

19. "A fire will occur when a flammable material is sufficiently heated in the presence of oxygen." It follows from this statement that a fire will occur whenever there is sufficient

    **(A)** heat, hydrogen, and fuel.

    **(B)** oxygen, hydrogen, and fuel.

    **(C)** heat, air, and fuel.

    **(D)** gasoline, heat, and fuel.

20. Suppose the vast majority of the fires that took place in 1980 in New York City occurred in a particular type of building structure. Before a firefighter could reasonably infer that this particular type of building structure was unduly susceptible to fires in New York City in 1980, he or she would have to know

    **(A)** the precise incidence, in terms of percent, represented by the somewhat loose phrase "vast majority."

    **(B)** the frequency of this type of building structure as compared with other types of building structures.

    **(C)** whether 1980 is validly to be taken as the fiscal year or the calendar year.

    **(D)** whether New York City can legitimately be considered as representative of the country as a whole in regard to the statistical frequency of this type of building structure.

21. "Every fire is a potential conflagration." Of the following, the most valid inference that can be drawn from this statement is that

    **(A)** no matter how insignificant a fire might appear to be, a first effort should be to isolate it.

    **(B)** the method of fighting a potential fire must be adapted to the unique circumstances surrounding that fire.

    **(C)** the apparatus and firefighters sent immediately to a fire should be sufficient to handle practically any conflagration that might be found.

    **(D)** the full potentialities of a fire are usually realized.

22. Suppose that, for a certain period of time studied, the percentage of telephone alarms that were false alarms was less than the percentage of fire box alarms that were false alarms. Of the following, the one that is most accurate, based solely on the preceding statement, is that during the period studied

    **(A)** more alarms were transmitted by telephone than by fire box.

    **(B)** more alarms were transmitted by fire box than by telephone.

    **(C)** relatively fewer false alarms were transmitted by telephone than by fire box.

    **(D)** relatively fewer false alarms were transmitted by fire box than by telephone.

23. "There are more engine companies than hook and ladder companies. However, to conclude that the number of firefighters assigned to engine companies exceeds the number of firefighters assigned to hook and ladder companies is to make a basic assumption." Of the following, the most accurate statement of the basic assumption referred to in this quotation is that

    **(A)** the number of hook and ladder companies does not differ greatly from the number of engine companies.

    **(B)** about the same number of firefighters, on the average, are assigned to each type of company.

    **(C)** an engine company, on the average, has fewer firefighters than a hook and ladder company.

    **(D)** the largest engine company is no larger than the largest hook and ladder company.

24. If the fire company due to arrive first arrived 1 minute before the company due second, and the company due third arrived 4 minutes after the alarm was sent, then the most accurate of the following statements is that the

    **(A)** second company due arrived 3 minutes before the third company due.

    **(B)** third company due arrived 3 minutes after the first company due.

    **(C)** second company due arrived 2 minutes after the alarm was sent.

    **(D)** second company due arrived 1 minute after the first company due.

25. "During June, 25 percent of all fires in a city were in buildings of Type A, 40 percent were in buildings of Type B, and 15 percent were in buildings of Type C." Of the following, the most accurate statement is that the total number of fires during June was

    (A) equal to the sum of the percentages of fires in the three types of buildings, divided by 100.

    (B) less than 100 percent.

    (C) one fourth the number of fires in buildings of Type B.

    (D) four times the number of fires in buildings of Type A.

26. "One or more public ambulances can be called to a street box by sending on the Morse key therein the following signals in the order given: the preliminary signal 777, the number of the street box, and the number to indicate how many ambulances are wanted." In accordance with these instructions, the proper signal for calling four ambulances to box 423 is

    (A) 4-777-423

    (B) 423-777-4

    (C) 7777-423

    (D) 777-423-4

27. The fire alarm box is an important element of the city's fire protection system. Of the following factors, the one that is of least value in helping a citizen send a fire alarm quickly by means of a fire alarm box is that

    (A) the mechanism of a fire alarm box is simple to operate.

    (B) fire alarm boxes are placed at very frequent intervals throughout the city.

    (C) there are several different types of fire alarm boxes in use.

    (D) specific directions for sending an alarm appear on each fire alarm box.

28. The number of fire companies in a city is chiefly determined by all of the following conditions EXCEPT

    (A) the number of hose streams likely to be required to handle such fires as might be expected.

    (B) the manpower and capacity of fire department pumping apparatus and of ladder service.

    (C) the type of fire alarm system in use.

    (D) the accessibility of various parts of the city to fire companies.

**29.** It is desirable for the fire department to have ladders of varying lengths mainly because some

    **(A)** fires occur at greater distances from the ground than others.

    **(B)** firefighters are more agile than others.

    **(C)** firefighters are taller than others.

    **(D)** fires are harder to extinguish than others.

**30.** The lever on fire alarm boxes for use by citizens should be

    **(A)** very easy to manipulate.

    **(B)** just a little difficult to manipulate.

    **(C)** very difficult to manipulate.

    **(D)** constructed without regard to ease of manipulation.

**31.** A large fire occurs that you, as a firefighter, are helping to extinguish. An emergency arises and you believe that a certain action should be taken. Your superior officer directs you to do something else that you consider to be undesirable. You should

    **(A)** take the initiative and follow what you originally thought to be the superior line of action.

    **(B)** think the matter over for a few minutes and weigh the virtues of the two lines of action.

    **(C)** waste no time but refer the problem immediately to another superior officer.

    **(D)** obey orders despite the fact that you disagree.

**32.** Suppose your company is extinguishing a very small fire in a parked automobile. Your commanding officer directs you to perform some act that, as far as you can see, is not going to help in any way to put out the fire. Of the following, the best reason for obeying the order instantly and without question is that

    **(A)** the fire department is a civil organization.

    **(B)** your officer, after all, has been in the service for a much longer period of time than you have been.

    **(C)** without discipline, the efficiency of your company would be greatly reduced.

    **(D)** the first duty of the commanding officer is to command.

33. Assume that you are a firefighter. While you are walking along in a quiet residential neighborhood at about 3 a.m. on a Sunday morning, another pedestrian calls your attention to smoke coming from several windows on the top floor of a three-story apartment house. Of the following, the best action for you to take immediately is to

    (A) race through the house, wake the tenants, and lead them to safety.

    (B) run to the nearest fire alarm box, turn in an alarm, and then run back into the house to arouse the tenants.

    (C) run to the nearest fire alarm box, turn in an alarm, and stay there to direct the fire apparatus when it arrives.

    (D) direct the other pedestrian to the nearest alarm box with directions to stay there after sending the alarm while you go to the apartment from which the smoke is issuing.

34. Assume that you are a firefighter, off duty, and in uniform in the basement of a department store. A large crowd is present. There are two stairways, 50 feet apart. A person you cannot see screams "Fire!" You should first

    (A) rush inside and sound the alarm at a fire alarm box.

    (B) find the person who shouted "Fire!," ascertain where the fire has occurred, and proceed to extinguish the fire.

    (C) jump on top of a nearby counter, order everyone to be quiet and not to move, find out who screamed "Fire!," and reprimand that person publicly.

    (D) jump on top of a counter, obtain the attention of the crowd, direct the crowd to walk to the nearest stairway, and announce that there is no immediate danger.

35. A number of prominent municipal officials are extremely interested in the operation of the local fire department. These officials frequently arrive at the scene of a fire almost as quickly as the firefighters. It is desirable that the fire department welcomes the interest of these municipal officials mainly because of the probability that

    (A) the officials will acquire considerable technical information.

    (B) the organization of the fire department will be modified.

    (C) fire administration will remain unaffected.

    (D) the work and problems of the fire department will receive due recognition.

**36.** The owner of a building at which you helped put out a fire complains bitterly to you that the firefighters broke a number of cellar windows even before setting out to extinguish the fire in the cellar. Of the following, the best action for you to take is to

**(A)** question the validity of the data as described by the owner.

**(B)** request that the owner put the statement in writing.

**(C)** explain the reason for breaking the windows.

**(D)** suggest that the owner have the cellar windows replaced with unbreakable glass.

**37.** An elderly man approaches you, as an officer on duty, to complain that the noise of apparatus responding to alarms wakes him in the middle of the night. Of the following, the best way for you to handle this situation is to

**(A)** quickly change the subject because the man obviously is a crank.

**(B)** ask the man for specific instances of apparatus making noise at night.

**(C)** explain the need for speed in the response of an apparatus.

**(D)** promise to avoid making unnecessary noise at night.

**38.** A firefighter assigned to deliver a talk before a civic organization first made a detailed outline of her talk. Then, with the help of two members of her company, she prepared several demonstrations and charts. The firefighter's procedure was

**(A)** good, chiefly because preparation increases the effectiveness of a talk.

**(B)** poor, chiefly because a talk should be guided by the questions and comments of the audience and should not follow a predetermined outline.

**(C)** good, chiefly because the audience will be impressed by the care the firefighter has taken in preparing the talk.

**(D)** poor, chiefly because the talk will appear rehearsed rather than spontaneous and natural.

**39.** Suppose you were assigned to be in charge of a new headquarters bureau that will have extensive correspondence with the public and very frequent mail contact with the other city departments. It is decided at the beginning that all communications from the headquarters officer are to go out over your signature. Of the following, the most likely result of this procedure is that

**(A)** the administrative head of the bureau will spend too much time preparing correspondence.

**(B)** execution of bureau policy will be unduly delayed.

**(C)** subordinate officers will tend to avoid responsibility for decisions based on bureau policy.

**(D)** uniformity of bureau policy as expressed in such communications will tend to be established.

40. Suppose you have been asked to answer a letter from a local board of trade requesting certain information. You find that you cannot grant this request. The best way to begin your answering letter is by

    **(A)** quoting the laws or regulations that forbid the release of this information.

    **(B)** stating that you are sorry the request cannot be granted.

    **(C)** explaining in detail the reasons for your decision.

    **(D)** commending the organization for its service to the community.

41. The relationship between the fire department and the press is like a two-way street. The press not only is a medium through which the fire department releases information to the public, the press also

    **(A)** is interested in the promotion of the department's program.

    **(B)** can teach the department good public relations.

    **(C)** makes the department aware of public opinion.

    **(D)** provides the basis for community cooperation with the department.

42. Assume that you have been asked to prepare an answer to a request from a citizen of the city addressed to the chief of the department. If this request cannot be granted, it is most desirable that the answering letter begin by

    **(A)** telling the person you were glad to receive the request.

    **(B)** listing the laws that make granting the request impossible.

    **(C)** saying that the request cannot be granted.

    **(D)** discussing the problem presented and showing why the fire department cannot be expected to grant the request.

43. Assume that you have been requested by the chief of the department to prepare for public distribution a statement dealing with a controversial matter. For you to present the department's point of view in a terse statement, making no reference to any other matter, is in general

    **(A)** undesirable; you should show all the statistical data you used, how you obtained the data, and how you arrived at the conclusions presented.

    **(B)** desirable; people will not read long statements.

    **(C)** undesirable; the statement should be developed from ideas and facts familiar to most readers.

    **(D)** desirable; the department's viewpoint should be made known in all controversial matters.

**44.** The most important function of good public relations in a fire department should be in

   **(A)** training personnel.

   **(B)** developing an understanding of fire dangers on the part of the public.

   **(C)** enacting new fire laws.

   **(D)** recruiting technically trained firefighters.

**45.** The primary purpose of the public relations program of an administrative organization should be to develop mutual understanding between the

   **(A)** public and those who benefit from organized service.

   **(B)** public and the organization.

   **(C)** organization and its affiliate organizations.

   **(D)** the personnel of the organization and the management.

**46.** You arrive in your office at 11 a.m., having been on an inspection tour since 8 a.m. A man has been waiting in your office for 2 hours. He is abusive because of his long wait, and he accuses you of sleeping off a hangover at the taxpayers' expense. You should

   **(A)** say that you have been working all morning and let him sit in the outer office a little longer until he cools off.

   **(B)** tell him you are too busy to see him and make an appointment for later in the day.

   **(C)** ignore his comments, courteously find out what his business is, and take care of him in a perfunctory manner.

   **(D)** explain briefly that your duties sometimes take you out of your office and that an appointment would have prevented the inconvenience.

**47.** During the routine inspection of a building, a citizen tells you very unfavorable personal comments concerning several top officials of the department. Of the following, it usually would be best for you to

   **(A)** try to change the subject as soon as possible.

   **(B)** attempt to convince the citizen that he is in error.

   **(C)** advise the citizen that your opinion might be like his but that you can't discuss it.

   **(D)** tell the citizen that it would be more proper for him to put his comments into writing.

48. "The fire officer should, as far as operations permit, answer any reasonable questions asked by the occupant of the premises involved in a fire." Of the following, however, the subject about which he should be most guarded in his comments is the

    (A) time of the arrival of companies.

    (B) methods used to control and extinguish the fire.

    (C) specific cause of the fire.

    (D) probable extent of the damage.

49. The public is most likely to judge personnel largely on the basis of their

    (A) experience.

    (B) training and education.

    (C) civic-mindedness.

    (D) manner and appearance while on duty.

50. When employees consistently engage in poor public relations practices, of the following, the cause is most often

    (A) disobedience of orders.

    (B) lack of emotional control.

    (C) bullheadedness.

    (D) poor supervision.

51. A citizen's support of a fire department program can best be enlisted by

    (A) minimizing the cost to the community.

    (B) telling him how backward the community is in its practices and why such a situation exists.

    (C) telling him that it is his civic duty to do all that he can to support the program.

    (D) informing him how the program will benefit him and his family.

52. The term "public relations" has been defined as the aggregate of every effort made to create and maintain goodwill and to prevent the growth of ill will. This concept assumes particular importance with regard to public agencies because

    (A) public relations become satisfactory in inverse ratio to the number of personnel employed in the agency.

    (B) legislators might react unfavorably to the public agency no matter how its public relations policy is developed.

    (C) they are much more dependent on public goodwill than commercial organizations are.

    (D) they are tax-supported and depend on the active and intelligent support of an informed public.

**53.** Of the following, the proper attitude for an administrative officer to adopt toward complaints from the public is that he or she should

(A) not only accept complaints but establish a regular procedure whereby they can be handled.

(B) avoid encouraging correspondence with the public on the subject of complaints.

(C) remember that it is his or her duty to get a job done, not to act as a public relations officer.

(D) recognize that complaints are rarely the basis for significant administrative action.

**54.** Of the following types of individuals who pull false alarms, reducing the number of false alarms by education alone is most effective for

(A) mental patients.

(B) drunks.

(C) children of school age.

(D) former firefighters.

**55.** Excluding direct enforcement measures, the most promising approach to relieve the municipal false alarm problem seems to be concerned with

(A) better location and distribution of alarm boxes.

(B) replacement of telegraph systems with telephone systems.

(C) improved electrical circuitry for verification.

(D) public relations and education.

**56.** An aspect of public relations that has been severely neglected by government agencies, but that has been stressed in business and industry, is

(A) use of radio.

(B) public speaking.

(C) advertising.

(D) training employees in personal contacts with the public.

**57.** In dealing with the public, it is helpful to know that most people generally are more willing to do something when they

(A) are not responsible.

(B) understand the reasons.

(C) will be given a little assistance.

(D) must learn a new skill.

58. A study shows that false alarms occur mostly between noon and 1 p.m. and between 3 and 10 p.m. The most likely explanation of these results is that many false alarms are sent by

    **(A)** schoolchildren.

    **(B)** drunks.

    **(C)** mental patients.

    **(D)** arsonists.

59. While visiting the lounge of a hotel, a firefighter discovers a fire that apparently has been burning for some time and is rapidly spreading. Of the following, the first action for her to take is to

    **(A)** find the nearest fire extinguisher and attempt to put out the fire.

    **(B)** notify the desk clerk of the fire and send an alarm from the nearest street alarm box.

    **(C)** send an alarm from the nearest street alarm box only.

    **(D)** run throughout the hotel and warn all occupants to evacuate the building.

60. When a fire occurs in the city of Merritt, the alarm transmitted over the fire alarm telegraph system sounds in every firehouse in Merritt. Of the following, the best justification for this practice is that

    **(A)** some companies are more efficient than others.

    **(B)** to listen to a large number of alarms keeps the firefighters alert.

    **(C)** for maximum effective disposition of fire apparatus, the staff at the central headquarters should be fully aware of the entire situation at any moment.

    **(D)** although certain companies are responding to one alarm, other companies must be prepared to respond to new alarms in the same or nearby districts.

61. Local firefighters have been called to the assistance of airport firefighters at the site of a major airplane crash. The primary concern of the firefighters must be

    **(A)** extinguishment of the fire.

    **(B)** protection of life.

    **(C)** protection of exposures.

    **(D)** overhauling.

exercises

62. Your fire company has just responded to a fire in the hold of a ship docked at a pier. Prior to your arrival, the ship's crew began to take certain actions to control the fire. Which of their following recent actions would prove least effective in controlling the fire?

    **(A)** They sealed the hatch covers on the hold with waterproof tape.

    **(B)** They shut down the ventilators from the hold and sealed them.

    **(C)** They flooded the hold with carbon dioxide.

    **(D)** They placed thermostats throughout the vessel to detect any increase in temperature.

63. Experience in fighting fires in high-rise structures that don't have sprinklers has created a new body of knowledge for firefighters. Which of the following statements is most correct with regard to lessons learned from such fires?

    **(A)** Central air conditioning, if allowed to continue to supply fresh air in the vicinity of the fire, will cool down the fire.

    **(B)** New lightweight construction helps restrict vertical extension of the fire to higher floors.

    **(C)** Extreme temperatures have no significant impact on a firefighter's effective work time.

    **(D)** Synthetic materials used in furnishings increase the threat to life and safety.

64. Your fire company responds to a fire involving drums of hazardous waste materials. Of the following actions, which would be the least correct for you to take?

    **(A)** Treat empty drums carefully because they might be more hazardous than full drums.

    **(B)** Avoid using water on a fuming substance until you have identified the substance.

    **(C)** Smell or touch the substance involved to get an indication of just what is burning.

    **(D)** Treat small leaks in small containers very carefully.

65. Your ladder company is operating at a serious fire in an old loft building. During operations, you become aware of a large accumulation of water on the third floor. The depth of the water is greater in the center of the room than on the sides. In addition, large slabs of plaster are falling from the walls. It would be most reasonable for you to

    **(A)** continue with your assigned firefighting duties. Large amounts of water always are necessary to extinguish a large fire.

    **(B)** take care to avoid being hit by the falling plaster, a normal occurrence in building fires.

(C) skirt the deep water in the middle of the room. Old buildings generally sag in the middle.

(D) give this information to your superior officer so he or she can assess the possibility of building collapse.

66. As a firefighter, you are summoned to a hazardous waste emergency involving radioactive isotopes. Of the following, the factor least likely to present a health hazard is

(A) atmospheric conditions.

(B) length of exposure.

(C) nature of the radiation.

(D) intensity of the radiation.

67. Your fire company responds to the site of a serious auto accident and finds the driver unconscious in the car. The victim's position severely limits your survey of the injuries involved. The vehicle itself, however, might give some clues. Of the following, which would be the least correct assumption?

(A) A cracked windshield or rearview mirror might indicate injury to the head and spine.

(B) A damaged steering wheel or column might indicate chest, head, or abdominal injury.

(C) Dents on the dashboard might indicate leg, knee, or pelvis trauma.

(D) Because the driver's seat belt is still in position, injuries to spine or abdomen are not likely.

68. The firefighters in a ladder company are required to search for victims in apartments above the site of a store fire. Of the following, the least reasonable procedure is to

(A) have a plan and move toward windows and secondary means of egress.

(B) in each bedroom, flip a mattress into a "U" position to indicate that the area has been searched.

(C) pause from time to time to listen for crying, moaning, or coughing.

(D) have the firefighters, after completing their first search for victims, repeat their search a second time to make sure all victims have been located.

**69.** In the case of an early morning major explosion in which a number of blocks are threatened with being engulfed by a fire, the structure that should demand the least attention from firefighters is a(n)

**(A)** hotel.

**(B)** theater.

**(C)** single-family dwelling.

**(D)** apartment house.

**70.** The fire department has criticized the management of several hotels for failure to call the fire department promptly when fires are discovered. The most probable reason for this delay by the management is that

**(A)** fire insurance rates are affected by the number of fires reported.

**(B)** most fires are extinguished by the hotel's staff before the fire department arrives.

**(C)** hotel guests frequently report fires erroneously.

**(D)** it is feared that hotel guests will be alarmed by the arrival of fire apparatus.

## ANSWERS AND EXPLANATIONS

| | | | | | | | |
|---|---|---|---|---|---|---|---|
| 1. | C | 19. | C | 37. | C | 55. | D |
| 2. | D | 20. | B | 38. | A | 56. | D |
| 3. | A | 21. | A | 39. | D | 57. | B |
| 4. | C | 22. | C | 40. | B | 58. | A |
| 5. | B | 23. | B | 41. | C | 59. | B |
| 6. | B | 24. | D | 42. | C | 60. | D |
| 7. | D | 25. | D | 43. | C | 61. | B |
| 8. | D | 26. | D | 44. | B | 62. | D |
| 9. | A | 27. | C | 45. | B | 63. | D |
| 10. | D | 28. | C | 46. | D | 64. | C |
| 11. | B | 29. | A | 47. | A | 65. | D |
| 12. | B | 30. | B | 48. | C | 66. | A |
| 13. | D | 31. | D | 49. | D | 67. | D |
| 14. | C | 32. | C | 50. | D | 68. | D |
| 15. | B | 33. | D | 51. | D | 69. | B |
| 16. | A | 34. | D | 52. | D | 70. | D |
| 17. | D | 35. | D | 53. | A | | |
| 18. | B | 36. | C | 54. | C | | |

1. **The correct answer is (C).** The rationale behind the civil service testing program is that the person with the higher score will become a better firefighter. If this is correct, then a firefighting force made up of high scorers on civil service exams will be more efficient than one made up of people appointed on the basis of friendships or political connections.

2. **The correct answer is (D).** A fire might get out of control because of an inadequate supply of water, but the lack of water could not be its cause.

**Completed Tables**
**(Questions 3–5)**

|  | Last Year | This Year | Increase |
| --- | --- | --- | --- |
| Firefighters | 9326 | 9744 | 418 |
| Lieutenants | 1355 | 1417 | 62 |
| Captains | 326 | 433 | 107 |
| Others | 462 | 505 | 43 |
| TOTAL | 11,469 | 12,099 | 630 |

3. **The correct answer is (A).** To calculate the number of people in the *Others* group last year, you must first find the number of *Captains* last year. Because the increase from last year to this year among *Captains* was 107, subtract 107 from 433 to find out the number of *Captains* last year. By adding the number of *Firefighters*, *Lieutenants*, and *Captains* and then subtracting that sum from last year's Total, we find that the number of *Others* last year was 462. 462 is closest to 450, so the answer to the question is choice (A).

4. **The correct answer is (C).** To determine percent of change, find the difference between the numbers and divide that difference by the original number: $107 \div 326 = 33\%$. Try the others and you will find that the *Captains* had the greatest percentage of increase by far.

5. **The correct answer is (B).** There are 9,744 *Firefighters* and 2,355 people of all other ranks. This constitutes a ratio of approximately 4:1.

6. **The correct answer is (B).** In firefighting, time is crucial. If a fire alarm box can be found at once, no time is lost in sounding the alarm.

7. **The correct answer is (D).** Obviously, the more information available to firefighters, the less time they need to spend finding things out for themselves and the more quickly they can attack the fire.

8. **The correct answer is (D).** If Mr. Smart had known how to use a fire alarm box, he might have alerted the firefighters far more quickly, and his house might have been saved. The public must be educated to use all resources.

9. **The correct answer is (A).** Presumably, a modern fire alarm system enables alarms to be transmitted and received most rapidly and accurately.

answers

The more quickly the firefighters can respond, the more quickly they can control the fire and the smaller it will be.

10. **The correct answer is (D).** It is impossible to overemphasize the need for speed in all aspects of firefighting.

11. **The correct answer is (B).** Obviously, the number of industrial fires can be reduced by proper fire safety education of workers in fire-prone industries.

12. **The correct answer is (B).** Experience is the best teacher. If a few firefighters drive frequently, they will become experts. If firefighters rotate the driving responsibility, each will drive less often, and no one will have the opportunity to develop expertise.

13. **The correct answer is (D).** The need for greater lengths of hose is less serious than problems in reaching the site of the fire or in gaining access to the building.

14. **The correct answer is (C).** A better-trained firefighter is more likely to recognize the best way to approach a problem.

15. **The correct answer is (B).** If fire inspections are made in conjunction with other safety inspections, remedial actions and follow-up also can be coordinated for efficiency and guarantees of compliance.

16. **The correct answer is (A).** Fireproof schools do not have fireproof contents. In the case of a fire, evacuation of students is a must. Fire drills are necessary so that teachers and students become familiar with procedures and will not panic in the case of a fire.

17. **The correct answer is (D).** The heat produced by all fires is not alike.

18. **The correct answer is (B).** If one length of hose is required for each story of a building, it must be assumed that the height of a story is similar in most buildings.

19. **The correct answer is (C).** Heat always is necessary for fire. Flammable material is the fuel; oxygen is the component of air that promotes combustion.

20. **The correct answer is (B).** If most of the buildings are of a particular construction, it stands to reason that most of the fires would occur in that kind of building. The answer to this question is entirely based on the frequency of that type of building as compared to all others.

21. **The correct answer is (A).** In other words, even a tiny fire can grow. Take all fires seriously.

22. **The correct answer is (C).** This is a combination reading comprehension and reasoning question. The correct answer is simply a restatement of the first sentence.

23. **The correct answer is (B).** If the same number of firefighters are assigned to engine companies as to hook and ladder companies, and if there are more engine companies, then there would be more firefighters assigned to engine companies.

24. **The correct answer is (D).** Because the first company arrived 1 minute before the second, the second company arrived 1 minute after the first. This is the only certain answer.

25. **The correct answer is (D).** If 25 percent of all fires were in buildings of Type A, the total number of fires was four times the number of fires in Type A buildings.

26. **The correct answer is (D).** The preliminary signal is the first signal (here 777), followed by box number 423 and the four ambulances wanted.

27. **The correct answer is (C).** Quite to the contrary, if directions for using alarm boxes vary from one to the next, the sending of alarms will be delayed while citizens study the directions. It is important for the directions to be uniform throughout the city.

28. **The correct answer is (C).** The type of alarm system has little bearing on the number of fire companies. What is important is that there should be sufficient manpower and sufficient equipment to reach all parts of the city efficiently.

29. **The correct answer is (A).** Longer ladders are more difficult to maneuver, so it is good to have short ladders for low fires. Long ladders, however, must be available to fight fires on higher floors.

30. **The correct answer is (B).** The logic of this answer is that slight difficulty will discourage accidental false alarms.

31. **The correct answer is (D).** There would be chaos if each firefighter were to create his or her own plan of action and follow it. You MUST obey orders. Your superior officer has had greater experience in fighting fires. At the station house after the fire, you might discuss the orders with your superior officer and learn about the reasoning behind them.

32. **The correct answer is (C).** Teamwork is vital in firefighting. Discipline coordinates teamwork.

33. **The correct answer is (D).** Sending in the alarm is the first priority, but the other pedestrian is perfectly qualified to do this chore. Assign the pedestrian to send the alarm and direct the firefighters while you, as a trained firefighter, begin to rouse and rescue.

34. **The correct answer is (D).** The fire is not visible and is not yet a clear and present danger. Of far greater concern is panic and stampede. Deal with the crowd first and, if possible, direct someone to turn in an alarm at the same time.

35. **The correct answer is (D).** As long as municipal officials do not get in the way of firefighters, their interest and appreciation should be most welcome. Recognition is good for morale, and appreciation of the firefighters' work can do no harm at budget time.

36. **The correct answer is (C).** The best course in dealing with irate citizens is patient explanation.

37. **The correct answer is (C).** Again, explanation is the best policy.

38. **The correct answer is (A).** Obviously, preparation makes for an organized, useful talk. Although the audience might be impressed with the firefighter's careful presentation, impressing the public must not be the goal of the firefighter.

39. **The correct answer is (D).** If one person oversees and signs all correspondence, that person has control over the content of the correspondence and can ensure consistency of policy and tactfulness of expression.

40. **The correct answer is (B).** This is a matter of tact and public relations. Begin with your regrets, and then explain the reasons for your refusal.

41. **The correct answer is (C).** A free press reports what it hears. The press reflects public opinion; the fire department must listen.

42. **The correct answer is (C).** A business letter, which is what this is, should come straight to the point and then explain in greater detail.

43. **The correct answer is (C).** You must not appear to be arbitrary in dealing with controversial matters with which the public is genuinely concerned. Explain in a way that is clear and reasonable.

44. **The correct answer is (B).** Public relations entails relating to the public. Of the choices, only explaining fire dangers involves dealing with the public.

45. **The correct answer is (B).** The public relations program of an organization has to do with the relations between that organization and the public.

46. **The correct answer is (D).** There is no point in further antagonizing the man. On the other hand, by all means, let him know that you were involved in fire department business. Suggesting that he might have made an appointment is constructive criticism.

47. **The correct answer is (A).** This might be a hard spot to get out of. No professional should discuss his or her superiors with the public. Do your best to find another topic of conversation. Avoid expressing any opinion.

48. **The correct answer is (C).** The firefighter is not qualified to determine the cause of a fire. That is the role of investigators. It is best to refer questions about the cause of a fire to the experts.

49. **The correct answer is (D).** The public bases judgments first on results (not offered here as an answer choice) and then on appearances.

50. **The correct answer is (D).** Occasional poor public relations practices might stem from lack of self-control or disobedience of orders. If this is a consistent problem, it more likely stems from poor supervision and lack of instruction in more desirable practices.

51. **The correct answer is (D).** Self-interest is a powerful stimulus.

52. **The correct answer is (D).** By definition, public agencies are supported by the public in terms of finances and cooperation. Goodwill is vital.

53. **The correct answer is (A).** A regular procedure for handling complaints leads to quick disposition of complaints and might even minimize their number. A public that knows its concerns are taken seriously is less likely to complain.

54. **The correct answer is (C).** It is impossible to change the behavior of mentally ill persons or drunks through education alone. If former firefighters are turning in false alarms, they have severe psychological problems; they already are fully aware of the dangers of false alarms. Schoolchildren, on the other hand, can be directed away from undesirable behavior through education and explanation.

55. **The correct answer is (D).** The public must be made aware of the dangers of false alarms and the expense to the taxpayers.

56. **The correct answer is (D).** Government agencies tend to neglect employee training in nonperformance areas. Some training in dealing with the public might lead to a better image and greater cooperation for government agencies.

57. **The correct answer is (B).** People always are more willing to cooperate with requests when they understand the reasons. Education is the key.

58. **The correct answer is (A).** From noon to 1 p.m. is the lunch hour; from 3 p.m. to 10 p.m., school is out. The culprits are schoolchildren.

59. **The correct answer is (B).** The alarm must be sounded immediately for a well-developed fire. The desk clerk is best equipped to notify hotel occupants of the fire and should do that while the firefighter sends in the alarm. The firefighter should then return to assist with evacuation.

60. **The correct answer is (D).** All fire companies must be aware of the location and involvement of all other companies at all times. This way, they know which companies are available to respond to subsequent alarms, and the entire city is certain to be protected at all times.

61. **The correct answer is (B).** The main objective of all firefighting at all times is to save the lives of civilians and other firefighters.

62. **The correct answer is (D).** Thermostats, strategically placed long before a fire, might have given advance warning and might have been useful in preventing the development of the fire. After the ship is actually on fire, however, placing thermostats is akin to closing the barn door after the horses have escaped. All the other actions should be useful in controlling the fire.

63. **The correct answer is (D).** Synthetic materials produce highly toxic products of combustion that seriously increase the threat to life and safety. All of the other statements are untrue. Introduction of fresh air contributes to the intensity of a fire. Fire spreads upward rapidly in the new lightweight high-rise construction. At extreme temperatures, a firefighter's effective work time is approximately 5 minutes.

64. **The correct answer is (C).** Under no circumstances should you touch, smell, or taste any suspected hazardous substance or any unknown substance involved in a fire. All of the other actions would be correct.

65. **The correct answer is (D).** Collapse of the building may be imminent. The weight of water causes beams to sag in the middle. This, in turn, pulls them from their supports on the sides of the building. The large slabs of falling plaster are a further indication of the shifting of floors and walls. The situation in the old loft building is by no means normal or routine.

66. **The correct answer is (A).** Dangers of radiation come from length of exposure, the nature of the radiation, and the intensity of the radiation. Atmospheric conditions have little influence on the hazards from radiation exposure.

67. **The correct answer is (D).** Although the use of seat belts does save lives and can limit many injuries, it is not a 100-percent effective measure. Use of seat belts, especially without the shoulder harness in place, does not always prevent injury to the abdomen, spine, and pelvis. The other assumptions all deserve consideration.

68. **The correct answer is (D).** A second search is very much in order, but it should be made by a different group of firefighters. If the first group has not found victims, it makes sense to send another group that might have different, more effective methods. The other procedures are all proper and helpful.

69. **The correct answer is (B).** In the early morning, a theater is unlikely to be occupied. Because lifesaving is the paramount concern, an unoccupied structure should not divert attention from buildings where lives are in danger.

70. **The correct answer is (D).** Your knowledge of human nature should tell you that the hotel's management is concerned with public relations. Fear of panic among guests and of publicity that the hotel is fire-prone can combine to discourage hotel management from reporting fires promptly.

# Spatial Orientation

······························································································

## OVERVIEW

- What do spatial orientation questions measure?
- Tips for answering spatial orientation questions

## WHAT DO SPATIAL ORIENTATION QUESTIONS MEASURE?

Spatial orientation questions measure your ability to keep a clear idea of where you are in relation to the space in which you happen to be. The ability to orient yourself in space is very important to a firefighter who must reach a fire in a minimum amount of time and in the safest and most effective manner. Spatial orientation also is vital to firefighters in a smoky environment. Knowing the direction in which you are facing after a number of turns with no visual clues might be a skill that will save your life.

Your firefighting training gives you clues and instructions to orient yourself in space. You will have lots of practice before you are put into a position in which you must rely on this skill. The spatial orientation questions on the exam are meant to test your native aptitude in this regard and to measure how carefully you read and how logically you follow through.

Firefighter exam spatial orientation questions tend to emphasize either where you are in a diagram or how to go from one spot to another on the diagram or map. Because these are not memorization questions, you probably will be allowed to use your pencil to write on the diagrams or maps as a way of testing your answer choices. When using your pencil to write on a diagram or map, be sure to write lightly. Erase any of your jottings that do not work out or are no longer needed. If several questions are based on the same diagram or map and you have made pencil markings for them, the diagram can get quite confusing with markings from prior questions. Hence, if there is another question to be answered on the basis of the same diagram, you should erase your markings as soon as you are done with them.

Many diagrams or maps use symbols, also called legends. Look at the whole page to see if there is a key to the legend. A dotted line might indicate movement. An arrow might indicate direction of movement or of permitted movement. An important feature of many diagrams and maps is the symbol indicating the directions of north, south, east, and west.

## chapter 6

## TIPS FOR ANSWERING SPATIAL ORIENTATION QUESTIONS

Questions often are based on phrases such as "turn left," "to the right," or "to the left of the rear entrance." The test-maker often approaches a diagram or map from the side or from the top, however, so that "left" and "right" do not correspond to where you are sitting in relation to the diagram. Just turn the test booklet sideways or upside-down to view the map or diagram from the standpoint of the question. Turn it so that "left" or "right" on the diagram or map is in the same direction as your left or right hand.

Occasionally, a question gives you information in verbal form and asks you to relate the information to a map or diagram. The question might say that a fire engine went north two blocks, for example, then turned east for two blocks, turned south for one block, and then went east one more block. The answer choices might consist of just lines showing the directions of movement. The correct line will go up, go left, come down halfway, and then go left again. In a question of this sort, you might find it helpful to make a diagram of your own in the margin of the test booklet. If you diagram the problem yourself, you are less likely to be confused by wrong answers.

## EXERCISE: SPATIAL ORIENTATION

### QUESTION 1 IS BASED ON THE FOLLOWING STREET MAP:

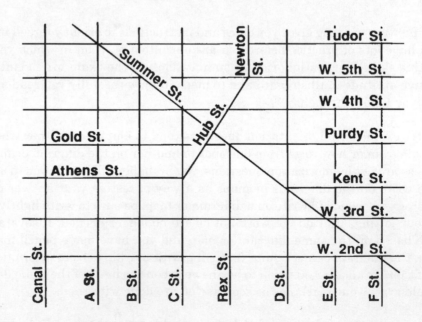

This portion of a city street map shows an area that is divided into four fire control sectors as follows:

—Sector Adam: Bounded by Tudor Street, Newton Street, Hub Street, Athens Street, and Canal Street

—Sector Boy: Bounded by Tudor Street, F Street, West 4th Street, Hub Street, and Newton Street

—Sector Charles: Bounded by West 4th Street, F Street, West 2nd Street, C Street, and Hub Street

—Sector David: Bounded by Athens Street, C Street, West 2nd Street, and Canal Street

1. There is a fire in the block bounded by West 4th Street, Summer Street, and Hub Street. The fire is in sector

   **(A)** David.

   **(B)** Charles.

   **(C)** Boy.

   **(D)** Adam.

**QUESTION 2 IS BASED ON THE FOLLOWING DIAGRAM AND PASSAGE:**

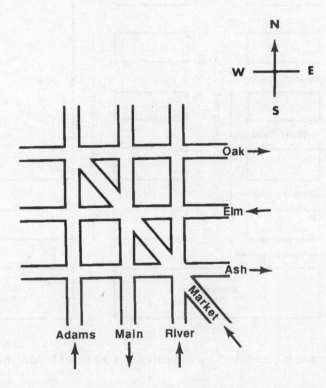

**As indicated by arrows on the accompanying street map, Adams and River Streets are one-way going north. Main is one-way going south, and Market is one-way going northwest. Oak and Ash are one-way streets going east, and Elm is one-way going west.**

exercises

2.  A fire engine heading north on River Street between Ash and Elm Streets receives a call to proceed to the intersection of Adams and Oak. To travel the shortest distance and not break any traffic regulations, the engine should turn left on

    **(A)** Elm and right on Market.

    **(B)** Market and proceed directly to Oak.

    **(C)** Elm and right on Adams.

    **(D)** Oak and proceed directly to Adams.

### QUESTIONS 3–5 ARE BASED ON THE FOLLOWING MAP:

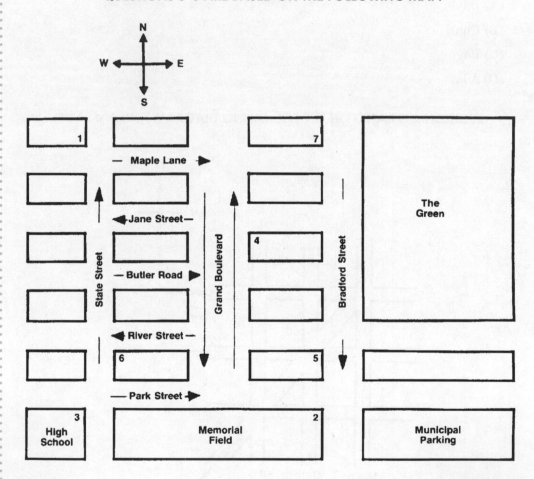

The flow of traffic is indicated by the arrows. You must follow the flow of the traffic.

3. If you are located at point (1) and travel east one block, turn right and travel four blocks, and then turn left and travel one block, you will be closest to point

   **(A)** 6

   **(B)** 5

   **(C)** 3

   **(D)** 2

4. You have responded to an alarm sent from the box at the corner of Butler Road and Bradford Street, and you learn that the fire is at Butler Road and Grand Boulevard. Which is the best route to take to the scene?

   **(A)** Go one block south on Bradford Street, go one block west on River Street, and then go one block north on Grand Boulevard.

   **(B)** Go one block south on Bradford Street, go two blocks west on River Street, go one block north on State Street, and then go one block east on Butler Road.

   **(C)** Go one block north on Bradford Street, go one block west on Jane Street, and then go one block south on Grand Boulevard.

   **(D)** Go two blocks south on Bradford Street, go one block west on Park Street, and then go two blocks north on Grand Boulevard.

5. You have just checked out and dismissed a false alarm at the high school when you receive word of a trash can fire on The Green near the corner of Maple Lane and Bradford Street. The best route to take would be to

   **(A)** go two blocks east on Park Street and then go four blocks north on Bradford Street.

   **(B)** go two blocks north on State Street, go two blocks east on Butler Road, and then go two blocks north on Bradford Street.

   **(C)** go four blocks north on State Street and then go two blocks east on Maple Lane.

   **(D)** go one block east on Park Street, go two blocks north on Grand Boulevard, go one block east on Butler Road, and then go two blocks north on Bradford Street.

## QUESTIONS 6–12 ARE BASED ON THE FOLLOWING DIAGRAM:

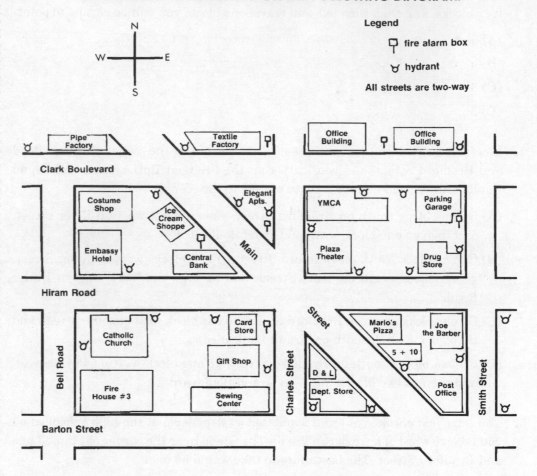

6. The parade down Main Street has cleared Clark Boulevard and is crossing Hiram Road when an alarm is received of a fire at the Post Office. The best hydrant for the firefighters to attach their first hose to should be the hydrant

   **(A)** at the corner of Barton Street and Main Street.

   **(B)** mid-block on the northeast side of Main Street.

   **(C)** mid-block on Smith Street between Hiram Road and Barton Street.

   **(D)** at the corner of Hiram Road and Smith Street.

7. The best route for the firefighters to take to the Post Office fire in question 6 is

   **(A)** straight down Barton Street.

   **(B)** north on Bell Road, right on Clark Boulevard, and then south on Smith Street.

   **(C)** north on Bell Road, right onto Hiram Road, and right again onto Main Street.

   **(D)** east on Barton Street, north onto Charles Street, right onto Hiram Road, and right again onto Smith Street.

8. At 5 p.m., word is received at the firehouse of a fire at the YMCA. The best route to this fire is

    **(A)** east one block on Barton Street and then left into Charles Street.

    **(B)** north on Bell Road to Clark Boulevard and then right onto Clark Boulevard to Charles Street.

    **(C)** east on Barton Street, left onto Smith Street, and then west onto Clark Boulevard.

    **(D)** north on Bell Road to Hiram Road, east onto Hiram Road, and then north onto Charles Street.

9. While the fire at the YMCA is raging, a strong wind is blowing from the southeast. Firefighters should train their hoses to avoid having the fire spread to the

    **(A)** Plaza Theater.

    **(B)** Parking Garage.

    **(C)** Elegant Apartments.

    **(D)** Textile Factory.

10. A fire alarm sent from the alarm box on Main Street between Clark Boulevard and Hiram Road is least likely to be reporting a fire

    **(A)** in an Office Building.

    **(B)** at the Pipe Factory.

    **(C)** at the Elegant Apartments.

    **(D)** at the Catholic Church.

11. On a hot summer evening, a firefighter is sent to make a routine check to ensure that no fire hydrants have been opened to reduce water pressure in the area. The firefighter leaves the firehouse by way of the Barton Street entrance and heads east. He takes the second left turn, proceeds one block, and makes another left, goes one block, and makes three successive right turns. At the next intersection, he turns left, then makes his first right, and stops the car. The firefighter is now

    **(A)** back at the firehouse.

    **(B)** at the corner of Hiram Road and Main Street facing southeast on Main Street.

    **(C)** at the corner of Charles Street and Clark Boulevard facing east on Clark Boulevard.

    **(D)** on Bell Road at the side entrance of the Embassy Hotel.

**12.** The most poorly protected block in terms of a pedestrian's ease in reporting a fire is the block bounded by

**(A)** Charles Street, Main Street, and Barton Street.

**(B)** Hiram Road, Main Street, Barton Street, and Smith Street.

**(C)** Clark Boulevard, Charles Street, Hiram Road, and Smith Street.

**(D)** Hiram Road, Bell Road, Barton Street, and Charles Street.

## QUESTIONS 13–16 ARE BASED ON THE FOLLOWING FLOOR PLAN AND INFORMATION:

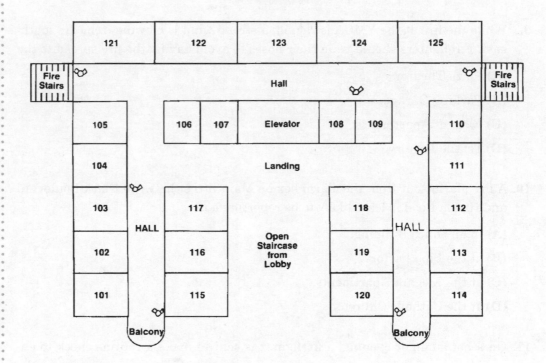

## Central Hotel

**Central Hotel is an elegant, old, five-story structure. The plan of each residential floor, with four floors above the lobby, is identical. ⬅️⬜➡️ represents a standpipe connection.**

**13.** In the event of a fire in the hotel lobby, the best means by which to evacuate the tenant in room 124 is

**(A)** via the elevator.

**(B)** down the fire stairs next to room 125.

**(C)** across the hall and down the wide open staircase.

**(D)** across the hall and down the corridor to the balcony.

14. In the same lobby fire, the tenant in room 217 should be evacuated by

    **(A)** running down the hall to the balcony.

    **(B)** taking the elevator.

    **(C)** whipping around the corner and down the open staircase.

    **(D)** using the fire stairs between rooms 205 and 221.

15. In the case of a fire in room 406, firefighters should

    **(A)** take the elevator to the roof, enter the fifth floor from the fire stairs, and connect to the standpipe at room 421.

    **(B)** run up the open staircase and connect hoses to the standpipe between rooms 403 and 404.

    **(C)** enter the building by way of a ladder placed in front and up to the balcony, then connect to the standpipe at rooms 403 and 404.

    **(D)** run up the fire stairs and connect to the standpipe at room 421.

16. Aside from the danger to the tenant of the involved room, the greatest danger from a fire in room 309 is to the tenants of rooms

    **(A)** 310 and 318.

    **(B)** 209 and 409.

    **(C)** 308 and 409.

    **(D)** 308 and 209.

## ANSWERS AND EXPLANATIONS

| | | | | | | | |
|---|---|---|---|---|---|---|---|
| 1. | D | 5. | C | 9. | D | 13. | B |
| 2. | A | 6. | C | 10. | A | 14. | A |
| 3. | D | 7. | B | 11. | C | 15. | D |
| 4. | A | 8. | A | 12. | B | 16. | C |

1. **The correct answer is (D).** Sector Adam is bounded by heavy black lines. The block with the fire is marked with an X.

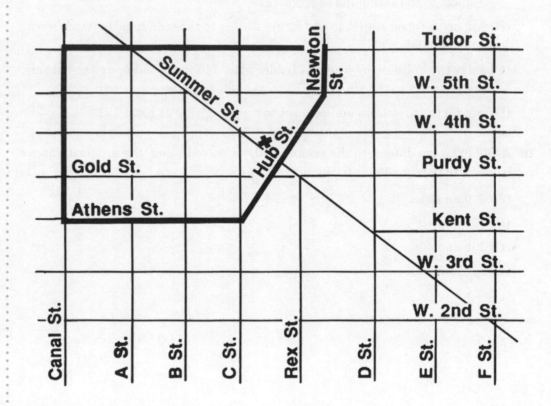

2. **The correct answer is (A).** The route is marked on the following diagram. Choice (C) would be slightly longer. Choices (B) and (D) are impossible from the given starting point.

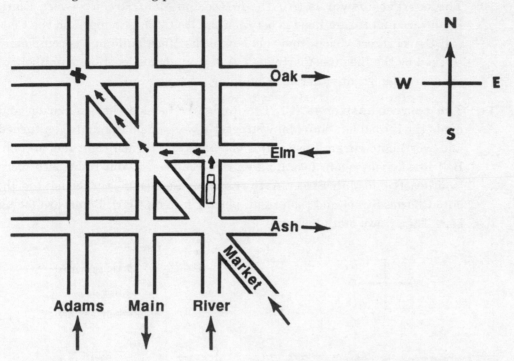

3. **The correct answer is (D).** From point (1), east one block is right on Maple Lane to Grand Boulevard, right four blocks is south on Grand Boulevard to Park Street, left is east onto Park Street, and one block brings us closest to point (2).

4. **The correct answer is (A).** Bradford Road is one-way southbound, so immediately eliminate choice (C). Park Street is one-way eastbound, so eliminate choice (D). Choice (B) is possible, but it is longer than choice (A) to no advantage.

5. **The correct answer is (C).** All other choices send you north on Bradford Street.

6. **The correct answer is (C).** With a parade coming down Main Street, it would be difficult to maneuver fire trucks and hoses on either side of the parade route. The hydrant at choice (C) is closer than the one at choice (D).

7. **The correct answer is (B).** Certainly, the route straight down Barton Street is the most direct, but remember the parade. Under the special circumstances of the day, the best route is to entirely skirt the parade. The parade has cleared Clark Boulevard, and the detour around the parade route is not really far.

8. **The correct answer is (A).** Look at the time of day. Clark Boulevard is clogged with workers on their way home and should be avoided if at all possible. Choice (D) is acceptable, but choice (A) is faster.

9. **The correct answer is (D).** Wind from the southeast is blowing toward the northwest. The Textile Factory is directly northwest from the YMCA.

10. **The correct answer is (A).** The fire box on Main Street between Clark Boulevard and Hiram Road is between the Ice Cream Shoppe and the Central Bank. Direct vision from this box to the office buildings is completely blocked by the Elegant Apartments. A person at this location is unlikely to be aware of a fire at either office building.

11. **The correct answer is (C).** The firefighter headed east on Barton and made the second left onto Main Street. He went one block and then turned left (west) onto Hiram Road. After one block, he turned right (north) onto Bell Road, right again (east) onto Clark Boulevard, and right once more (southeast) onto Main Street. At the next intersection, he turned left (north) onto Charles Street and then right, placing him on Clark Boulevard facing east. The arrows mark his route:

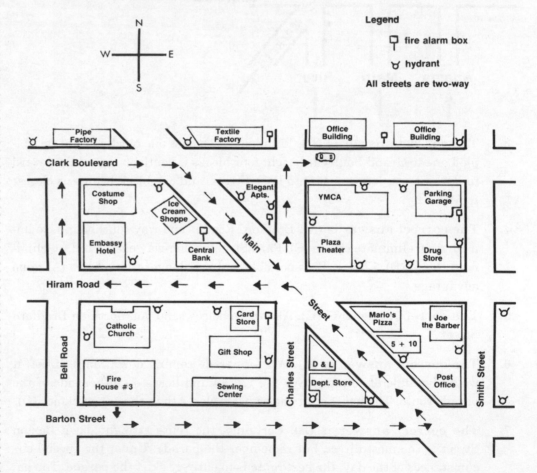

12. **The correct answer is (B).** There is no fire alarm box on this block. It is quite far from the nearest box on the next block, and it is not close to the firehouse. The block in choice (A) also has no box, but it is closer to the nearest box and to the firehouse.

13. **The correct answer is (B).** Fire, heat, and smoke will rise rapidly up the open staircase from the lobby and will fill the halls of the floors above. Tenants being evacuated should travel the shortest possible distances in the halls. The fire stairs are close and are safe because they are enclosed. Elevators are never safe in a fire. Further, because the fire is in the lobby, patrons should be kept away from the lobby.

14. **The correct answer is (A).** The reasoning is the same as in question 13. This tenant will reach fresh air and safety most quickly and with least exposure to smoke by running to the balcony for rescue from outside.

15. **The correct answer is (D).** As previously mentioned, the elevator and open staircase present hazards to all people—firefighters as well as civilians. The safely enclosed fire stairs are a much quicker and safer route by which to carry hoses to the upper floor.

16. **The correct answer is (C).** Room 308 shares a common wall with room 309 and is clearly in jeopardy. Common walls tend to be flimsier than floors and ceilings. Because heat and smoke rise, the danger is greater to the room above room 309 than to the room below, though it too should be evacuated.

answers

# Observation and Memory

## OVERVIEW

- How are observation and memory tested?

## HOW ARE OBSERVATION AND MEMORY TESTED?

Firefighters do not work alone. Whatever it is they are doing—fire inspections, equipment maintenance, firefighting—they always do their work in pairs, groups, or teams. Much of a firefighter's work is done under direction and supervision. It is important, however, that the firefighter keep alert at all times and not rely solely on the observations and judgments of others.

Fire inspections provide the ideal opportunity for firefighters to learn neighborhood and building layouts, locations of firefighting aids, and possible hazardous situations. It is vital for a firefighter out on inspection to observe everything about the premises. It is to be hoped that observation of problem situations during inspections will lead to corrective and preventive measures so that fires never occur at the premises.

Fires do occur, however. In the event of a fire, the firefighter who has noticed and remembered that chemicals stored near the site of the fire might give off toxic fumes when burned will be the better prepared and more effective firefighter. The firefighter who has noticed and remembered the locations of fire stairs and firehose connectors will be on the scene more quickly and will attack the fire with the best efficiency. Should a firefighter find him- or herself inside a premises engulfed in dense smoke, memory of layout and points of exit can save his or her life.

Because observation and memory are so important to firefighters, many firefighter exams try to measure this ability in applicants. The usual method of measuring this skill is to distribute a diagram or a picture (or perhaps two diagrams or pictures) to applicants to study for a given period of time. After the study period, usually 5 minutes per picture or diagram, the pictorial materials are collected. The applicant then must answer a series of multiple-choice questions based on information in the pictures or diagrams.

chapter 7

The pictures and diagrams on firefighter exams usually are not too complex. Five minutes is adequate for absorbing the information, provided that the 5 minutes are used systematically. Your problem is that you cannot write anything down. All notes must be mental notes. It is worthwhile, therefore, to establish in advance the categories into which you will fit information and then to approach the pictorial material to fill in the blanks of each category.

Here are some specifics to look for as you examine a pictorial:

❶ **People.** Are there any people in the picture or diagram? If not, proceed to the next category. If so, where are they located? What landmark objects (windows, doors, fire escapes, fire sources) are they near? Are they in danger?

❷ **Fire activity.** Is there fire? Is there smoke? Where is the activity located? Again, where is the fire activity located with respect to landmarks? How does it endanger life and property? Is there a predictable direction in which it might spread? (In connection with this last point, note whether there is a directional indicator—N, S, E, W—anywhere on the page.)

❸ **Layout.** If the picture is an outdoor scene, look for fire equipment such as alarm boxes and hydrants. Note what these items are nearest to and farthest from. Study heights of buildings, natures of buildings, and fire escapes. If you have in hand an indoor scene or floor plan, focus on adjacent rooms, means of entering one room from the next, and locations of doors, windows, and fire escapes. Count and try to remember numbers of doors and doorways and numbers of windows.

❹ **Special details.** Is there a smoke alarm? Where? Is there an obstruction noted? Is there an indication that doors are open or closed? If so, which ones? What is unique to this particular scene or diagram? What feature would likely lead to a question on an exam?

## EXERCISE 1: OBSERVATION AND MEMORY

**Directions:** Study the diagram for 5 minutes, set it aside, answer the multiple-choice questions, and then study the answer discussions.

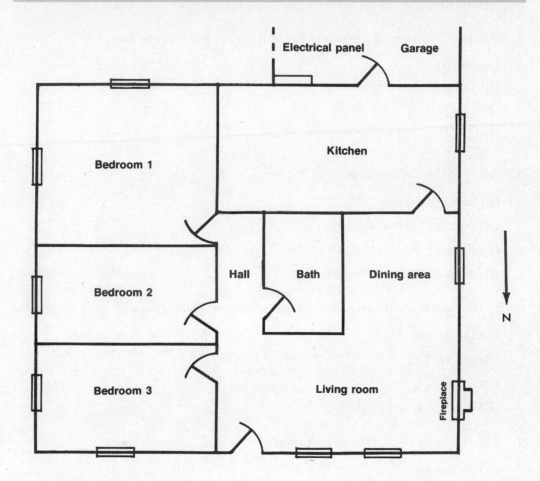

1. The main entry to these premises is the

   (A) door to the kitchen.

   (B) front vestibule.

   (C) door to the living room.

   (D) door to the dining area.

2. Direct access to the bathroom can be made from

   (A) bedroom 1.

   (B) the living room.

   (C) bedroom 2.

   (D) None of the above

3. The room with the fewest windows is

    (A) bedroom 2.

    (B) the bathroom.

    (C) the dining area.

    (D) the kitchen.

4. The exposure with the greatest possible number of exits is the

    (A) north.

    (B) south.

    (C) east.

    (D) west.

5. Entry to the kitchen is possible from the

    (A) dining area only.

    (B) dining area and the hall.

    (C) dining area and the garage.

    (D) dining area, hall, and garage.

6. In the case of an electrical fire, power should be shut off in

    (A) the garage.

    (B) the kitchen.

    (C) the hall.

    (D) There is no way to tell.

7. The stairway goes upstairs by way of

    (A) the garage.

    (B) the hall.

    (C) bedroom 2.

    (D) There is no stairway.

8. If there were an out-of-control fire spreading from the fireplace, a person's best escape from the bathroom would be

    (A) out the bathroom door to the hall, into bedroom 1, and out the south window.

    (B) out the bathroom door to the hall, through the living room and dining area, through the kitchen, and out by way of the garage.

    (C) out the bathroom door, down the hall, and out the front door.

    (D) across the hall into bedroom 2 and out the window.

9. Connecting rooms are

   **(A)** bedroom 1 and bedroom 2.

   **(B)** bedroom 2 and bedroom 3.

   **(C)** bedroom 3 and the dining area.

   **(D)** the dining area and the kitchen.

10. The premises diagrammed here most probably are a(n)

    **(A)** single-family detached house.

    **(B)** semi-detached house.

    **(C)** apartment in a low-rise building.

    **(D)** apartment in a high-rise building.

## ANSWERS AND EXPLANATIONS

| 1. | C | 3. | B | 5. | C | 7. | D | 9. | D |
|----|---|----|---|----|---|----|---|----|---|
| 2. | D | 4. | A | 6. | A | 8. | C | 10. | A |

1.  **The correct answer is (C).** The front door enters directly into the living room. The door to the kitchen from the garage is a secondary entrance. There is no anteroom indicated in front of the living room.

2.  **The correct answer is (D).** The bathroom door opens only into the hallway.

3.  **The correct answer is (B).** The bathroom has no windows at all. Each of the other areas has one window.

4.  **The correct answer is (A).** Most directional markers indicate North as up. This marker is unusual in that North is pointing down. Reverse the diagram to make North point up and get your bearings. The front of the house is the northern exposure. There are three windows and a door along the front (north) of the house.

5.  **The correct answer is (C).** The hall does not go through to the kitchen; however, entry to the kitchen can be made through either the dining area or the garage.

6.  **The correct answer is (A).** The electrical panel is in the garage.

7.  **The correct answer is (D).** There is no stairway.

8.  **The correct answer is (C).** This question requires both memory of the layout and common sense. Exit is always faster and safer by way of a door than by way of a window if it is possible to reach a door. The fire is at the end of the living room farthest from the door. The short run out the door makes most sense.

9. **The correct answer is (D).** Of these, only the dining area and kitchen connect.

10. **The correct answer is (A).** This takes some inference in conjunction with memory. The presence of a garage entry directly to the kitchen rules out an apartment house of any height. Windows on all four sides indicate that the house is freestanding and fully detached. There are no windows in a party wall.

In studying this floor plan, you should have noticed:

➊ **People.** None.

➋ **Fire activity.** None.

➌ **Layout.** The northern orientation of the house. Front entry to the living room. L-shaped living/dining area. Access to the bathroom from the hall only. Three nonconnecting bedrooms in a row. Corner bedrooms with two windows each. The hallway does not go through to the kitchen. The door from the dining area to the kitchen (as opposed to a possible open doorway without a door). Access to the kitchen through the garage. Fully interior bathroom.

➍ **Special details.** Electrical panel in the garage. Fireplace in the living room. No fire-protective devices.

# EXERCISE 2: OBSERVATION AND MEMORY

**Directions:** Study the diagram for 5 minutes, set it aside, answer the multiple-choice questions, and then study the answer discussions.

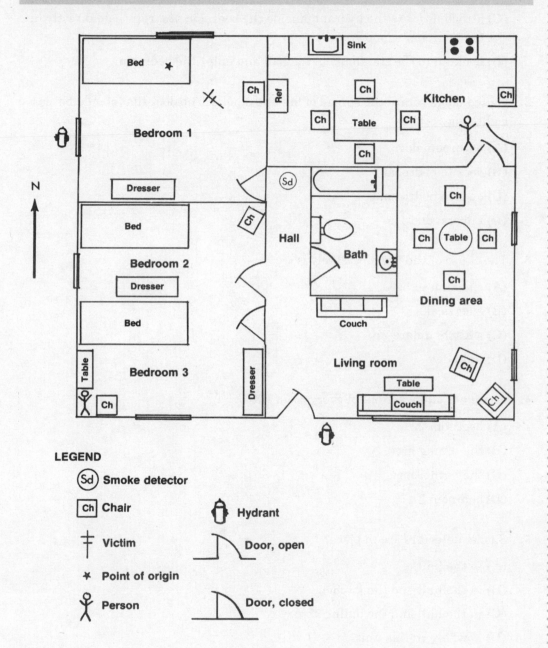

**LEGEND**

(Sd) Smoke detector

Ch Chair

✝ Victim

✷ Point of origin

Person

Hydrant

Door, open

Door, closed

1. To directly attack the fire, firefighters should

    **(A)** attach hoses to the hydrant next to the front door, enter through the living room, and go down the hall.

    **(B)** attach hoses to the hydrant outside the kitchen door, running them out the kitchen window and into the bedroom window.

    **(C)** attach hoses to the hydrant outside the bedroom and run them directly into the bedroom window.

    **(D)** hook up to the standpipe in the hall and enter the bedroom.

2. Closed doors retard the spread of fire and smoke. The door that should be closed right away is the

    **(A)** bathroom door.

    **(B)** door to bedroom 2.

    **(C)** door to bedroom 3.

    **(D)** kitchen door.

3. The cause of the fire most likely is

    **(A)** smoking in bed.

    **(B)** electrical.

    **(C)** a kitchen flare-up.

    **(D)** unknown.

4. There are two ways out of every room EXCEPT

    **(A)** bedroom 2.

    **(B)** the dining area.

    **(C)** the bathroom.

    **(D)** bedroom 3.

5. Smoke detectors are in place

    **(A)** in the hall.

    **(B)** in the hall and the kitchen.

    **(C)** in the hall and the dining area.

    **(D)** nowhere in this unit.

6. In a search for people who might need rescue, firefighters should especially concentrate in

    (A) the kitchen.

    (B) bedroom 3.

    (C) the bathroom.

    (D) the dining area.

7. Doors to the outside of this unit can be found on the

    (A) south and east walls.

    (B) north and east walls.

    (C) south and west walls.

    (D) north, south, and east walls.

8. The total number of people in the unit is

    (A) 0

    (B) 1

    (C) 2

    (D) 3

9. A room with one window is

    (A) bedroom 1.

    (B) bedroom 3.

    (C) the living room.

    (D) the bathroom.

10. If the fire were to extend throughout the unit, the risk of involving furnishings would be greatest in

    (A) bedroom 3.

    (B) the living room.

    (C) the dining area.

    (D) the kitchen.

## ANSWERS AND EXPLANATIONS

| 1. | C | 3. | A | 5. | A | 7. | A | 9. | B |
|----|---|----|---|----|---|----|---|----|---|
| 2. | B | 4. | C | 6. | B | 8. | D | 10. | B |

1. **The correct answer is (C).** For the most direct attack on the fire, attach to the hydrant directly outside the involved room. The hydrant outside the front door should be utilized immediately thereafter for protection of the other rooms in the unit. With careful observation and memory, you should be aware that there is no hydrant outside the kitchen door and that the device in the hall is a smoke detector, not a standpipe.

2. **The correct answer is (B).** The bathroom door and the door to bedroom 3 already are closed. Danger to the immediately adjacent bedroom 2 is far greater than danger to the remote kitchen.

3. **The correct answer is (A).** The point of origin of the fire is a bed, and the victim is in the bedroom. Suspicion of smoking in bed as the cause of the fire is legitimate.

4. **The correct answer is (C).** Only the bathroom has just one exit. Bedrooms 2 and 3 each have a door and a window, and the dining area is wide open to the living room and the kitchen and has window access as well.

5. **The correct answer is (A).** There is one smoke detector at the end of the hall.

6. **The correct answer is (B).** The victim on the floor in bedroom 1 and the person in the kitchen are quite obvious. Note the person hiding behind a chair in the corner of bedroom 3. Of course, the bathroom and dining area must be checked as well, but they offer fewer hiding places and require less concentrated attention.

7. **The correct answer is (A).** Look at the directional marker. The north wall of the unit is at the top of the page, so the doors are on the south and east walls.

8. **The correct answer is (D).** The three people are the victim in bedroom 1, the person standing in the kitchen, and the person hiding in bedroom 3.

9. **The correct answer is (B).** Even though it is a corner bedroom, bedroom 3 has only one window. Bedroom 1 and the living room both have two windows, and the bathroom has none.

10. **The correct answer is (B).** The living room is full of upholstered furniture. The other rooms are more sparsely furnished.

In studying this diagram, you should have noticed:

**❶ People.** Three—one each in bedroom 1, bedroom 3, and the kitchen.

**❷ Fire activity.** Point of origin in the bedroom.

**❸ Layout.** Exit doors from the living room and kitchen. Three nonconnecting bedrooms in a row. Fully interior bathroom with exit to the hallway only. L-shaped living/dining area. The hallway does not go through to the kitchen. Locations of windows. General volume and placement of furniture. Open and closed doors.

**❹ Special details.** Hydrants at the front door and outside the window of bedroom 1. Smoke detector in the hall. Position of the victim.

answers

# EXERCISE 3: OBSERVATION AND MEMORY

**Directions:** Study the sketch for 5 minutes, set it aside, answer the multiple-choice questions, and then study the answer discussions.

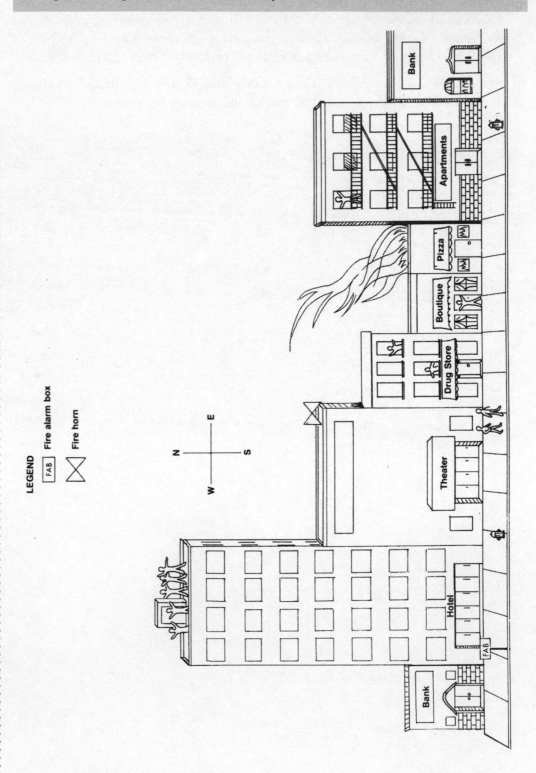

1.  The fire is in

    **(A)** the building between the apartments and the boutique.

    **(B)** the building next to the drug store.

    **(C)** a bank.

    **(D)** the building between the pizza parlor and the apartments.

2.  Of the people who can be seen, those in the most immediate danger are the

    **(A)** people on the roof of the hotel.

    **(B)** residents of the apartment house.

    **(C)** people who live above the drug store.

    **(D)** person in the doorway of the boutique.

3.  If the hour were 8 a.m., the least peril to persons would occur at the

    **(A)** hotel.

    **(B)** building between the hotel and the drug store.

    **(C)** sidewalk.

    **(D)** building housing the drug store.

4.  The best spot from which the firefighters should train their hoses on the fire is the

    **(A)** hydrant in front of the theater.

    **(B)** hydrant in front of the pizza parlor.

    **(C)** roof of the boutique.

    **(D)** hydrant in front of the apartment house.

5.  The building least endangered by this fire is

    **(A)** the apartment house.

    **(B)** a bank.

    **(C)** the theater.

    **(D)** the pizza parlor.

6.  The people on the roof of the hotel should be told to

    **(A)** stay there.

    **(B)** go inside and stay inside.

    **(C)** take the elevator down to the lobby and leave the hotel.

    **(D)** take the stairs down to the lobby and leave the hotel.

7.  The total number of people visible in the drawing is

    **(A)** 7
    **(B)** 9
    **(C)** 11
    **(D)** 12

8.  If firefighters want to get to the roof of the burning building after the flames and smoke have been brought under control, their best route is

    **(A)** directly from the roof of the apartment house.
    **(B)** from the roof of the hotel to the roof of the drug store to the roof of the burning building.
    **(C)** from the roof of the building housing the drug store to the roof of the building next door to the roof of the burning building.
    **(D)** from the roof of the bank to the roof of the theater to the roof of the burning building.

9.  The fire probably was reported from the fire alarm box

    **(A)** in front of the hotel.
    **(B)** on the wall of the theater.
    **(C)** in front of the pizza parlor.
    **(D)** on the fire escape of the apartment house.

10. The wind is blowing from the

    **(A)** west.
    **(B)** east.
    **(C)** north.
    **(D)** south.

## ANSWERS AND EXPLANATIONS

| | | | | | | | | | |
|---|---|---|---|---|---|---|---|---|---|
| 1. | A | 3. | B | 5. | B | 7. | C | 9. | A |
| 2. | C | 4. | D | 6. | D | 8. | C | 10. | B |

1.  **The correct answer is (A).** The fire is in the pizza parlor, which is between the apartments and the boutique.

2.  **The correct answer is (C).** The people who live in the apartments over the drug store are separated from the fire by only one building. With the wind blowing in their direction, they clearly are in greater danger than the people in the apartment house next to the fire but upwind of it. The person at street level can just walk away; the people on the hotel roof are quite a bit removed.

3.  **The correct answer is (B).** The building between the hotel and the drug store is a theater. At 8 a.m., a theater is likely to be deserted, creating peril to property but not to life.

4.  **The correct answer is (D).** From the hydrant in front of the apartment house, the firefighters can move in quite close behind the fire. This is an ideal location. They can approach the fire directly without having to force their way through heat, flame, and smoke. The hydrant in front of the theater is farther away, and the approach is from a less favorable direction. There is no hydrant in front of the pizza parlor itself. The roof of the boutique is too dangerous a spot from which to fight the fire at this active stage.

5.  **The correct answer is (B).** The two banks are the two least endangered buildings. The bank beside the apartment house is separated from the fire by the apartment house, and it is away from the path of the fire. The bank next to the hotel is the building farthest from the fire and is separated from it by a couple of large, sturdy buildings.

6.  **The correct answer is (D).** This judgment question must be based on your recall of the whole scene. The people on the hotel roof (and the people inside the hotel) are in no immediate danger, but smoke and fire can spread and precautions must be taken. With the emergency in the area, electrical interruption could occur at any time. People should avoid the elevator and evacuate the hotel by way of the stairs. The roof of the hotel offers a wonderful vantage point for watching the fire, but a fire is not a show, and the smoke eventually will reach them. They must go.

7.  **The correct answer is (C).** 5 people on the hotel roof + 2 pedestrians + 2 in the apartments over the drug store + 1 in the doorway of the boutique + 1 on the apartment house fire escape = 11.

answers

8. **The correct answer is (C).** Consider relative heights of adjacent buildings in answering this question. The drop from the apartment house is too steep, especially onto a roof weakened by fire. The route from the drug store allows for a gradual, cautious approach.

9. **The correct answer is (A).** Did you notice the legend on the page? The fire alarm box in front of the hotel is the only one in the picture. The device on the theater is a fire horn.

10. **The correct answer is (B).** The flame and smoke, and therefore the wind, are blowing from the east toward the west.

In studying this sketch, you should have noticed:

❶ **People.** Numbers, locations, and locations with reference to the fire.

❷ **Fire activity.** Location and location with reference to landmarks. Direction. Extent of involvement.

❸ **Layout.** Types of buildings. Sizes of buildings. Arrangement of buildings. Relationships between buildings in terms of size and types of occupancy.

❹ **Special details.** Legend on page. Number and location of hydrants, fire alarm boxes, and fire horns.

## EXERCISE 4: OBSERVATION AND MEMORY

**Directions:** Study the diagram for 5 minutes, set it aside, answer the multiple-choice questions, and then study the answer discussions.

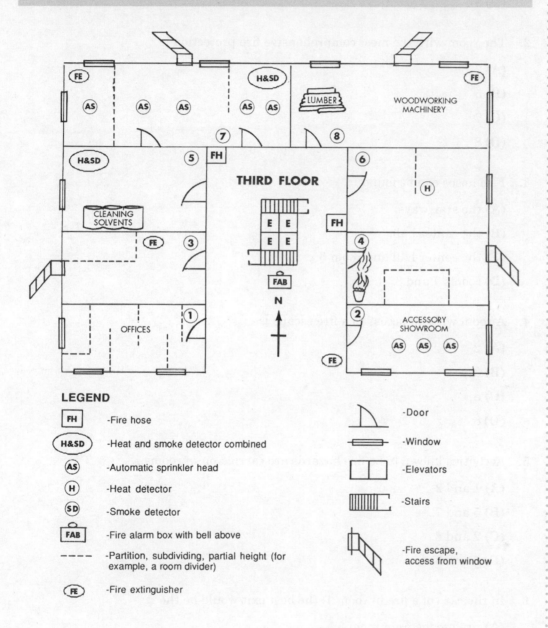

**LEGEND**

| FH | -Fire hose |
| H&SD | -Heat and smoke detector combined |
| AS | -Automatic sprinkler head |
| H | Heat detector |
| SD | -Smoke detector |
| FAB | -Fire alarm box with bell above |
| - - - - | -Partition, subdividing, partial height (for example, a room divider) |
| FE | -Fire extinguisher |

- Door
- Window
- Elevators
- Stairs
- Fire escape, access from window

exercises

1. The room most poorly protected in case of fire is

    **(A)** 1

    **(B)** 4

    **(C)** 6

    **(D)** 8

2. The room with the most comprehensive fire protection is

    **(A)** 2

    **(B)** 5

    **(C)** 7

    **(D)** 8

3. Fire hoses can be found in

    **(A)** the stairways.

    **(B)** the center hall.

    **(C)** the center hall and room 6.

    **(D)** rooms 7 and 8.

4. A room without access to a fire escape is

    **(A)** 2

    **(B)** 3

    **(C)** 5

    **(D)** 8

5. Activities known to be fire hazards are carried on in rooms

    **(A)** 1 and 2.

    **(B)** 5 and 7.

    **(C)** 2 and 8.

    **(D)** 5 and 8.

6. In the case of a fire in room 1, the best exit would be the

    **(A)** fire escape from room 3.

    **(B)** elevators.

    **(C)** north stairs.

    **(D)** south stairs.

7. The room in which a dangerous situation exists right now is

   **(A)** 2

   **(B)** 4

   **(C)** 6

   **(D)** 8

8. In the case of a fire, an alert would be sounded from

   **(A)** the east fire escape.

   **(B)** room 3.

   **(C)** the hall.

   **(D)** the elevators.

9. The intricate pattern of partitions would make quick exit most difficult from room

   **(A)** 1

   **(B)** 2

   **(C)** 6

   **(D)** 7

10. The total number of windows on this floor is

    **(A)** 14

    **(B)** 16

    **(C)** 18

    **(D)** 20

## ANSWERS AND EXPLANATIONS

| 1. | A | 3. | B | 5. | D | 7. | B | 9. | A |
|----|---|----|---|----|---|----|---|----|---|
| 2. | C | 4. | A | 6. | D | 8. | C | 10. | C |

1.  **The correct answer is (A).** Room 1 has no fire-protective device of any kind and has no independent exit to the outside. Room 6 is not much better off, but at least it has a heat detector. Room 4 has a fire escape but no internal protective device. Room 8 is well-protected.

2.  **The correct answer is (C).** Room 7 has automatic sprinklers, a heat and smoke detector, a fire extinguisher, and a fire escape. Room 2 has only automatic sprinklers, and room 5 has only a heat and smoke detector. Room 8 has a smoke detector, a fire extinguisher, and a fire escape, but it lacks a sprinkler system.

3.   **The correct answer is (B).** Fire hoses are only in the hallway.

4.   **The correct answer is (A).** Room 2 has no access to a fire escape. Rooms 3 and 8 have fire escapes, and room 5 has a door connecting it with room 7, which has a fire escape.

5.   **The correct answer is (D).** Use and storage of cleaning solvents and woodworking are activities that can pose fire hazards. The usual activities in offices and accessory showrooms do not present fire hazards. The nature of the business in the other areas is unspecified.

6.   **The correct answer is (D).** Room 1 is not connected to a fire escape, so its occupants would have to go into the hall to escape the fire. Once in the hall, the quickest way out would be via the nearest (the south) stairway. Entering room 3 to go out the window and down the fire escape would be slower and more cumbersome. In addition, room 3 is the next room to become involved in the fire. Elevators should be avoided in a fire situation.

7.   **The correct answer is (B).** A wastebasket is on fire in room 4.

8.   **The correct answer is (C).** There is a fire alarm box topped by a bell in the hall next to the south staircase. The device in room 3 is a fire extinguisher, not an alarm.

9.   **The correct answer is (A).** Room 1 is full of half partitions.

10.   **The correct answer is (C).** Be sure to count the windows connected to fire escapes. There are 18 windows in all.

In studying this diagram, you should have noticed:

❶ **People.** None.

❷ **Fire activity.** Wastebasket fire in room 4.

❸ **Layout.** Center core building with four elevators and two sets of stairs. All rooms have doors to the hallway and at least one window. Corner rooms have multiple windows. Rooms 5 and 7 connect. Rooms 3, 4, 7, and 8 have fire escapes.

❹ **Special details.** Two fire hoses, a fire extinguisher, and a fire alarm box with bell in the hall. Fire-hazardous activities in rooms 5 and 8. Fire extinguishers in rooms 3, 7, and 8 and in the hall. Sprinkler systems in rooms 2 and 7. Heat detector in room 6. Smoke detector in room 8. Heat and smoke detectors in rooms 5 and 7. Diagram is of the third floor of the building. Extensive legend.

## EXERCISE 5: OBSERVATION AND MEMORY

**Directions:** Study the sketch for 5 minutes, set it aside, answer the multiple-choice questions, and then study the answer discussions.

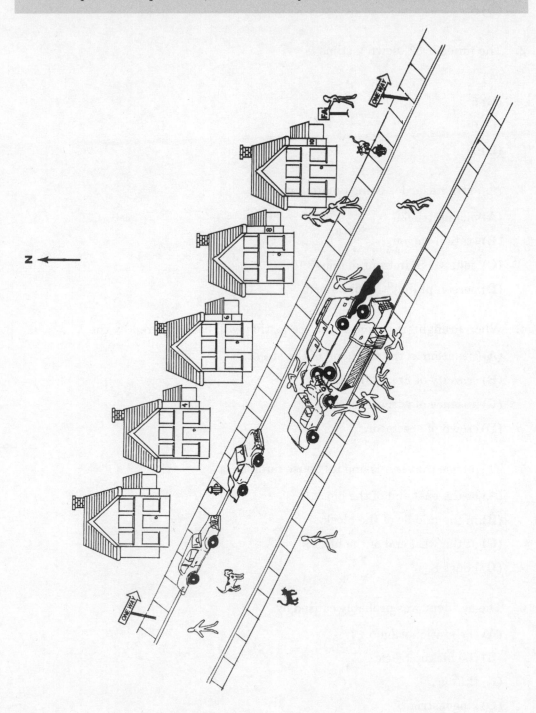

1. The number of vehicles involved in the accident is

   **(A)** 2

   **(B)** 3

   **(C)** 4

   **(D)** 5

2. The number of known victims is

   **(A)** 4

   **(B)** 5

   **(C)** 6

   **(D)** 7

3. An extra hazard is being created by the

   **(A)** pickup truck.

   **(B)** station wagon.

   **(C)** pedestrian in front of house 10.

   **(D)** person pulling the fire alarm.

4. When firefighters arrive at the scene, they will be hampered by the

   **(A)** situation at the west end of the street.

   **(B)** scarcity of fire hydrants.

   **(C)** absence of witnesses.

   **(D)** crowd of spectators.

5. The house that is missing its house number is

   **(A)** at the east end of the block.

   **(B)** in the middle of the block.

   **(C)** at the west end of the block.

   **(D)** house 8.

6. The accident was probably caused by

   **(A)** the station wagon.

   **(B)** the pickup truck.

   **(C)** the car.

   **(D)** a pedestrian.

7. The animals on the scene are

   (A) two dogs and a cat.

   (B) two cats and a dog.

   (C) three cats and a dog.

   (D) two cats and two dogs.

8. The house that is different from the others is house

   (A) 4

   (B) 6

   (C) 10

   (D) All are exactly alike.

9. The vehicle that is totally overturned is the

   (A) station wagon.

   (B) sedan.

   (C) pickup truck.

   (D) convertible.

10. When firefighters arrive on the scene, they must first direct their attention to the

    (A) victims.

    (B) houses in danger of catching fire.

    (C) cars on fire.

    (D) bystanders.

## ANSWERS AND EXPLANATIONS

| 1. | B | 3. | B | 5. | C | 7. | B | 9. | C |
|----|---|----|---|----|---|----|---|----|---|
| 2. | C | 4. | A | 6. | A | 8. | D | 10. | A |

1. **The correct answer is (B).** Three vehicles are piled up in the middle of the street. The other two cars are parked.

2. **The correct answer is (C).** Six people are lying in the street. We have no way of knowing how many are in the cars.

3. **The correct answer is (B).** The puddle forming from the back end of the station wagon most likely is gasoline. With the three vehicles actively on fire, a gasoline puddle creates a serious explosion threat.

4. **The correct answer is (A).** At the west end of the street, two cars are parked directly in front of and completely blocking access to the fire hydrant. This is both dangerous and illegal.

5. **The correct answer is (C).** House 2 at the far west end of the block is missing its house number.

6. **The correct answer is (A).** Note the one-way signs at each end of the street. It is possible that one or more of the vehicles spun around on impact, but on the face of the situation, it appears that the station wagon caused the accident by driving the wrong way on a one-way street.

7. **The correct answer is (B).** There are two cats and a dog in the scene. The animals have no bearing on the fire problem, but an observant test-taker must notice them.

8. **The correct answer is (D).** All the houses are exactly alike.

9. **The correct answer is (C).** The pickup truck has completely overturned and is resting on the top of its cab. The other two vehicles are more or less upright.

10. **The correct answer is (A).** Saving lives is more important than saving property. The bystanders must get out of the way by themselves; the victims need assistance.

In studying this sketch, you should have noticed:

❶ **People.** Six lying in the street, one turning in a fire alarm at the street box, five approaching the scene.

❷ **Fire activity.** Three vehicles apparently on fire after a collision.

❸ **Layout.** Three cars in a pileup in the middle of street, one upside-down. Five identical houses in a row. Two parked cars blocking one hydrant. Second hydrant at the end of the street. Fire alarm box on corner.

❹ **Special details.** Puddle coming from the rear end of the station wagon. An overturned pickup truck. One-way street, eastbound. Two cats and one dog.

# Mechanical Reasoning

## OVERVIEW

- What do mechanical reasoning questions measure?

## WHAT DO MECHANICAL REASONING QUESTIONS MEASURE?

The following questions evaluate your understanding of some basic mechanical principles. Some questions simply ask you to identify a pictured tool or to specify the tool's function. Other questions ask you to predict the outcomes of mechanical activities or to reason backward from an event to a probable mechanical cause.

The firefighter exam does not assume that you have knowledge about specific firefighting equipment, tools, or techniques. It does assume, however, that you have a basic understanding of things that are mechanical in nature. Firefighters rely heavily on the machinery, tools, and safety equipment that help them perform their job successfully and safely. Therefore, it is important for a firefighter to be able to predict the outcome of a mechanical activity or to find the mechanical cause of an event.

The following questions provide you with a broad selection of the types of mechanical reasoning questions you might encounter on the firefighter exam. A full set of explanations follows the questions.

chapter 8

# EXERCISE: MECHANICAL REASONING

> **Directions:** Choose the best answer and circle the corresponding letter.

1. The main reason a fiberglass ladder would be used in place of an aluminum ladder is when there are concerns about the

   **(A)** length of the ladder.

   **(B)** conductivity of the ladder.

   **(C)** flammability of the ladder.

   **(D)** "rust-resistance" of the ladder.

2. The main advantage of four-wheel drive is

   **(A)** higher speed.

   **(B)** better traction.

   **(C)** better gas mileage.

   **(D)** greater durability of the vehicle.

3. One complete revolution of the sprocket wheel below will bring weight W2 higher than weight W1 by

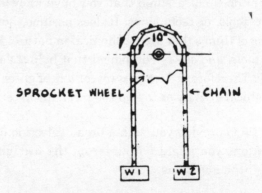

   **(A)** 20 inches.

   **(B)** 40 inches.

   **(C)** 30 inches.

   **(D)** 50 inches.

**4.** If you are cooking on your stove at home and the grease in your frying pan catches fire, the method that is NOT proper for putting out the fire is

**(A)** extinguishing the fire with carbon dioxide.

**(B)** covering the pan with its lid.

**(C)** pouring water on the fire.

**(D)** extinguishing the fire with a noncombustible foam.

**5.** In the figure shown here, assume that all valves are closed. For air to flow from R through G and then through S to M, open valves

**(A)** 1, 2, 6, and 4.

**(B)** 7, 3, and 4.

**(C)** 7, 6, and 4.

**(D)** 7, 3, and 5.

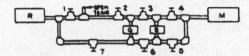

**6.** Which of the following would be best for cutting wire?

(A)

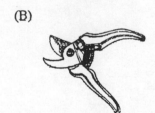

(B)

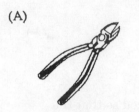

(C)

(D)

7. The function of the generator or alternator is to

   **(A)** start the engine.

   **(B)** carry electricity from the battery to the engine.

   **(C)** keep the battery charged.

   **(D)** control the production of hydrocarbons.

8. The principal objection to using water from a hose to put out a fire involving electrical equipment is that

   **(A)** serious shock might result.

   **(B)** it might spread the fire.

   **(C)** metal parts might rust.

   **(D)** fuses might blow out.

9. The following figure shows four gears. If gear 1 turns as shown, the gears turning in the same direction are

   **(A)** 2, 3, and 4.

   **(B)** 2 and 4.

   **(C)** 2 and 3.

   **(D)** 3 and 4.

   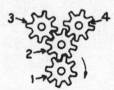

10. Light fixtures suspended from chains should be wired so that the

    **(A)** wires do not support the fixture.

    **(B)** wires help support the fixture.

    **(C)** chains have an insulated link.

    **(D)** chain is not grounded to prevent short circuits.

11. Hose X is 50 feet long and $\frac{1}{2}$ inch wide, and hose Y is 50 feet long and 1 inch wide. If they begin filling at the same time and an equal amount of water pressure is used to fill each hose,

    **(A)** hose X will fill with water first.

    **(B)** hose Y will fill with water first.

    **(C)** they both will fill with water at the same time.

    **(D)** there is no way of knowing which hose will fill first.

**12.** In the figure on the right, all four springs are identical. In case 1 with the springs end to end, the stretch of each spring caused by the 5-pound weight is

**(A)** half as much as in case 2.

**(B)** the same as in case 2.

**(C)** twice as much as in case 2.

**(D)** four times as much as in case 2.

**13.** You might use this instrument if you wanted to

**(A)** pitch a tent.

**(B)** poke holes in a fabric.

**(C)** locate studs in a wall.

**(D)** drill holes at equal short distances along a board.

**14.** After using cotton rags to clean a spill with paint thinner, you should

**(A)** leave them in a pile.

**(B)** stuff them all in a metal can.

**(C)** stuff them all in a plastic container.

**(D)** hang them separately to dry.

**15.** One complete revolution of the following windlass drum will move the weight up

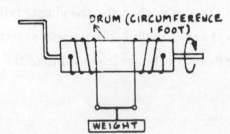

**(A)** $\frac{1}{2}$ foot.

**(B)** $1\frac{1}{2}$ feet.

**(C)** 1 foot.

**(D)** 2 feet.

**16.** The frequency of oiling and greasing bearings and other moving parts of machinery depends mainly on the

**(A)** size of the parts requiring lubrication.

**(B)** speed at which the parts move.

**(C)** ability of the operator.

**(D)** amount of use of the equipment.

17. The tool shown here is a

    **(A)** screw.

    **(B)** screwdriver.

    **(C)** drill bit.

    **(D)** corkscrew.

18. If pipes A and B are free to move back and forth but are held so they cannot turn and the coupling is turned four revolutions with a wrench, the overall length of the pipes and coupling will

    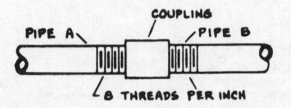

    **(A)** decrease $\frac{1}{2}$ inch.

    **(B)** remain the same.

    **(C)** increase or decrease 1 inch, depending on the direction of turning.

    **(D)** increase $\frac{1}{2}$ inch.

19. The plug of a portable tool should be removed from the convenience outlet by grasping the plug, not by pulling on the cord. This is because

    **(A)** the plug is easier to grip than the cord.

    **(B)** pulling on the cord might cause the plug to fall on the floor and break.

    **(C)** pulling on the cord might break the wires off the plug terminals.

    **(D)** the plug generally is better insulated than the cord.

20. The following figure is a device that is attached to the top of a ladder. Its purpose is to

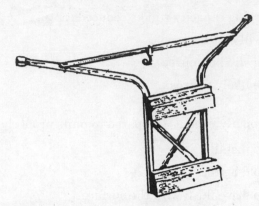

   **(A)** allow one to hang the ladder for easier storage.

   **(B)** increase the strength of the ladder.

   **(C)** give the ladder a greater amount of stability.

   **(D)** allow the ladder to be used upside down.

21. The figure represents an enclosed water chamber that is partially filled with water. The number 1 indicates air in the chamber, and 2 indicates a pipe by which water enters the chamber. If the water pressure in the pipe (2) increases, then the

   **(A)** water pressure in the chamber will be decreased.

   **(B)** water level in the chamber will fall.

   **(C)** air in the chamber will be compressed.

   **(D)** air in the chamber will expand.

22. Products stored in aerosol cans often have warnings against storing such containers in areas that might exceed 120°F. The main reason for these warnings is that high temperatures might cause the contents of the container to

   **(A)** shrink.

   **(B)** shift.

   **(C)** separate.

   **(D)** expand.

23. When removing the insulation from a wire before making a splice, care should be taken to avoid nicking the wire mainly because the

    (A) current carrying capacity will be reduced.

    (B) resistance will be increased.

    (C) wire tinning will be injured.

    (D) wire is more likely to break.

24. Wood screws properly used as compared to nails properly used

    (A) are easier to install.

    (B) generally hold better.

    (C) are easier to drive flush with the surface.

    (D) are more likely to split the wood.

25. If a hose releases water at 2,500 psi, the pressure of the water coming out of the hose is

    (A) greater than if it were 1,500 psi.

    (B) less than if it were 1,500 psi.

    (C) about the same as if it were 1,500 psi.

    (D) close to that found in a household sink.

26. The figure shows a cutter and a steel block. For proper cutting, they should move respectively in directions

    (A) 1 and 4.

    (B) 2 and 3.

    (C) 1 and 3.

    (D) 2 and 4.

27. The reading shown on the gauge is

    (A) 10.35

    (B) 13.5

    (C) 10.7

    (D) 17.0

28. An electric light bulb operated at more than its rated voltage will result in a

    (A) longer life and dimmer light.

    (B) longer life and brighter light.

    (C) shorter life and brighter light.

    (D) shorter life and dimmer light.

**29.** The wrench used principally for pipe work is

(A)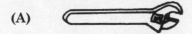

(B)

(C)

(D)

**30.** A slight coating of rust on small tools is best removed by

**(A)** applying a heavy coat of petroleum jelly.

**(B)** rubbing with kerosene and fine steel wool.

**(C)** scraping with a sharp knife.

**(D)** rubbing with a dry cloth.

**31.** Which of the following is most likely to have difficulty stopping at high speeds?

(A)

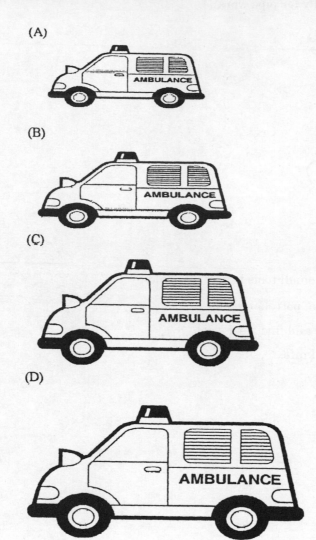

(B)

(C)

(D)

**32.** To bring the level of the water in the tanks to a height of $2\frac{1}{2}$ feet, the quantity of water to be added is

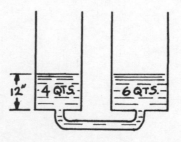

**(A)** 10 quarts.

**(B)** 20 quarts.

**(C)** 15 quarts.

**(D)** 25 quarts.

**33.** The outlet that will accept the plug is

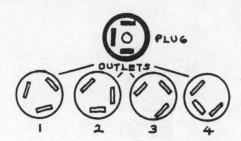

(A) 1

(B) 2

(C) 3

(D) 4

**34.** The micrometer shown here reads

(A) 0.2270

(B) 0.2120

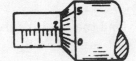

(C) 0.2252

(D) 0.2020

**35.** When a spark plug ignites fuel in the combustion chamber,

(A) you should stop the car immediately.

(B) the spark plug is functioning properly.

(C) the battery is not working.

(D) the vehicle is out of gas.

**36.** Which of the following statements is true?

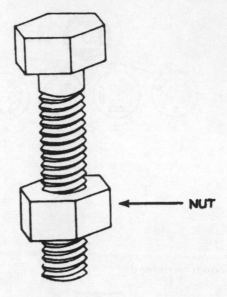

NUT

(A) If the nut is held stationary and the head is turned clockwise, the bolt will move up.

(B) If the head of the bolt is held stationary and the nut is turned clockwise, the nut will move down.

(C) If the head of the bolt is held stationary and the nut is turned clockwise, the nut will move up.

(D) If the nut is held stationary and the bolt is turned counterclockwise, the nut will move up.

**37.** A good lubricant for locks is

(A) graphite.

(B) grease.

(C) mineral oil.

(D) motor oil.

**38.** The best way to neutralize an acid is to mix it with a(n)

(A) acid of a weaker strength.

(B) neutral substance.

(C) base of equal strength.

(D) stronger acid.

**39.** The tool shown here is a

(A) punch.

(B) drill holder.

(C) Phillips-type screwdriver.

(D) socket wrench.

**40.** Wood ladders should not be painted because

(A) the paint will wear off rapidly due to the conditions under which ladders are used.

(B) ladders are slippery when painted.

(C) it is more effective to store the ladder in a dry place.

(D) paint will hide defects in the ladder.

**41.** If pipe A is held in a vise and pipe B is turned 10 revolutions with a wrench, the overall length of the pipes and coupling will decrease

(A) $\frac{5}{8}$ inch.

(B) $2\frac{1}{2}$ inches.

(C) $1\frac{1}{4}$ inches.

(D) $3\frac{3}{4}$ inches.

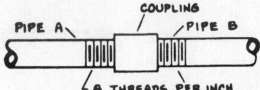

**42.** Which of the following saws is used to make curved cuts?

(A)

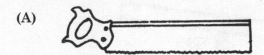

(B)

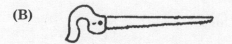

(C)

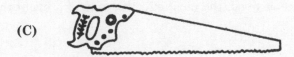

(D)

43. A method that can be used to prevent the formation of "skin" on a partially used can of oil paint is to

    **(A)** turn the can upside down every few months.

    **(B)** pour a thin layer of solvent over the top of the paint.

    **(C)** store the paint in a well-ventilated room.

    **(D)** avoid shaking the can after it has been sealed.

44. The reason that a lubricant prevents rubbing surfaces from becoming hot is that the oil

    **(A)** is cold and cools off the rubbing metal surfaces.

    **(B)** is sticky, preventing the surfaces from moving over each other too rapidly.

    **(C)** forms a smooth layer between the two surfaces, preventing them from coming into contact.

    **(D)** makes the surfaces smooth so they move easily over each other.

45. The tool shown here is used to

    **(A)** set nails.

    **(B)** drill holes in concrete.

    **(C)** cut a brick accurately.

    **(D)** centerpunch for holes.

46. If the block on which the lever is resting is moved closer to the brick, the brick will be

    **(A)** easier to lift and will be lifted higher.

    **(B)** harder to lift and will be lifted higher.

    **(C)** easier to lift but will not be lifted as high.

    **(D)** harder to lift and will not be lifted as high.

47. When driving a nail into a piece of wood, the most effective way to prevent the wood from splitting is to

    **(A)** use a carpenter's hammer.

    **(B)** first drill a small hole in the wood.

    **(C)** drive the nail into the wood slowly.

    **(D)** drive the nail into the wood quickly.

**48.** If the head of a hammer has become loose on the handle, it should be properly tightened by

(A) driving the handle further into the head.

(B) driving a nail alongside the present wedge.

(C) using a slightly larger wedge.

(D) soaking the handle in water.

**49.** The following figure represents a water tank containing water. The number 1 indicates an intake pipe, and the number 2 indicates a discharge pipe. Of the following, the least accurate statement is that the

(A) tank will eventually overflow if water flows through the intake pipe at a faster rate than it flows through the discharge pipe.

(B) tank will empty completely if the intake pipe is closed and the discharge pipe is allowed to remain open.

(C) water in the tank will remain at a constant level if the rate of intake is equal to the rate of discharge.

(D) water in the tank will rise if the intake pipe is operating when the discharge pipe is closed.

**50.** The tool shown here is most often used for cutting

(A) metal.

(B) wood.

(C) tile.

(D) glass.

**51.** The tool shown here is a(n)

(A) Allen-head wrench.

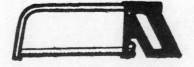

(B) double scraper.

(C) offset screwdriver.

(D) nail puller.

**52.** Wires often are spliced using a fitting like the one shown here. The use of this fitting does away with the need for

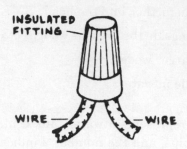

INSULATED FITTING

WIRE — — WIRE

(A) skinning.

(B) cleaning.

(C) twisting.

(D) soldering.

**53.** With which of the following screw heads do you use an Allen wrench?

(A)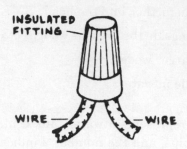

(B)

(C)

(D)

**54.** A trowel is a tool used for

(A) clearing out drain pipes.

(B) digging holes for footings.

(C) carving notches into wood.

(D) smoothing out cement.

55. The main reason some nails are galvanized is to make

   **(A)** them stronger.

   **(B)** the point sharper.

   **(C)** them smoother.

   **(D)** them rust resistant.

56. When driving in extremely low temperatures, a car's engine might overheat if the

   **(A)** heater is on for too long.

   **(B)** coolant freezes.

   **(C)** gas tank is not full enough.

   **(D)** oil recently was changed.

57. Which of the following would be best for digging into the ground?

   (A)

   (B)

   (C)

   (D)

**exercises**

**58.** The easiest way to chop through wood with an ax is to align the blade

    **(A)** with the grain.

    **(B)** across the grain at 90°.

    **(C)** across the grain at 45°.

    **(D)** around the grain.

**59.** The convenience outlet known as a polarized outlet is number

    **(A)** 1

    **(B)** 2

    **(C)** 3

    **(D)** 4

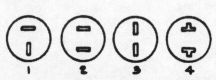

**60.** The tool used to measure the depth of a hole is

    **(A)**

    **(B)**

    **(C)**

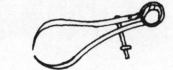

**61.** Neutral wire can quickly be recognized by the

    **(A)** greenish color.

    **(B)** bluish color.

    **(C)** natural or whitish color.

    **(D)** black color.

62. If a wrench were attached to the top of a fire hydrant, as pictured here, how would you turn the water on?

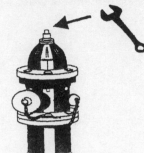

    **(A)** Pull the wrench up.

    **(B)** Turn the wrench counterclockwise.

    **(C)** Push the wrench down.

    **(D)** Turn the wrench clockwise.

63. If the flush tank of a water-closet fixture overflows, the fault is likely to be

    **(A)** failure of the ball to seat properly.

    **(B)** excessive water pressure.

    **(C)** a defective trap in the toilet bowl.

    **(D)** a waterlogged float.

64. When using a hammer with one hand, to use the greatest amount of force, you should hold the hammer

    **(A)** near the end of the handle.

    **(B)** in the middle of the handle.

    **(C)** near the head of the hammer.

    **(D)** by the head of the hammer.

**65.** Study the gear wheels in the figure shown here and then determine which of the following statements is true.

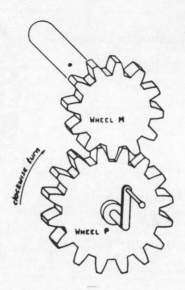

**(A)** If you turn wheel M clockwise by means of the handle, wheel P also will turn clockwise.

**(B)** It will take the same time for a tooth of wheel P to make a full turn as it will for a tooth of wheel M.

**(C)** It will take less time for a tooth of wheel P to make a full turn than it will take a tooth of wheel M.

**(D)** It will take more time for a tooth of wheel P to make a full turn than it will for a tooth of wheel M.

**66.** Locknuts frequently are used in making electrical connections on terminal boards. The purpose of the locknuts is to

**(A)** eliminate the use of flat washers.

**(B)** prevent unauthorized personnel from tampering with the connections.

**(C)** keep the connections from loosening through vibration.

**(D)** increase the contact area at the connection point.

**67.** When dragging a slightly leaky, cloth-covered hose filled with water across a gravel surface, the element most likely to make the task difficult is the

**(A)** cloth hose covering.

**(B)** gravel surface.

**(C)** leaking water.

**(D)** weight of the water.

**68.** The tool shown here is used to

(A) ream holes in wood.

(B) countersink holes in soft metals.

(C) turn Phillips-head screws.

(D) drill holes in concrete.

**69.** The tool shown here is used for

(A) soldering.

(B) caulking.

(C) shooting.

(D) scoring.

**70.** A squeegee is a tool used in

(A) drying windows after washing.

(B) cleaning inside boiler surfaces.

(C) the central vacuum cleaning system.

(D) clearing stoppage in waste lines.

**71.** Boxes and fittings intended for outdoor use should be

(A) of weatherproof type.

(B) stamped steel of not less than No. 16.

(C) standard gauge.

(D) stamped steel plated with cadmium.

**72.** The device used to change AC to DC is a

(A) frequency changer.

(B) regulator.

(C) transformer.

(D) rectifier.

**73.** The figure shown here shows a governor on a rotating shaft. As the shaft speeds up, the governor balls will move

(A) down.

(B) upward and inward.

(C) upward.

(D) inward.

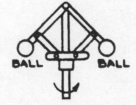

74. The purpose of an air valve in a heating system is to

    **(A)** prevent pressure from building up in a room due to the heated air.

    **(B)** relieve the air from steam radiators.

    **(C)** allow excessive steam pressure in the boiler to escape to the atmosphere.

    **(D)** control the temperature in the room.

75. If a fuse of higher than the required current rating is used in an electrical circuit,

    **(A)** better protection will be afforded.

    **(B)** the fuse will blow more often because it carries more current.

    **(C)** serious damage might result to the circuit from overload.

    **(D)** maintenance of the large fuse will be higher.

76. If the following dark lines represent different ways a ladder can be set up, which line represents the safest way to set up a ladder?

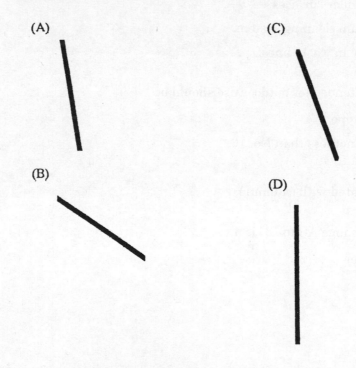

77. Paint is "thinned" with

(A) linseed oil.

(B) varnish.

(C) turpentine.

(D) gasoline.

78. The tool shown here is a(n)

(A) offset wrench.

(B) box wrench.

(C) spanner wrench.

(D) open end wrench.

79. If gear A makes one clockwise revolution per minute, which one of the following is true?

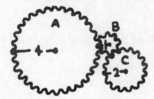

(A) Gear B makes one counterclockwise revolution every 4 minutes.

(B) Gear C makes two clockwise revolutions every minute.

(C) Gear B makes four clockwise revolutions every minute.

(D) Gear C makes one counterclockwise revolution every 8 minutes.

80. Referring to the figure, which one of the following statements is true?

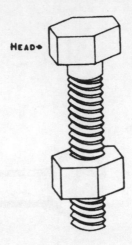

HEAD→

(A) If the nut is held stationary and the head is turned clockwise, the bolt will move down.

(B) If the head of the bolt is held stationary and the nut is turned clockwise, the nut will move down.

(C) If the head of the bolt is held stationary and the nut is turned clockwise, the nut will move up.

(D) If the nut is held stationary and the head is turned counterclockwise, the bolt will move up.

81. If the float of a flush tank leaks and fills with water, the most probable result is that

(A) there will be no water in the tank.

(B) the ball cock will remain open.

(C) water will flow over the tank rim onto the floor.

(D) the flush ball will not seat properly.

82. If the temperature gauge indicates that the engine is getting overheated,

(A) allow it to cool down.

(B) pour in cold water immediately.

(C) pour in hot water immediately.

(D) pour in a cooling antifreeze at once.

83. A crack in the distributor cap might cause problems by allowing

(A) oil to leak out.

(B) electricity to leak out.

(C) fuel to seep in.

(D) moisture to seep in.

**84.** The sketch here shows a head-on view of a three-pronged plug used with portable electrical power tools. Considering the danger of shock when using such tools, it is evident that the function of the U-shaped prong is to

**(A)** ensure that the other two prongs enter the outlet with the proper polarity.

**(B)** provide a half-voltage connection when doing light work.

**(C)** prevent accidental pulling of the plug from the outlet.

**(D)** connect the metallic shell of the tool motor to ground.

**85.** The object pictured here is a

**(A)** spark plug.

**(B)** lug nut.

**(C)** drill bit.

**(D)** crank shaft.

## ANSWERS AND EXPLANATIONS

| | | | | | | | |
|---|---|---|---|---|---|---|---|
| 1. | B | 23. | D | 45. | B | 67. | D |
| 2. | B | 24. | B | 46. | C | 68. | C |
| 3. | B | 25. | A | 47. | B | 69. | B |
| 4. | C | 26. | C | 48. | C | 70. | A |
| 5. | D | 27. | D | 49. | B | 71. | A |
| 6. | A | 28. | C | 50. | A | 72. | D |
| 7. | C | 29. | C | 51. | C | 73. | C |
| 8. | A | 30. | B | 52. | D | 74. | B |
| 9. | D | 31. | D | 53. | C | 75. | C |
| 10. | A | 32. | C | 54. | D | 76. | C |
| 11. | A | 33. | C | 55. | D | 77. | C |
| 12. | C | 34. | A | 56. | B | 78. | D |
| 13. | D | 35. | B | 57. | C | 79. | B |
| 14. | D | 36. | B | 58. | A | 80. | C |
| 15. | C | 37. | A | 59. | A | 81. | B |
| 16. | D | 38. | C | 60. | B | 82. | A |
| 17. | C | 39. | D | 61. | C | 83. | D |
| 18. | B | 40. | D | 62. | B | 84. | D |
| 19. | C | 41. | C | 63. | D | 85. | A |
| 20. | C | 42. | B | 64. | A | | |
| 21. | C | 43. | B | 65. | D | | |
| 22. | D | 44. | C | 66. | C | | |

1. **The correct answer is (B).** The main reason a fiberglass ladder would be used in place of an aluminum ladder is if there were concerns about conductivity or the risk of electrical shock from nearby electrical wires. A fiberglass ladder does not have any advantages in length, choice (A), over an aluminum ladder. Fiberglass and aluminum are not flammable, choice (C), nor are they susceptible to rusting, choice (D).

2. **The correct answer is (B).** Because power from the engine is transmitted to both the front wheels and the back wheels, a four-wheel drive vehicle has much better traction. Four-wheel drive is especially useful in mud, sand, and very uneven terrain. Although four-wheel drive tends to be a feature of durable vehicles, choice (D), such as Jeeps, it is not the four-wheel drive that makes them durable.

3. **The correct answer is (B).** One half the circumference of the sprocket wheel is 10 inches; therefore, the entire circumference is 20 inches. In one complete revolution of the wheel, the chain will move 20 inches. As weight

answers

2 moves up 20 inches, weight 1 will move down 20 inches. The difference between the heights of the two weights will be 40 inches.

4. **The correct answer is (C).** Water and grease do not mix. Therefore, if you pour water on a grease fire, it might cause the grease to splatter and actually spread the fire. Fire feeds off oxygen, so extinguishing it with carbon dioxide would work, choice (A). Covering the pan would cut off the oxygen supply, choice (B), and a noncombustible foam also would cut off the oxygen supply, choice (D).

5. **The correct answer is (D).** The air from R must follow a route down through valve 7, up through G, through valve 3, down through S, to the right through 5, and then up and over to M. The air could not pass through valves 1 and 2 to G because it would escape through the opening between valves 1 and 2. If either valve 4 or valve 6 were to be opened, the air would be diverted from the appointed route.

6. **The correct answer is (A).** This is a wire cutter. Choice (B) is a pruning clipper. Choice (C) is a hedge clipper. Choice (D) is a pair of pliers.

7. **The correct answer is (C).** The generator or alternator is operated by the car's engine. It produces electricity that flows to the battery and keeps the battery charged.

8. **The correct answer is (A).** Water is not likely to spread an electrical fire, choice (B), though it might be ineffective at putting it out. It is true that metal parts might rust if soaked with water, choice (C), and that fuses might blow, choice (D), but these are not serious considerations. There is a real danger from fatal shock if water is used on an electrical fire. The danger is even greater if that water comes from a hose because there would then be a great deal of water on the floor and the firefighters would be standing in it. Electricity travels rapidly through water, and the people standing in the water would be its targets.

9. **The correct answer is (D).** A turning gear always turns the gear with which it interlocks in the opposite direction. If gear 1 turns clockwise, gear 2 must turn counterclockwise. In turn, gears 3 and 4, because they are both turned by gear 2, must both turn clockwise.

10. **The correct answer is (A).** The answer to this question is pure common sense. Electrical wires should serve only one purpose—to supply electricity. If electrical wires are required to support weight, there is danger of breakage in the wires and of damage to the fixture itself caused by tension at the connections.

11. **The correct answer is (A).** Hose X and hose Y are the same length and are filled starting at the same time with the same amount of water pressure. Hose X, however, is only $\frac{1}{2}$ inch wide; hose Y is 1 inch wide. This means that hose X has less space inside it and will fill up faster than hose Y.

12. **The correct answer is (C).** In case 2, each spring bears one half the weight of the 5-pound weight. Each spring in case 2 is therefore stretched by $2\frac{1}{2}$ pounds. In case 1, each spring bears a full 5-pound load. Each spring in case 1 must be stretched twice as much as each spring in case 2.

13. **The correct answer is (D).** The compass shown would be very useful in marking out equal short distances on a board. The compass would not, of course, be of use in the actual drilling.

14. **The correct answer is (D).** Rags used to clean a spill with paint thinner should always be allowed to dry after use. Leaving them in a pile, choice (A), stuffing them in a metal can, choice (B), or stuffing them in a plastic container, choice (C), might cause the rags to catch fire due to the paint thinner on them.

15. **The correct answer is (C).** Because the circumference of the drum is 1 foot, one complete revolution of the drum will take up 1 foot of each rope. As each of the separate ropes supporting the weight is shortened by 1 foot, the weight will move up 1 foot.

16. **The correct answer is (D).** Lubrication of machinery is scheduled according to time elapsed and amount of use. All the other reasons offered are irrelevant.

17. **The correct answer is (C).** The tool is a drill bit. The smooth end is inserted into the drill chuck and is tightened so it is very secure. When the drill motor goes, the bit turns very fast and drills holes wherever it is applied.

18. **The correct answer is (B).** If the coupling is turned but the pipes are held firm so they cannot turn, the coupling will move along the length of one or the other of the pipes, but the overall length of the three pieces will remain the same.

19. **The correct answer is (C).** Yanking at an electric cord can cause hidden damage inside the plug. The force can cause terminals to loosen or bits of wire to break off and fray. Frayed wires can, in turn, come into contact with the opposite pole, causing a fire or short circuit in the plug at some later date.

20. **The correct answer is (C).** The figure is a picture of a stabilizer. This device is attached to the top of a ladder to give it a greater amount of stability by reducing the likelihood that the ladder will slide to the left or the right.

21. **The correct answer is (C).** If water pressure in the pipe is increased, more water will flow into the water chamber. Because the chamber is enclosed, the air will be unable to escape. As more water enters the chamber, the existing air must be compressed into a smaller space.

22. **The correct answer is (D).** The main reason for storing aerosol cans in temperatures that do not exceed 120°F is because high temperatures might cause the contents of the can to expand. This might then cause the can to burst.

23. **The correct answer is (D).** A nick in a wire can be dangerous because it weakens the wire at that point and can lead to breakage. If the wire is nicked during stripping, you should cut off the weakened portion and begin again. Later breakage from an unnoticed weakness can lead to a short circuit.

24. **The correct answer is (B).** Wood screws are usually more difficult to install than nails, but they often are preferable because they generally hold better and are less likely to split the wood.

25. **The correct answer is (A).** Psi refers to how much water pressure is present in pounds per square inch. Therefore, 2,500 psi is much greater than 1,500 psi. This amount of water pressure is far greater than that coming out of a household sink, choice (D).

26. **The correct answer is (C).** Common sense should give you the answer to this question. For the cutter to cut the block, the two must be in contact. Obviously, the block must move in direction 3 to make contact with the cutter. For a cutter to cut, it must move in a direction that enables the sharp edge of its teeth to bite into the object being cut. In direction 1, the teeth will bite into the block. In direction 2, the back edge of the teeth will just slide off the block.

27. **The correct answer is (D).** Each division marks 2 units: 20 units/10 divisions = 2 units/division. The pointer is $3\frac{1}{2}$ divisions above 10: 2 units/division $\times 3\frac{1}{2}$ divisions = 7 + 10 = 17.

28. **The correct answer is (C).** If an electric light bulb is operated at more than its rated voltage, the extra surge of electricity will cause the bulb to burn more brightly. The same excess electrical force, however, will weaken the filament and will cause the bulb to burn out more quickly.

29. **The correct answer is (C).** This is a pipe wrench. Choice (A) is a crescent or expandable wrench. Choice (B) is a ratchet wrench. Choice (D) is an open end wrench.

30. **The correct answer is (B).** Kerosene and fine steel wool will effectively remove a thin layer of rust. Petroleum jelly, choice (A), can prevent rust but not remove it. A sharp knife, choice (C), might chip off thick crusty rust but not a thin coat. A dry cloth, choice (D), alone will not remove rust.

31. **The correct answer is (D).** A larger vehicle is likely to be heavier and therefore will have more difficulty stopping at high speeds.

32. **The correct answer is (C).** Ten quarts of water have brought the water level to 1 foot. An additional 15 quarts would raise the water level by $1\frac{1}{2}$ feet to a total height of $2\frac{1}{2}$ feet.

33. **The correct answer is (C).** This shape plug will only go into a similarly shaped socket. The other outlets are of different shapes.

34. **The correct answer is (A).** The measurements that can be made on the micrometer are: (1) 2 major divisions and 1 minor division on the ruler-type scale: .2 + .025 = .225; and (2) 2 minor divisions above 0 on the rotating scale: .002. Summing, we find the final measurement is .225 + .002 = .227.

35. **The correct answer is (B).** The main purpose of a spark plug is to ignite fuel vapors in the combustion chamber. The battery must be working for this to occur, choice (C); the vehicle must have gas in it for this to occur, choice (D); and this is definitely not a reason to stop the car, choice (A).

36. **The correct answer is (B).** Clockwise is left to right. If the nut moves, it follows the threads of the bolt downward.

37. **The correct answer is (A).** Graphite, which is powdered carbon, is very slippery and will not bind the small springs and metal parts of a lock.

38. **The correct answer is (C).** A base is a substance that is the "opposite" of an acid. The best way to neutralize an acid is to mix it with a base of equal strength. An acid of weaker strength will not have as good of an effect, choice (A). A neutral substance will work better than an acid, but it will not completely neutralize the acid, choice (B). A stronger acid will only make the acid stronger.

39. **The correct answer is (D).** Although this tool looks like a screwdriver, the head will fit into a hex nut and works like a socket wrench.

40. **The correct answer is (D).** You cannot see a defect in a painted ladder such as a knot or a split in the wood. A ladder should never be painted.

41. **The correct answer is (C).** The overall length of the pipes and coupling could decrease or increase, depending on the direction in which pipe B is turned. As stated in this question, however, pipe B is turned to disappear into the coupling. Because there are eight threads to the inch, eight complete revolutions of the pipe would shorten the pipes and coupling by 1 inch. An additional two turns, for a total of ten, would shorten the pipes and coupling by an additional $\frac{2}{8}$ or $\frac{1}{4}$ of an inch.

42. **The correct answer is (B).** This is a keyhole saw used to make curved cuts. Choice (A) is a backsaw. Choice (C) is a rip or crosscut saw. Choice (D) is a hacksaw.

43. **The correct answer is (B).** "Skin" forms when air combines with paint. To stop this from happening, pour a thin layer of solvent over the paint. The air will be prevented from reaching the paint and forming a layer of skin.

44. **The correct answer is (C).** When two pieces of metal rub together, the friction causes a great deal of heat. Oil reduces the friction between the two pieces of metal.

45. **The correct answer is (B).** The tool shown is a star drill. It is hit with a hammer to make a hole in concrete.

46. **The correct answer is (C).** If the block is moved toward the brick, the moment for a given force exerted will increase (being further from the force), making it easier to lift. The height will be made smaller, hardly raising the brick when moved to the limit (directly underneath it).

47. **The correct answer is (B).** By first drilling a small hole into the wood, you can reduce the amount of stress the nail places on the wood and can therefore reduce the likelihood that the wood will split.

48. **The correct answer is (C).** If you look at the top of a hammer where it is joined to the handle, you will see the top of either a wooden wedge or a metal wedge. Driving another wedge into the handle will tighten the hammer.

49. **The correct answer is (B).** If pipe 2 is open while pipe 1 is closed, the level will drop to the lowest level of pipe 2. This leaves the volume below pipe 2 still filled, having no way to discharge. All other statements are true.

50. **The correct answer is (A).** The tool shown is a hacksaw, which is used to cut metal.

51. **The correct answer is (C).** This tool is an offset screwdriver. It is used for tightening screws in hard-to-reach places where a regular screwdriver cannot turn in a complete revolution.

52. **The correct answer is (D).** This is a mechanical or solderless connector. It does away with the need to solder wires and is found in house wiring.

53. **The correct answer is (C).** An Allen wrench is hexagon-shaped and will fit into screw C.

54. **The correct answer is (D).** A trowel is a tool with a flat face that often is used to smooth out cement.

55. **The correct answer is (D).** A galvanized nail is covered with a protective coating that keeps the nail from rusting.

56. **The correct answer is (B).** It seems strange to think that a car could overheat when it is cold outside; however, extremely cold temperatures can freeze the engine coolant. This makes the coolant useless and leads to an overheated engine.

57. **The correct answer is (C).** This type of shovel is best for digging into the ground. The other shovels are used for shoveling snow, choices (A) and (D), and for shoveling grain, choice (B).

58. **The correct answer is (A).** Wood is weakest when a cut is made with the grain. Any angle that is across the grain will not be as effective, choices (B) and (C). There is no way to cut around the grain, choice (D).

59. **The correct answer is (A).** The plug can go into the outlet in only one way in a polarized outlet. In the other outlets, the plug can be reversed.

60. **The correct answer is (B).** The flattened part rests at the top of the hole, and the ruler is pushed down into the hole until it reaches the bottom. The depth of the hole is then read from the ruler.

61. **The correct answer is (C).** Neutral wire is whitish in color. The hot lead is black, and the ground wire is green.

62. **The correct answer is (B).** Just as with any water valve, you should turn it counterclockwise to turn the water on. Pulling the wrench up, choice (A), and pushing it down, choice (C), will have no effect. Turning the wrench clockwise will actually tighten the valve.

63. **The correct answer is (D).** The water shutoff valve on a flush tank is closed by the force of a lightweight ball rising inside the tank. If this float becomes waterlogged, it will not rise and shut off the water.

64. **The correct answer is (A).** Holding the hammer near the end of the handle enables you to take advantage of the weight of the hammer with each swing and will result in a more forceful blow.

65. **The correct answer is (D).** Wheel P has 16 teeth; wheel M has 12 teeth. When wheel M makes a full turn, wheel P will still have four more teeth to turn. Therefore, wheel P is slower and takes more time to turn.

66. **The correct answer is (C).** Locknuts are bent so that their metal edges will bite into the terminal board and will require the use of a wrench to loosen them.

67. **The correct answer is (D).** The weight of the water is most likely to make the task difficult because it makes the hose extremely heavy and hard to drag. The cloth hose covering will have little effect, choice (A); the gravel surface might actually make the task easier by providing a movable surface over which to drag the hose, choice (B); and the leaking water will have little consequence, choice (C).

68. **The correct answer is (C).** This is a Phillips-head screwdriver. It will turn Phillips-head screws with this shape.

69. **The correct answer is (B).** The tool pictured is a caulk gun and is used for caulking.

70. **The correct answer is (A).** A squeegee is a rubber wiper that removes water from a wet window.

71. **The correct answer is (A).** Outdoor boxes and fittings must be weatherproof to withstand any problems caused by moisture.

72. **The correct answer is (D).** A rectifier, or diode, is a device that changes AC to DC.

73. **The correct answer is (C).** The centrifugal force acts to pull the balls outward. Because the two balls are connected to a yolk around the center bar, this outward motion pulls the balls upward.

74. **The correct answer is (B).** An air valve on a radiator removes air from the steam pipes. If air is trapped in the pipes, it prevents the steam from going to the radiator. This prevents the radiator from producing heat.

75. **The correct answer is (C).** Never use a fuse with a rating higher than the one specifically called for in the circuit. A fuse is a safety device used to protect a circuit from serious damage caused by too high a current.

76. **The correct answer is (C).** This is the best angle for a ladder setup. Both choices (A) and (D) might cause a person standing at the top to fall over backward. Choice (B) might cause the bottom of the ladder to slip backward, causing the ladder to fall.

77. **The correct answer is (C).** Paint is made thinner or easier to apply by diluting it with turpentine. Linseed oil and varnish are not used as paint thinners.

78. **The correct answer is (D).** The opened face on this tool shows that it is an open end wrench.

79. **The correct answer is (B).** Gear A turns in the opposite direction from gear B. A clockwise turn of A results in a counterclockwise revolution of B. Because the distance traversed by A (perimeter = $\pi \times$ diameter = $\pi \times 4$) is twice that of C (perimeter = $\pi \times 2$), the speed of C is doubled.

80. **The correct answer is (C).** To tighten the bolt, turn it counterclockwise. To tighten the nut on the bolt, the reverse is true—turn it clockwise.

81. **The correct answer is (B).** The ball cock will remain open, the float will not rise with the incoming water, and the tank will continue to fill because the shutoff valve will not be shut off.

82. **The correct answer is (A).** If an engine starts to overheat, you must stop the car immediately. Otherwise, the metal will expand and damage the engine. If you pour anything on an overheated engine, the rapid cooling might cause the block to crack.

83. **The correct answer is (D).** A crack in the distributor cap might cause moisture to seep in and cause a short in the system. This might prevent the car from starting.

84. **The correct answer is (D).** The third prong in the plug is the grounding wire.

85. **The correct answer is (A).** The object pictured is a spark plug.

# PART III
## PRACTICE TESTS

# Practice Test 1

**ANSWER SHEET**

| | | | |
|---|---|---|---|
| 1. Ⓐ Ⓑ Ⓒ Ⓓ Ⓔ | 26. Ⓐ Ⓑ Ⓒ Ⓓ Ⓔ | 51. Ⓐ Ⓑ Ⓒ Ⓓ Ⓔ | 76. Ⓐ Ⓑ Ⓒ Ⓓ Ⓔ |
| 2. Ⓐ Ⓑ Ⓒ Ⓓ Ⓔ | 27. Ⓐ Ⓑ Ⓒ Ⓓ Ⓔ | 52. Ⓐ Ⓑ Ⓒ Ⓓ Ⓔ | 77. Ⓐ Ⓑ Ⓒ Ⓓ Ⓔ |
| 3. Ⓐ Ⓑ Ⓒ Ⓓ Ⓔ | 28. Ⓐ Ⓑ Ⓒ Ⓓ Ⓔ | 53. Ⓐ Ⓑ Ⓒ Ⓓ Ⓔ | 78. Ⓐ Ⓑ Ⓒ Ⓓ Ⓔ |
| 4. Ⓐ Ⓑ Ⓒ Ⓓ Ⓔ | 29. Ⓐ Ⓑ Ⓒ Ⓓ Ⓔ | 54. Ⓐ Ⓑ Ⓒ Ⓓ Ⓔ | 79. Ⓐ Ⓑ Ⓒ Ⓓ Ⓔ |
| 5. Ⓐ Ⓑ Ⓒ Ⓓ Ⓔ | 30. Ⓐ Ⓑ Ⓒ Ⓓ Ⓔ | 55. Ⓐ Ⓑ Ⓒ Ⓓ Ⓔ | 80. Ⓐ Ⓑ Ⓒ Ⓓ Ⓔ |
| 6. Ⓐ Ⓑ Ⓒ Ⓓ Ⓔ | 31. Ⓐ Ⓑ Ⓒ Ⓓ Ⓔ | 56. Ⓐ Ⓑ Ⓒ Ⓓ Ⓔ | 81. Ⓐ Ⓑ Ⓒ Ⓓ Ⓔ |
| 7. Ⓐ Ⓑ Ⓒ Ⓓ Ⓔ | 32. Ⓐ Ⓑ Ⓒ Ⓓ Ⓔ | 57. Ⓐ Ⓑ Ⓒ Ⓓ Ⓔ | 82. Ⓐ Ⓑ Ⓒ Ⓓ Ⓔ |
| 8. Ⓐ Ⓑ Ⓒ Ⓓ Ⓔ | 33. Ⓐ Ⓑ Ⓒ Ⓓ Ⓔ | 58. Ⓐ Ⓑ Ⓒ Ⓓ Ⓔ | 83. Ⓐ Ⓑ Ⓒ Ⓓ Ⓔ |
| 9. Ⓐ Ⓑ Ⓒ Ⓓ Ⓔ | 34. Ⓐ Ⓑ Ⓒ Ⓓ Ⓔ | 59. Ⓐ Ⓑ Ⓒ Ⓓ Ⓔ | 84. Ⓐ Ⓑ Ⓒ Ⓓ Ⓔ |
| 10. Ⓐ Ⓑ Ⓒ Ⓓ Ⓔ | 35. Ⓐ Ⓑ Ⓒ Ⓓ Ⓔ | 60. Ⓐ Ⓑ Ⓒ Ⓓ Ⓔ | 85. Ⓐ Ⓑ Ⓒ Ⓓ Ⓔ |
| 11. Ⓐ Ⓑ Ⓒ Ⓓ Ⓔ | 36. Ⓐ Ⓑ Ⓒ Ⓓ Ⓔ | 61. Ⓐ Ⓑ Ⓒ Ⓓ Ⓔ | 86. Ⓐ Ⓑ Ⓒ Ⓓ Ⓔ |
| 12. Ⓐ Ⓑ Ⓒ Ⓓ Ⓔ | 37. Ⓐ Ⓑ Ⓒ Ⓓ Ⓔ | 62. Ⓐ Ⓑ Ⓒ Ⓓ Ⓔ | 87. Ⓐ Ⓑ Ⓒ Ⓓ Ⓔ |
| 13. Ⓐ Ⓑ Ⓒ Ⓓ Ⓔ | 38. Ⓐ Ⓑ Ⓒ Ⓓ Ⓔ | 63. Ⓐ Ⓑ Ⓒ Ⓓ Ⓔ | 88. Ⓐ Ⓑ Ⓒ Ⓓ Ⓔ |
| 14. Ⓐ Ⓑ Ⓒ Ⓓ Ⓔ | 39. Ⓐ Ⓑ Ⓒ Ⓓ Ⓔ | 64. Ⓐ Ⓑ Ⓒ Ⓓ Ⓔ | 89. Ⓐ Ⓑ Ⓒ Ⓓ Ⓔ |
| 15. Ⓐ Ⓑ Ⓒ Ⓓ Ⓔ | 40. Ⓐ Ⓑ Ⓒ Ⓓ Ⓔ | 65. Ⓐ Ⓑ Ⓒ Ⓓ Ⓔ | 90. Ⓐ Ⓑ Ⓒ Ⓓ Ⓔ |
| 16. Ⓐ Ⓑ Ⓒ Ⓓ Ⓔ | 41. Ⓐ Ⓑ Ⓒ Ⓓ Ⓔ | 66. Ⓐ Ⓑ Ⓒ Ⓓ Ⓔ | 91. Ⓐ Ⓑ Ⓒ Ⓓ Ⓔ |
| 17. Ⓐ Ⓑ Ⓒ Ⓓ Ⓔ | 42. Ⓐ Ⓑ Ⓒ Ⓓ Ⓔ | 67. Ⓐ Ⓑ Ⓒ Ⓓ Ⓔ | 92. Ⓐ Ⓑ Ⓒ Ⓓ Ⓔ |
| 18. Ⓐ Ⓑ Ⓒ Ⓓ Ⓔ | 43. Ⓐ Ⓑ Ⓒ Ⓓ Ⓔ | 68. Ⓐ Ⓑ Ⓒ Ⓓ Ⓔ | 93. Ⓐ Ⓑ Ⓒ Ⓓ Ⓔ |
| 19. Ⓐ Ⓑ Ⓒ Ⓓ Ⓔ | 44. Ⓐ Ⓑ Ⓒ Ⓓ Ⓔ | 69. Ⓐ Ⓑ Ⓒ Ⓓ Ⓔ | 94. Ⓐ Ⓑ Ⓒ Ⓓ Ⓔ |
| 20. Ⓐ Ⓑ Ⓒ Ⓓ Ⓔ | 45. Ⓐ Ⓑ Ⓒ Ⓓ Ⓔ | 70. Ⓐ Ⓑ Ⓒ Ⓓ Ⓔ | 95. Ⓐ Ⓑ Ⓒ Ⓓ Ⓔ |
| 21. Ⓐ Ⓑ Ⓒ Ⓓ Ⓔ | 46. Ⓐ Ⓑ Ⓒ Ⓓ Ⓔ | 71. Ⓐ Ⓑ Ⓒ Ⓓ Ⓔ | 96. Ⓐ Ⓑ Ⓒ Ⓓ Ⓔ |
| 22. Ⓐ Ⓑ Ⓒ Ⓓ Ⓔ | 47. Ⓐ Ⓑ Ⓒ Ⓓ Ⓔ | 72. Ⓐ Ⓑ Ⓒ Ⓓ Ⓔ | 97. Ⓐ Ⓑ Ⓒ Ⓓ Ⓔ |
| 23. Ⓐ Ⓑ Ⓒ Ⓓ Ⓔ | 48. Ⓐ Ⓑ Ⓒ Ⓓ Ⓔ | 73. Ⓐ Ⓑ Ⓒ Ⓓ Ⓔ | 98. Ⓐ Ⓑ Ⓒ Ⓓ Ⓔ |
| 24. Ⓐ Ⓑ Ⓒ Ⓓ Ⓔ | 49. Ⓐ Ⓑ Ⓒ Ⓓ Ⓔ | 74. Ⓐ Ⓑ Ⓒ Ⓓ Ⓔ | 99. Ⓐ Ⓑ Ⓒ Ⓓ Ⓔ |
| 25. Ⓐ Ⓑ Ⓒ Ⓓ Ⓔ | 50. Ⓐ Ⓑ Ⓒ Ⓓ Ⓔ | 75. Ⓐ Ⓑ Ⓒ Ⓓ Ⓔ | 100. Ⓐ Ⓑ Ⓒ Ⓓ Ⓔ |

practice test

# PRACTICE TEST 1

## 100 Questions • 210 Minutes

**Directions:** Study and memorize the details of the buildings on Pine Street. Questions 1–13 are based on the picture and the legend. You may not refer back to either the picture or the legend while answering these questions. Mark your answer on the answer sheet.

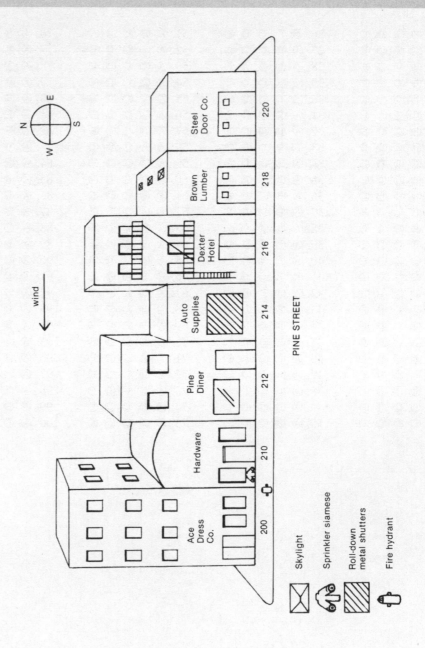

1. The building that would require the longest ladder to enable you to climb to the top floor would most likely be

   (A) 212 Pine Street.
   (B) 216 Pine Street.
   (C) 200 Pine Street.
   (D) 220 Pine Street.

2. If a fire occurred in a building on Pine Street at 3 a.m., the greatest life hazard would most likely be in the building at

   (A) 212 Pine Street.
   (B) 218 Pine Street.
   (C) 220 Pine Street.
   (D) 216 Pine Street.

3. For a building fire on Pine Street, if the fire department pumper is connected to the hydrant, the firefighter would need the longest stretch of hose to reach the

   (A) Ace Dress Co.
   (B) Pine Diner.
   (C) Auto Supplies Store.
   (D) Steel Door Co.

4. A firefighter is opening the skylights to remove smoke and heat from the building. You would be correct to assume that he is on the roof of the

   (A) lumber company.
   (B) dress company.
   (C) hardware store.
   (D) auto supply store.

5. As a firefighter, you are ordered by Lieutenant Brown to connect a hoseline to the automatic sprinkler siamese connection. You would find this connection in front of

   (A) 220 Pine Street.
   (B) 216 Pine Street.
   (C) 200 Pine Street.
   (D) 210 Pine Street.

6. Roll-down metal shutters are installed on storefronts to deter burglars. They might cause a fire to burn undetected, however, until it has gained considerable headway. The tenant on this block with roll-down metal shutters is the

   (A) hardware store.
   (B) diner.
   (C) lumber company.
   (D) auto supply store.

7. A firefighter is cutting a hole in the roof close to the front wall of 218 Pine Street. Severe conditions on the roof require that the firefighters make a fast exit from the roof. The safest and quickest exit can be made by

   (A) walking across the roof to 216 Pine Street.
   (B) stepping over the parapet to the roof of 220 Pine Street.
   (C) jumping across to the building behind 218 Pine Street.
   (D) waiting until a ladder is placed against the building.

8. The best protection against fire in its beginning stage is the automatic wet sprinkler system. The tenant on Pine Street that has the automatic wet sprinkler system is the

(A) dress company.

(B) hardware store.

(C) lumber company.

(D) Dexter Hotel.

9. When extinguishing a smoky fire, it often is advantageous to operate from the windward side of the fire (with the wind at your back). From the diagram of Pine Street, you have observed that the wind is blowing from the

(A) north.

(B) south.

(C) east.

(D) west.

10. Hose stream penetration is important when extinguishing fires. The building on Pine Street that would require the deepest penetration of a hose stream is occupied by

(A) the Steel Door Co.

(B) Brown Lumber.

(C) Pine Diner.

(D) the auto supply store.

11. Generally speaking, a curved roof is a trussed roof. Trussed roofs are notorious for rapid failure when subjected to high heat and fire for a short period of time. The building on Pine Street with a trussed roof is the

(A) hardware store.

(B) lumber company.

(C) Steel Door Co.

(D) Ace Dress Co.

12. Baled material and fibers become extremely heavy when wet. This causes overloading of floors that can result in collapse of the building. Effective use of water would be necessary for a fire at

(A) 200 Pine Street.

(B) 212 Pine Street.

(C) 214 Pine Street.

(D) 218 Pine Street.

13. Protecting exposures is an important part of firefighting strategy. If a heavy fire condition exists in Brown Lumber and your strategy is to protect the adjoining building, you most likely would have your firefighters stretch a hoseline into the

(A) Dexter Hotel.

(B) Ace Dress Co.

(C) Pine Diner.

(D) hardware store.

14. A problem likely to be found in a hotel fire that is not usually found in fires involving other types of residences, such as apartment houses, is

(A) obstructions in hallways and other passageways.

(B) large numbers of persons in relation to the number of rooms.

(C) delay in the transmission of the fire alarm.

(D) many occupants who don't know the location of exits.

15. On the way to work one morning, a firefighter notices a high-tension wire that has blown down and is lying across the sidewalk and into the road. The most appropriate course of action for the firefighter to take would be to

(A) move the wire to one side by means of a stock or branch of a tree and continue on her way.

(B) continue on her way to the firehouse and report the situation to the officer on duty.

(C) call the public utility company from the first public telephone she passes on her way to the firehouse.

(D) stand by the wire to warn away passersby and ask one of them to call the public utility company.

16. In general, firefighters on prevention inspection duty do not inspect the living quarters of private dwellings unless the occupants agree to the inspection. The best explanation why private dwellings are excluded from compulsory inspections is that

(A) private dwellings seldom catch fire.

(B) fires in private dwellings are more easily extinguished than other types of fires.

(C) people might resent such inspections as an invasion of privacy.

(D) the monetary value of private dwellings is lower than that of other types of occupancies.

17. Fire lines usually are established by the police to keep bystanders out of the immediate vicinity while firefighters are fighting the fire. Of the following, the best justification for the establishment of these fire lines is to

(A) prevent theft of property from partially destroyed apartments or stores.

(B) prevent interference with firefighter operations.

(C) give privacy to the victims of the fire.

(D) help apprehend arsonists in the crowd.

18. What is 12 squared?

(A) 144

(B) 169

(C) 196

(D) 225

(E) 256

19. At a large fire outside the city, three water tankers have arrived to supply water to the firefighters. Tanker A holds 1,500 gallons, Tanker B holds 2,500 gallons, and Tanker C holds 3,200 gallons of water. How many gallons of water are available to the firefighters?

(A) 4,000 gallons

(B) 4,700 gallons

(C) 5,700 gallons

(D) 6,400 gallons

(E) 7,200 gallons

**20.** A fire engine is required to carry six hoses. The lengths of the hoses are 100 feet, 100 feet, 150 feet, 150 feet, 250 feet, and 300 feet. What is the average hose length?

(A) 300 feet

(B) 275 feet

(C) 200 feet

(D) 175 feet

(E) 150 feet

**21.** A firefighter needs to crawl through a window that is 5 feet high and 2 feet wide. What is the area of the opening?

(A) 30 square feet

(B) 25 square feet

(C) 15 square feet

(D) 10 square feet

(E) 5 square feet

## QUESTIONS 22–23 ARE BASED ON THE FOLLOWING TABLE:

Four different categories of fire engine carry ladders or platforms. Each of these has a specific range when the ladder or platform is extended.

| Category | Range |
| --- | --- |
| Aerial ladder | 50 to 135 feet |
| Aerial ladder platform | 85 to 110 feet |
| Telescoping aerial platform | 50 to 100 feet |
| Articulating aerial platform | 55 to 102 feet |

**22.** Which of the four categories has the largest range?

(A) Aerial ladder platform

(B) Aerial ladder

(C) Articulating ladder platform

(D) Telescoping aerial platform

(E) Articulating aerial platform

**23.** If the articulating aerial platform were extended to 75 feet, what percentage of its range would be in use?

(A) 42.5%

(B) 43.5%

(C) 45.3%

(D) 46.2%

(E) 47.6%

**24.** Visitors near patients in oxygen tents are not permitted to smoke. The best of the following reasons for this prohibition is that

(A) the flame of the cigarette or cigar might flare dangerously.

(B) smoking tobacco is irritating to persons with respiratory diseases.

(C) smoking in bed is one of the major causes of fire.

(D) diseases can be transmitted by means of tobacco smoke.

25. At a hot and smoky fire, a lieutenant orders firefighters to work in pairs. Of the following, the best justification for this order is that

   (A) better communications result because one person can bring messages back to the lieutenant.

   (B) more efficient operation results because many vital activities require two people.

   (C) better morale results because firefighters are more willing to face danger in pairs.

   (D) safer operations result because one firefighter can help if the other becomes disabled.

26. Firefighters frequently open windows, doors, and skylights of a building on fire in a planned or systematic way to ventilate the building. The least likely to be accomplished by ventilation is the

   (A) increase in visibility of firefighters.

   (B) slowdown in the rate of burning of the building.

   (C) reduction in the danger from toxic gases.

   (D) control of the direction of travel of the fire.

27. At the first sign of a fire, the manager of a movie theater had the lights turned on and made the following announcement: "Ladies and gentlemen, the management has found it necessary to dismiss the audience. Please remain seated until it is time for your aisle to file out. In leaving the theater, follow the directions of the ushers. There is no danger involved." The manager's action in this situation was

   (A) proper.

   (B) improper, chiefly because the manager did not tell the audience the reason for the dismissal.

   (C) improper, chiefly because the manager did not permit all members of the audience to leave at once.

   (D) improper, chiefly because the manager misled the audience by saying that there was no danger.

28. In general, sprinkler heads must be replaced each time they are used. The best explanation for this is that the sprinkler heads

   (A) are subject to rusting after discharging water.

   (B) might become clogged after discharging water.

   (C) have a distorted pattern of water discharge after use.

   (D) are set off by the effect of heat on metal and cannot be reset.

29. After a fire in an apartment had been brought under control, firefighters were engaged in extinguishing the last traces of fire. A firefighter who noticed an expensive vase in the room in which activities were concentrated moved it to an empty closet in another room. The firefighter's action was

   (A) proper, chiefly because the owners would realize that extreme care was taken to avoid damage to their possessions.

   (B) improper, chiefly because disturbance of personal possessions should be kept to a minimum.

   (C) proper, chiefly because the chance of damage is reduced.

   (D) improper, chiefly because the firefighter should have devoted this time to putting out the fire.

30. While standing in front of a firehouse, a firefighter is approached by a woman with a baby carriage. The woman asks the firefighter if he will "keep an eye" on the baby while she visits a doctor in a nearby building. In this situation, the best course of action for the firefighter to take is to

   (A) agree to the woman's request but warn her that it might be necessary to leave to answer an alarm.

   (B) refuse politely after explaining that he is on duty and cannot become involved in other activities.

   (C) refer the woman to the officer on duty.

   (D) ask the officer on duty for permission to grant the favor.

31. Suppose that, while cooking, a pan of grease catches fire. The method that would be most effective, if available, in putting out the fire is to

   (A) dash a bucket of water on the fire.

   (B) direct a stream of water on the fire from a fire extinguisher.

   (C) pour a bottle of household ammonia water on the fire.

   (D) empty a box of baking soda on the fire.

32. Persons engaged in certain hazardous activities are required to obtain a fire department permit or certificate for which a fee is charged. The main reason for requiring permits or certificates is to

   (A) obtain revenue for the city government.

   (B) prevent unqualified persons from engaging in these activities.

   (C) obtain information about these activities to plan for fire emergencies.

   (D) warn the public of the hazardous nature of these activities.

33. A firefighter traveling to work is stopped by a citizen who complains that the employees of a nearby store frequently pile empty crates and boxes in a doorway, blocking passage. The most appropriate action for the firefighter to take is to

(A) assure the citizen that the fire department's inspection activities will eventually catch up with the store.

(B) obtain the address of the store and investigate to determine whether the citizen's complaint is justified.

(C) obtain the address of the store and report the complaint to a superior officer.

(D) ask the citizen for specific dates on which this practice occurred to determine whether the complaint is justified.

34. The crime of arson is defined as the willful burning of a house or other property. The best illustration of arson is a fire in a(n)

(A) apartment started by a 4-year-old boy playing with matches.

(B) barn started by a drunken man who overturns a lantern.

(C) store started by the bankrupt owner to collect insurance.

(D) house started by a neighbor who carelessly burned leaves in the garden.

35. The fire department now uses companies on fire duty with their apparatus for fire prevention inspection in commercial buildings. The change most important in making this inspection procedure practicable was the

(A) reduction of hours of work for firefighters.

(B) use of two-way radio equipment.

(C) use of enclosed cabs on fire apparatus.

(D) increase in property value in the city.

36. Many fires are caused by improper use of oxyacetylene torches. The main cause of such fires is the

(A) high pressure under which the gases are stored.

(B) failure to control or extinguish sparks.

(C) high temperatures generated by the equipment.

(D) explosive nature of the gases used.

37. The most important reason for having members of the fire department wear uniforms is to

(A) indicate the semimilitary nature of the fire department.

(B) build morale and esprit de corps of members.

(C) identify members on duty to the public and other members.

(D) provide clothing suitable for the work performed.

**38.** Of the following types of fires, the one that is likely to have the least amount of damage from water used in extinguishing it is a fire in a(n)

(A) rubber toy factory.

(B) retail hardware store.

(C) outdoor lumberyard.

(D) furniture warehouse.

**39.** "In a case of suspected arson, it is important for firefighters engaged in fighting the fire to remember conditions that existed at the time of their arrival. Particular attention should be given to all doors and windows." The main justification for this statement is that knowledge of the condition of the doors and windows might indicate

(A) where the fire started.

(B) who set the fire.

(C) the best way to ventilate the building.

(D) that someone wanted to prevent it from being extinguished.

**40.** "After a fire has been extinguished, every effort should be made to determine how the fire started." Of the following, the chief reason for determining the origin of the fire is to

(A) reduce the amount of damage caused by the fire.

(B) determine how the fire should have been fought.

(C) eliminate causes of fire in the future.

(D) explain delays in fighting the fire.

**41.** A firefighter inspecting buildings in a commercial area came to one building whose outside surface appeared to be of natural stone. The owner told the firefighter that it was not necessary to inspect the building because it was "fireproof." The firefighter, however, completed inspection of the building. Of the following, the best reason for continuing the inspection is that

(A) stone buildings catch fire as readily as wooden buildings.

(B) the fire department cannot make exceptions in its inspection procedures.

(C) the building might have been built of imitation stone.

(D) interiors and contents of stone buildings can catch fire.

**42.** The least valid reason for the fire department to investigate the causes of fire is to

(A) determine whether the fire was the result of arson.

(B) estimate the amount of loss for insurance purposes.

(C) gather information useful in fire prevention.

(D) discover violations of the Fire Prevention Code.

43. While on duty at a fire, a probationary firefighter receives an order from the lieutenant that appears to conflict with the principles of firefighting taught at the fire school. Of the following, the best course of action for the firefighter to take is to follow the order and, at a convenient time after the fire, to

(A) discuss the apparent inconsistency with the lieutenant.

(B) discuss the apparent inconsistency with another officer.

(C) mention this apparent inconsistency in an informal discussion.

(D) ask a more experienced firefighter about the apparent inconsistency.

44. When fighting fires on piers, the fire department frequently drafts salt water from the harbor. The chief advantage of using harbor water instead of relying on water from street mains is that harbor water is

(A) less likely to cause water damage.

(B) available in unlimited quantities.

(C) more effective in extinguishing fires due to its salt content.

(D) less likely to freeze in low temperatures due to its salt content.

**QUESTIONS 45–48 ARE BASED ON THE FOLLOWING PASSAGE:**

The canister type of gas mask consists of a tight-fitting facepiece connected to a canister containing chemicals that filter toxic gases and smoke from otherwise breathable air. These masks are of value when used with due regard to the fact that 2 or 3 percent of gas in air is about the highest concentration that the chemicals in the canister will absorb and that these masks do not provide the oxygen necessary for the support of life. In general, if flame is visible, there is sufficient oxygen for firefighters although toxic gases might be present. Where there is heavy smoke and no flame, an oxygen deficiency might exist. Fatalities have occurred where filter-type canister masks have been used in attempting rescue from manholes, wells, basements, or other locations deficient in oxygen.

45. If the mask described in the paragraph is used in an atmosphere containing oxygen, nitrogen, and carbon monoxide, we would expect the mask to remove from the air breathed the

(A) nitrogen only.

(B) carbon monoxide only.

(C) nitrogen and the carbon monoxide.

(D) None of the above

46. According to this paragraph, when firefighters are wearing these masks at a fire in which flame is visible, the firefighters generally can feel that, as far as breathing is concerned, they are

   (A) safe, because the mask will provide them with sufficient oxygen to live.

   (B) unsafe, unless the gas concentration is below 2 or 3 percent.

   (C) safe, provided the gas concentration is above 2 or 3 percent.

   (D) unsafe, because the mask will not provide them with sufficient oxygen to live.

47. According to this paragraph, fatalities have occurred to persons using this type of gas mask in manholes, wells, and basements because

   (A) the supply of oxygen provided by the mask ran out.

   (B) the air in those places did not contain enough oxygen to support life.

   (C) heavy smoke interfered with the operation of the mask.

   (D) the chemicals in the canister did not function properly.

48. The following formula can be used to show, in general, the operation of the gas mask described in the preceding paragraph:

(Chemicals in canister) → (air + gases) = breathable air

The arrow in the formula, when expressed in words, most nearly means

   (A) replace.

   (B) are changed into.

   (C) act upon.

   (D) give off.

**QUESTIONS 49–51 ARE BASED ON THE FOLLOWING PASSAGE:**

The only openings permitted in fire partitions, except openings for ventilating ducts, shall be those required for doors. There shall be but one such door opening unless the provision of additional openings would not exceed in total width of all doorways 25 percent of the length of the wall. The minimum distance between openings shall be 3 feet. The maximum area for such a door opening shall be 80 square feet, except that such openings for the passage of trucks may be a maximum of 140 square feet.

**49.** According to the paragraph, openings in fire partitions are permitted only for

(A) doors.

(B) doors and windows.

(C) doors and ventilation ducts.

(D) doors, windows, and ventilation ducts.

**50.** In a fire partition 22 feet long and 10 feet high, the maximum number of doors 3 feet wide and 7 feet high is

(A) 1

(B) 2

(C) 3

(D) 4

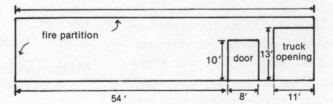

**51.** The most accurate statement about the preceding layout is that the

(A) total width of the openings is too large.

(B) truck opening is too large.

(C) truck and door openings are too close together.

(D) layout is acceptable.

**52.** Of the following, the basic purpose of fire department inspection of private property is to

(A) make sure that fire department equipment is properly maintained.

(B) secure proper maintenance of all entrance and exit facilities.

(C) obtain correction of conditions creating undue fire hazards.

(D) make sure that walls are properly plastered and painted to prevent the spread of fire.

**53.** Of the following items commonly found in a household, the one that uses the most electric current is a(n)

(A) 150-watt light bulb.

(B) toaster.

(C) door buzzer.

(D) 8-inch electric fan.

**QUESTIONS 54–56 ARE BASED ON THE FOLLOWING PASSAGE:**

The average daily flow of water through public water systems in American cities ranges generally between 40 and 250 gallons per capita, depending on the underground leakage in the system, the amount of waste in domestic premises, and the quantity used for industrial purposes. The problem of supplying this water has become serious in many cities. Supplies, which were once adequate, in many cases have become seriously deficient due to greater demands with increased population and growing industrial use of water. Waterworks, operating on fixed schedules of water charges, in many cases have not been able to afford the heavy capital expenditures necessary to provide adequate supply, storage, and distribution facilities. Thus, the adequacy of a public water supply for fire protection in any given location cannot properly be taken for granted.

**54.** The following four programs are possible ways in which American communities might try to reduce the seriousness of the water shortage problem. The one program that does NOT directly follow from the paragraph is that of

(A) regular replacement of old street water mains by new ones.

(B) inspection and repair of leaky plumbing fixtures.

(C) fire prevention inspection and education to reduce the amount of water used to extinguish fires.

(D) research into industrial processes to reduce the amount of water used in those processes.

**55.** The main conclusion reached by the author of the paragraph is that

(A) there is a waste of precious natural resources in America.

(B) communities have failed to control the industrial use of water.

(C) a need exists for increasing the revenue of waterworks to build up adequate supplies of water.

(D) fire departments cannot assume they will always have the necessary supply in water available to fight fires.

**56.** Per capita consumption of water in a community is determined by which of the following formulas?

**(A)** Population ÷ Total consumption in gallons = Per capita consumption in gallons

**(B)** Total consumption in gallons ÷ Population = Per capita consumption in gallons

**(C)** Total consumption in gallons × Population = Per capita consumption in gallons

**(D)** Total consumption in gallons − Population = Per capita consumption in gallons

**QUESTIONS 57–60 ARE BASED ON THE FOLLOWING PASSAGE:**

Most accidents that lead to firefighter injuries or fatalities occur at the scene of an emergency. To avoid accidents, it is important to have a system that clarifies the roles and responsibilities of personnel.

The Incident Command System (ICS) is a system that was adopted by the National Fire Academy (NFA) to coordinate resources during a variety of emergency situations. The ICS assists different agencies in coordinating their efforts by using common terminology and operating procedures. The five major functional areas of the ICS are command, operations, planning, logistics, and finance.

Command—For a large-scale operation to run safely and effectively, it is necessary to have people in charge. Individuals occupying positions such as safety officer, liaison, and public information officer are responsible for the ordering and releasing of resources.

Operations—Once command has established the strategic goals, operations is responsible for managing all the operations that directly affect the primary mission. Operations is divided into branches that can be divided again into groups assigned to a particular task such as rescue or ventilation.

Planning—As the incident develops, this area is responsible for the collection, evaluation, dissemination, and use of relevant information. Command will adjust the ordering and releasing of resources according to the information compiled by planning.

Logistics—This area is responsible for supporting the incident through its support and service branches. The support branch maintains supplies, facilities, and vehicle services. The service branch maintains medical, communications, and food services.

Finance—This section takes care of all costs and financial aspects of the incident. Finance is usually only a concern, however, during large-scale incidents that last for a particularly long period of time.

57. The title that best describes the content of this passage is

(A) "The Purpose and Origin of the National Fire Academy (NFA)"

(B) "Why Most Injuries and Fatalities Occur at the Scene of an Emergency"

(C) "The Purpose and Functions of the Incident Command System (ICS)"

(D) "Why Command Is the Most Important Functional Area of the ICS"

(E) "The Different Branches of Operations"

58. It is necessary to have _____ _____ for a large-scale operation to run safely and effectively.

(A) people in charge

(B) adequate ventilation

(C) sufficient funds

(D) food services

(E) firefighter injuries

59. If an operation does not turn out to take much time to resolve, which functional area of the ICS is least likely to be necessary?

(A) Command

(B) Operations

(C) Planning

(D) Logistics

(E) Finance

60. Logistics is NOT in charge of maintaining what service?

(A) Medical

(B) Rescue

(C) Vehicle

(D) Supplies

(E) Communications

**QUESTIONS 61–64 ARE BASED ON THE FOLLOWING PASSAGE:**

When arson is the suspected cause of a fire, it is important to ensure the proper collection of evidence at the fire scene. The fire investigator typically is responsible for collecting evidence at the scene. It might be the case, however, that an investigator is not available immediately after a fire. In these situations, it is important for the firefighters on the scene to ensure the proper collection of evidence.

The fire department has the authority to control a fire scene for as long after a fire as is necessary. As soon as the last firefighter leaves the scene, however, the department's control of it is limited. Often times, a search warrant or written consent is necessary to make later visits to the scene. It is therefore important that all the evidence be tagged, marked, and photographed as soon as possible. If it is necessary to remove any evidence from the scene, firefighters should be careful not to damage any of the evidence in the process.

Until the investigation of the fire is complete, no one should be allowed to enter the scene unless accompanied by a fire officer. Any entry, even by the owner of the premises, should be carefully logged. The person's name, time of entry, time of departure, and descriptions of any items taken from the premises should be recorded.

If it is deemed necessary, fire scenes can be guarded to ensure the safety of anyone near the scene and to ensure that no one tampers with the evidence. Depending on the number of possible entrances to the scene and the severity of safety concerns, a fire scene could be watched by one person or a full-time guard force.

**61.** If an investigator is not available immediately after a fire, firefighters should

**(A)** destroy the evidence at the scene.

**(B)** tag, mark, and photograph the evidence.

**(C)** allow the owner to find the evidence alone.

**(D)** leave the fire scene unguarded.

**(E)** have no control over the evidence.

**62.** All of the following should be logged when someone enters the fire scene EXCEPT

**(A)** a description of the person.

**(B)** the person's name.

**(C)** time of entry.

**(D)** time of departure.

**(E)** a description of items taken.

63. The fire department has the authority to control a fire scene

    (A) for about two weeks.

    (B) after proper collection of evidence.

    (C) for as long after a fire as necessary.

    (D) when the last firefighter leaves.

    (E) if the scene is adequately guarded.

64. To ensure that the fire scene is not contaminated by the activity of the firefighters, they should NOT

    (A) secure the fire scene.

    (B) call the fire investigator.

    (C) damage any evidence.

    (D) tag any evidence.

    (E) be careful with equipment.

65. Sand and ashes frequently are placed on icy pavements to prevent skidding. The effect of the sand and ashes is to increase

    (A) inertia.

    (B) gravity.

    (C) momentum.

    (D) friction.

66. The air near the ceiling of a room usually is warmer than the air near the floor because

    (A) there is better air circulation at floor level.

    (B) warm air is lighter than cold air.

    (C) windows usually are nearer the floor than the ceiling.

    (D) heating pipes usually run along the ceiling.

**67.** It is safer to use the ladder positioned as shown in Diagram 1 than as shown in Diagram 2 because in Diagram 1

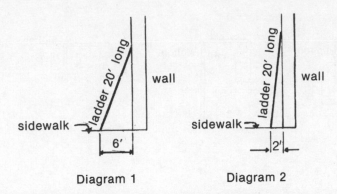

Diagram 1          Diagram 2

**(A)** less strain is placed on the center rungs of the ladder.

**(B)** it is easier to grip and stand on the ladder.

**(C)** the ladder reaches a lower height.

**(D)** the ladder is less likely to tip over backward.

**68.** A substance that is subject to spontaneous heating is one that

**(A)** is explosive when heated.

**(B)** is capable of catching fire without an external source of heat.

**(C)** acts to speed up the burning of material.

**(D)** liberates oxygen when heated.

**69.** "Chief officers shall instruct their firefighters that, when transmitting particulars of a fire, they shall include the fact that foods are involved if the fire involves premises where foodstuffs are sold or stored." Of the following, the best justification for this regulation is that, when foodstuffs are involved in a fire,

**(A)** the fire can reach serious proportions.

**(B)** police protection might be desirable to prevent looting.

**(C)** relatively little firefighting equipment may be needed.

**(D)** inspection to detect contamination may be desirable.

**70.** A fire can continue when fuel, oxygen from the air or another source, and a sufficiently high temperature to maintain combustion are present. The method most commonly used to extinguish fires is to

**(A)** remove the fuel.

**(B)** exclude the oxygen from the burning material.

**(C)** reduce the temperature of the burning material.

**(D)** smother the flames of the burning material.

**71.** Of the following, the siphon arrangement that would most quickly transfer a solution from the container on the left side to the one on the right side is numbered

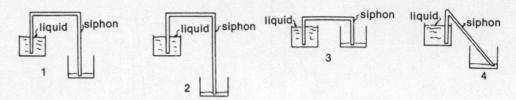

**(A)** 1

**(B)** 2

**(C)** 3

**(D)** 4

**72.** Static electricity is a hazard in industry chiefly because it can cause

**(A)** dangerous or painful burns.

**(B)** chemical decomposition of toxic elements.

**(C)** sparks that can start an explosion.

**(D)** overheating of electrical equipment.

**73.** A rowboat will float deeper in freshwater than in salt water because

**(A)** in the salt water, the salt will occupy part of the space.

**(B)** freshwater is heavier than salt water.

**(C)** salt water is heavier than freshwater.

**(D)** salt water offers less resistance than freshwater.

74. In the following diagram, it is easier to get the load onto the platform by using the ramp than by lifting it directly onto the platform. This is true because the effect of the ramp is to

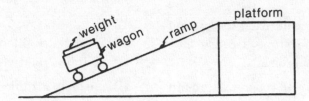

(A) reduce the amount of friction so that less force is required.

(B) distribute the weight over a larger area.

(C) support part of the load so that less force is needed to move the wagon.

(D) increase the effect of the moving weight.

75. More weight can be lifted by the method shown in diagram 2 than the method in diagram 1 because

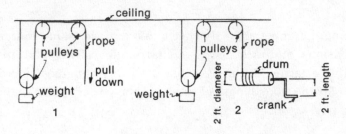

(A) it takes less force to turn a crank than it does to pull in a straight line.

(B) the drum will prevent the weight from falling by itself.

(C) the length of the crank is larger than the radius of the drum.

(D) the drum has more rope on it, easing the pull.

**76.** Two balls of the same size but different weights are dropped from a 10-foot height. The most accurate statement is that

**(A)** both balls will reach the ground at the same time because they are the same size.

**(B)** both balls will reach the ground at the same time because the effect of gravity is the same on both balls.

**(C)** the heavier ball will reach the ground first because it weighs more.

**(D)** the lighter ball will reach the ground first because air resistance is greater on the heavier ball.

**77.** "A partially filled gasoline drum is a more dangerous fire hazard than a full one." Of the following, the best justification for this statement is that

**(A)** a partially filled gasoline drum contains relatively little air.

**(B)** gasoline is difficult to ignite.

**(C)** when a gasoline drum is full, the gasoline is more explosive.

**(D)** gasoline vapors are more explosive than gasoline itself.

**78.** The following diagrams show flywheels made of the same material, with the same dimensions, and attached to similar engines. The solid areas represent equal weights attached to the flywheel. If all three engines are running at the same speed for the same length of time and the power to the engines is shut off simultaneously,

 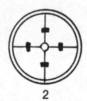 

1            2            3

**(A)** wheel 1 will continue turning longest.

**(B)** wheel 2 will continue turning longest.

**(C)** wheel 3 will continue turning longest.

**(D)** all three wheels will continue turning for the same time.

**79.** The substance that expands when freezing is

**(A)** alcohol.

**(B)** ammonia.

**(C)** mercury.

**(D)** water.

**FIRE CLASSIFICATION SYSTEM**

| Type of Fire | Combustible Materials | Extinguishing Process |
|---|---|---|
| Class A | paper, cloth, wood, rubber, plastics | water, foam |
| Class B | paints, oil, gasoline, mineral spirits | oxygen exclusion (the blanketing effect) |
| Class C | energized electrical equipment | dry chemicals, carbon dioxide |
| Class D | combustible metals (aluminum, titanium) | depends on the type of metal |

**80.** It would be dangerous to use water to extinguish a Class C fire because

(A) it might cause the fire to spread.

(B) there is a risk of electrical shock.

(C) water is only for Class B fires.

(D) the metals are combustible.

(E) water is not to be used for plastics.

**81.** If a painter is working near an open flame and the paint catches fire, the extinguishing process she should use

(A) depends on the type of paint.

(B) is dry chemicals or carbon dioxide.

(C) is water or a noncombustible foam.

(D) is oxygen exclusion (blanketing effect).

(E) is combustible metals (aluminum).

**82.** When drilling a hole through a piece of wood with an auger bit, it is considered good practice to clamp a piece of scrap wood to the underside of the piece through which the hole is being drilled. The main reason for this is to

(A) direct the auger bit.

(B) speed up the drilling operation.

(C) prevent the drill from wobbling.

(D) prevent splintering of the wood.

**83.** "Firefighters holding a life net should keep their eyes on the person jumping from a burning building." Of the following, the best justification for this statement is that

(A) a person attempting to jump into a life net might overestimate the distance of the net from the building.

(B) some people will not jump into a life net unless given confidence.

(C) a person jumping into a life net might be seriously injured if the net is not allowed to give slightly at the moment of impact.

(D) firefighters holding a life net should be evenly spaced around the net at the moment of impact to distribute the shock.

**84.** If boiling water is poured into a drinking glass, the glass is likely to crack. If a metal spoon is first placed in the glass, however, the glass is much less likely to crack. The reason that the glass with the spoon is less likely to crack is that the spoon

(A) distributes the water over a larger surface of the glass.

(B) quickly absorbs heat from the water.

(C) reinforces the glass.

(D) reduces the amount of water that can be poured into the glass.

**85.** It takes more energy to force water through a long pipe than through a short pipe of the same diameter. The principal reason for this is

(A) gravity.

(B) friction.

(C) inertia.

(D) cohesion.

**86.** A pump discharging at 300 pounds psi pressure delivers water through 100 feet of pipe laid horizontally. If the valve at the end of the pipe is shut so that no water can flow, the pressure at the valve is, for practical purposes,

(A) greater than the pressure at the pump.

(B) equal to the pressure at the pump.

(C) less than the pressure at the pump.

(D) greater or less than the pressure at the pump, depending on the type of pump used.

87. The leverage system in the following sketch is used to raise a weight. To reduce the amount of force required to raise the weight, it is necessary to

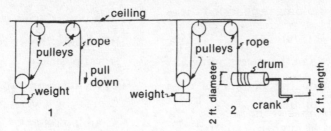

   **(A)** decrease the length of the lever.

   **(B)** place the weight closer to the fulcrum.

   **(C)** move the weight closer to the person applying the force.

   **(D)** move the fulcrum farther from the weight.

### QUESTIONS 88–91 ARE BASED ON THE FOLLOWING PASSAGE:

Old buildings are liable to possess a special degree of fire risk merely because they are old. Outdated electrical wiring systems and installation of new heating appliances for which the building was not designed can contribute to the increased hazard. Old buildings often have been altered many times, parts of the structure might antedate building codes, and dangerous defects may have been covered up. On the average, old buildings contain more lumber than comparable new buildings. This, in itself, makes old buildings more susceptible to fire. It is not true, though, that sound lumber in old buildings is drier than new lumber. Moisture content of lumber varies with that of the atmosphere to which it is exposed.

88. According to the paragraph, old buildings present a special fire hazard chiefly because of the

   **(A)** poor planning of the buildings when first designed.

   **(B)** haphazard alteration of the buildings.

   **(C)** dryness of old lumber.

   **(D)** inadequate enforcement of the building codes.

89. We can conclude from the paragraph that lumber

   **(A)** should not be used in buildings unless absolutely necessary.

   **(B)** should not be used near electrical equipment.

   **(C)** is more inflammable than newer types of building materials.

   **(D)** tends to lose its moisture at a constant rate.

**90.** According to the paragraph, the amount of moisture in the wooden parts of a building depends on the

(A) age of the building.

(B) moisture in the surrounding air.

(C) type of heating equipment used in the building.

(D) quality of lumber used.

**91.** In regard to building codes, the paragraph implies that

(A) old buildings are exempt from the provisions of building codes.

(B) some buildings now in use were built before building codes were adopted.

(C) building codes usually don't cover electrical wiring systems.

(D) building codes generally are inadequate.

**92.** A firefighter must become familiar with first-aid methods. Of the following, the best reason for this rule is that

(A) it saves expense to have someone take care of simple injuries not requiring a doctor.

(B) it can keep the fire department out of a lawsuit if a simple injury is taken care of promptly.

(C) it is of great advantage that a firefighter be prepared to act quickly in an emergency.

(D) in doing first aid, it is sometimes better to do nothing than to do the wrong thing.

**93.** The method that is NOT recommended procedure for treating shock is

(A) administering oxygen.

(B) placing the victim in a horizontal position.

(C) trying to prevent the loss of body heat by covering the victim with a blanket.

(D) having the victim drink warm water.

**94.** To control severe bleeding, a tourniquet should be applied only as a last resort. Once applied, it is left in place. Which of the following statements is NOT correct?

(A) The tourniquet is placed two inches from the injury between the heart and the location of the injury.

(B) Tighten the tourniquet only enough to stop the bleeding.

(C) Mark the person with the letters T–K and the time of day the tourniquet was applied.

(D) Loosen and retighten the tourniquet from time to time.

**95.** An open fracture generally means a fracture in which

(A) the broken bone has penetrated the skin.

(B) the bone is broken at more than one point.

(C) the bone not only has been broken but has been dislocated as well.

(D) more than one bone has been broken.

96. A newly appointed firefighter should realize that the principal function of an immediate superior officer is to

   (A) inflict discipline for rule infractions.

   (B) direct the activities of subordinates so they function in a proper and acceptable manner.

   (C) make recommendations to promote the most able subordinates under his or her command.

   (D) keep records that properly reflect the activities of subordinates.

**TO ANSWER QUESTIONS 97–98, CHOOSE THE CORRECT SPELLING OF THE WORD TO COMPLETE THE SENTENCE.**

97. A fire's need for oxygen to continue the burning process often results in an oxygen-deficient _____ in a structural fire.

   (A) atmosfere

   (B) atmosphere

   (C) atmosfear

   (D) atmosphear

   (E) atmospheir

98. Class A and Class B fires can be extinguished with an _____ film-forming foam.

   (A) aquius

   (B) aqwious

   (C) aqweus

   (D) aqueous

   (E) aquious

TO ANSWER QUESTIONS 99–100, CHOOSE THE APPROPRIATE WORD TO COMPLETE THE SENTENCE.

99. _____ they had left the fire scene, some smoldering ashes caused the fire to start up again.

(A) After

(B) While

(C) Until

(D) Whether

(E) Where

100. The fire was so big and lasted for such a long time that the _____ supplies were running low.

(A) firefighteres's

(B) firefighters's

(C) firefighters'

(D) firefighterz

(E) firefighteres

## ANSWERS AND EXPLANATIONS

| | | | | | | | | |
|---|---|---|---|---|---|---|---|
| 1. | C | 21. | D | 41. | D | 61. | B | 81. | D |
| 2. | D | 22. | B | 42. | B | 62. | A | 82. | D |
| 3. | D | 23. | A | 43. | A | 63. | C | 83. | A |
| 4. | A | 24. | A | 44. | B | 64. | C | 84. | B |
| 5. | D | 25. | D | 45. | B | 65. | D | 85. | B |
| 6. | D | 26. | B | 46. | B | 66. | B | 86. | B |
| 7. | B | 27. | A | 47. | B | 67. | D | 87. | B |
| 8. | B | 28. | D | 48. | C | 68. | B | 88. | B |
| 9. | C | 29. | C | 49. | C | 69. | D | 89. | C |
| 10. | B | 30. | B | 50. | A | 70. | C | 90. | B |
| 11. | A | 31. | D | 51. | B | 71. | B | 91. | B |
| 12. | A | 32. | B | 52. | C | 72. | C | 92. | C |
| 13. | A | 33. | C | 53. | B | 73. | C | 93. | D |
| 14. | D | 34. | C | 54. | C | 74. | C | 94. | D |
| 15. | D | 35. | B | 55. | D | 75. | C | 95. | A |
| 16. | C | 36. | B | 56. | B | 76. | B | 96. | B |
| 17. | B | 37. | C | 57. | C | 77. | D | 97. | B |
| 18. | A | 38. | C | 58. | A | 78. | C | 98. | D |
| 19. | E | 39. | D | 59. | E | 79. | D | 99. | A |
| 20. | D | 40. | C | 60. | B | 80. | B | 100. | C |

1.  **The correct answer is (C).** 200 Pine Street is the tallest building on the block.

2.  **The correct answer is (D).** The Dexter Hotel at 216 Pine Street has occupants who are likely to be asleep in the early morning hours. The area is commercial/industrial, and most businesses are not open at 3 a.m.

3.  **The correct answer is (D).** This is the building farthest from the hydrant.

4.  **The correct answer is (A).** Note the skylights on the diagram.

5.  **The correct answer is (D).** Refer to the legend and the diagram.

6.  **The correct answer is (D).** Refer to the legend and the diagram.

7.  **The correct answer is (B).** 218 and 220 Pine Street are both one-story buildings.

8.  **The correct answer is (B).** The siamese connection on the front wall of the building indicates that the building has an automatic wet sprinkler system.

9. **The correct answer is (C).** The direction of wind is indicated by the arrow and compass points at the top of the diagram.

10. **The correct answer is (B).** The building that houses Brown Lumber would require the deepest penetration of a hose stream because of the type of material stored there.

11. **The correct answer is (A).** Refer to the diagram and the definition of a trussed roof in the first sentence of the question.

12. **The correct answer is (A).** Baled material, fibers, and other materials are used in the manufacture of dresses and other clothing.

13. **The correct answer is (A).** The Dexter Hotel is immediately adjacent to the building afire. Fire burning through the roof of Brown Lumber would expose the hotel to flames, especially with the wind blowing from the east.

14. **The correct answer is (D).** Generally speaking, guests staying at a hotel are so preoccupied with their activities that they rarely take the time to familiarize themselves with the location of the fire exits and fire stairs.

15. **The correct answer is (D).** The firefighter knows the dangerous situation presented by the live wire and the tendency for it to recoil (wire has memory) without notice. The firefighter should protect the public until the fire department or utility company arrives.

16. **The correct answer is (C).** Entry into a private home or apartment is illegal unless permission to do so is given by the occupant. In the event of a fire, however, permission is not required.

17. **The correct answer is (B).** Bystanders in the immediate vicinity of the fire hinder effective firefighter operations. They impede movement of fire apparatus, stretching of hoselines, and rescue efforts. There is also the danger of bystanders being injured by falling debris.

18. **The correct answer is (A).** To square a number, just multiply the number by itself. In this case, $12 \times 12 = 144$.

19. **The correct answer is (E).** The amount of water available to the firefighters is the sum of all of the water in the three tankers. Thus, $1,500 + 2,500 + 3,200 = 7,200$ gallons.

20. **The correct answer is (D).** To find the average of a list of numbers, first add them up ($100 + 100 + 150 + 150 + 250 + 300 = 1,050$), and then divide the sum of the numbers by the number of lengths listed ($1,050 \div 6 = 175$).

21. **The correct answer is (D).** To find the area of a flat surface, multiply the length times the width. $5 \times 2 = 10$ square feet

22. **The correct answer is (B).** The range is equal to the largest number minus the smallest number. Thus, the largest range is $135 - 50 = 85$. This is the range of the aerial ladder.

23. **The correct answer is (A).** The range is equal to $102 - 55 = 47$. Because $75 - 55 = 20$, $20 \div 47 = 42.5\%$.

24. **The correct answer is (A).** Oxygen is a noncombustible gas; however, it will support combustion and will accelerate the fire in whatever is burning.

25. **The correct answer is (D).** Fires are always unpredictable. In hot, smoky conditions, the buddy system makes for greater firefighter safety.

26. **The correct answer is (B).** Unless the ventilation of a fire building is synchronized with advancing hoselines, the in-rush of fresh air can cause a fire to accelerate.

27. **The correct answer is (A).** Patrons in a theater generally are surprised when the lights are turned on suddenly. The manager calmed the patrons at once and informed them of how to make a safe and speedy exit. This was very well done.

28. **The correct answer is (D).** A sprinkler head has a fusible metal disc, which is used to hold back the water. The heat from the fire causes the disc to melt at a predetermined temperature and, in doing so, releases the water. It cannot be used again; therefore, it must be replaced.

29. **The correct answer is (C).** The firefighter's job is to extinguish fires, to save lives, and at every opportunity, to make an effort to hold property damage from fire or water to a minimum. With the fire out and no danger to life, the firefighter correctly saved property.

30. **The correct answer is (B).** The firefighter is on duty to respond to fires and emergencies. If he "kept an eye" on the child, he would not be available to respond with his unit.

31. **The correct answer is (D).** The baking soda would smother the fire. The other extinguishing agents would cause the grease to splatter and would spread the fire from the pan to surrounding areas.

32. **The correct answer is (B).** Before obtaining a certificate or permit from the fire department, people applying must show fire department examiners that they are trained and qualified to engage in the particular hazardous work.

33. **The correct answer is (C).** By reporting the complaint, the officer will then cause an official inspection to be made. If the doorway is blocked, a violation order will be issued to the person in charge.

34. **The correct answer is (C).** The bankrupt owner of the store deliberately set the fire in the store. The other three fires were started accidentally.

35. **The correct answer is (B).** The two-way radio enables the fire company to be away from the fire station and still be in radio contact with the fire alarm dispatcher and available to respond to alarms.

36. **The correct answer is (B).** Oxyacetylene torches generate a high-intensity flame that is used to cut metal. As the metal is cut, the hot pieces and sparks bounce from the metal. Numerous fires are started by sparks and hot metal falling on combustible material.

37. **The correct answer is (C).** The firefighter in uniform serves as a beacon to a person seeking aid. The uniform also identifies the firefighter as a person with authority. This may be a deterrent to anyone intent on committing a criminal act.

38. **The correct answer is (C).** The majority of wood and wood products at an outdoor lumberyard are stored outdoors and are exposed to all weather conditions. Wetting by hose streams should have little adverse effect on the lumber.

39. **The correct answer is (D).** The arsonist generally will open windows to allow air in to accelerate the fire. An open or unlocked door usually indicates that someone was in the building prior to the arrival of the firefighters.

40. **The correct answer is (C).** Fire prevention is a very important aspect of the firefighter's role. Study of the causes of fires can lead to steps to eliminate those causes at other locations, thus preventing future fires.

41. **The correct answer is (D).** The term "fireproof building" means that the walls, floors, stairways, partitions, and other structural components of the building are made of noncombustible material and will resist fire. The contents and furnishings, however, are combustible and are likely to burn.

42. **The correct answer is (B).** The fire department would investigate a fire for the reasons stated in choices (A), (C), and (D). The amount of loss for insurance purposes would be investigated by an adjuster hired by the insurance company or the victim of the fire.

43. **The correct answer is (A).** The lieutenant might have been aware of other circumstances or conditions at the fire that the probationary firefighter, because of inexperience, was not aware of. The officer who gave the order is in the best position to explain it.

44. **The correct answer is (B).** Pier fires usually are fires of long duration that require enormous quantities of water to extinguish. By using the salt water, we have an unlimited supply and we can conserve our freshwater resources.

45. **The correct answer is (B).** Carbon monoxide is a toxic gas. The atmospheric air we breathe is composed of 84 percent nitrogen and 16 percent oxygen; obviously, nitrogen is not a toxic gas.

46. **The correct answer is (B).** The flame indicates that there is sufficient oxygen. Gas concentration is low enough to be handled by the mask.

answers

47. **The correct answer is (B).** This is stated in the last sentence of the paragraph.

48. **The correct answer is (C).** The canister-type gas mask contains chemicals that filter toxic gases and smoke, resulting in breathable air.

49. **The correct answer is (C).** This is stated in the first sentence.

50. **The correct answer is (A).** This is stated in the second sentence. Doorway openings permitted in the wall are a maximum of 25 percent of 22 feet, or 5.5 feet. One opening would be 3 feet, and two openings would be a total of 6 feet. Therefore, choice (A) must be the answer.

51. **The correct answer is (B).** This is stated in the last sentence. The maximum opening for a motor vehicle is 140 square feet. The opening in the diagram is 13 feet × 11 feet, or 143 square feet. Therefore, it is too large.

52. **The correct answer is (C).** The purpose of fire department inspections of any and all property is to eliminate conditions creating undue fire hazards.

53. **The correct answer is (B).** The average household toaster uses 1,400 watts of energy, much more than the other items mentioned in choices (A), (C), and (D).

54. **The correct answer is (C).** Choices (A), (B), and (D) are programs that investigate the causes of water usage, as stated in the first sentence.

55. **The correct answer is (D).** This is stated in the last sentence.

56. **The correct answer is (B).** To obtain the amount used by each person, take the total number of gallons consumed and divide this number by the total number of people (population).

57. **The correct answer is (C).** This entire passage discusses the purpose and functions of the Incident Command System (ICS).

58. **The correct answer is (A).** Under the heading of command, the passage states that, for a large-scale operation to run safely and effectively, it is necessary to have people in charge.

59. **The correct answer is (E).** Under the heading of finance, the passage states that finance is usually only a concern during large-scale incidents that last for a particularly long period of time.

60. **The correct answer is (B).** The only thing that is not listed as a responsibility of logistics is rescue.

61. **The correct answer is (B).** The first paragraph states that, if an investigator is not available immediately after a fire, it is important for the firefighters on the scene to ensure proper collection of the evidence. The second paragraph then states that the firefighters should tag, mark, and photograph the evidence as soon as possible.

**62.** **The correct answer is (A).** The third paragraph indicates that only the person's name, time of entry, time of departure, and a description of the items taken should be logged.

**63.** **The correct answer is (C).** The first sentence of the second paragraph states that the fire department has the authority to control a fire scene for as long after a fire as necessary.

**64.** **The correct answer is (C).** The last sentence of the second paragraph states that it is important for firefighters to not damage any evidence.

**65.** **The correct answer is (D).** Friction opposes motion. By placing a coarse substance between two smooth surfaces, friction is generated and skidding is reduced.

**66.** **The correct answer is (B).** Heated air expands; it becomes lighter and rises. This is called convection.

**67.** **The correct answer is (D).** The ladder with the gentler grade is more stable and is less likely to tip over. The horizontal distance between the wall and the base of the ladder is determined by dividing the length of the ladder by 5 and adding 2. Example: For a 20-foot ladder, dividing by 5 equals 4; 4 plus 2 equals 6; 6 feet is the proper distance from the wall for the 20-foot ladder.

**68.** **The correct answer is (B).** Spontaneous heating is the increase in temperature of a material or substance without drawing heat from its surroundings.

**69.** **The correct answer is (D).** Food that has been subjected to unusual heat and perhaps also to water might well have undergone changes that render it unfit for human consumption. For the protection of the public, it is important for this possibility to be recognized and for the food to be inspected.

**70.** **The correct answer is (C).** The most common method of extinguishing a fire is to pour water on it. Water is a very effective cooling agent. The action of water is to reduce the temperature of the burning material.

**71.** **The correct answer is (B).** The principle that applies to this problem is as follows: Pressure in a liquid is equal to the height multiplied by the density, or $P = H \times D$. Because siphon 2 has a greater depth, the liquid at the bottom of the siphon 2 tube has a greater pressure than that of siphon 1. The greater pressure also produces greater velocity of the moving liquid. Therefore, the flow in siphon 2 would be faster, making it more efficient.

**72.** **The correct answer is (C).** Static electricity is the electrification of material through physical contact and separation. Sparks might result, constituting a fire or explosion hazard.

**73.** **The correct answer is (C).** Salt water has greater density than freshwater. For this reason, it is easier to float in salt water than in freshwater.

74. **The correct answer is (C).** The inclined ramp (or plane) is simply a sloping platform that enables a person to raise an object without having to lift it vertically. It is a simple machine that helps you perform work.

75. **The correct answer is (C).** A distinct mechanical advantage is gained by using a large lever (the crank) to move a small object (the drum).

76. **The correct answer is (B).** Both balls will reach the ground at the same time because the gravitational force on objects with the same surface size varies with the distance of the objects from the center of the earth. The weight of the objects does not matter.

77. **The correct answer is (D).** If you do not know the answer, you can use reasoning to figure out this question. Choices (B) and (C) are irrelevant, and choice (A) makes no sense. It is reasonable to assume that the space in a gasoline drum is filled with gasoline vapors.

78. **The correct answer is (C).** The explanation is based on a well-known principle of centrifugal force.

79. **The correct answer is (D).** When water freezes, it increases its volume by one eleventh. When 11 cubic inches of water freeze, for example, 12 cubic inches of ice form. This is why water pipes burst and auto radiators without antifreeze are severely damaged.

80. **The correct answer is (B).** The chart indicates that Class C fires involve energized electrical equipment. Because water is a good conductor of electricity, there is a serious risk of electrical shock if water is used on a Class C fire.

81. **The correct answer is (D).** The table indicates that, when paints are the combustible material, oxygen exclusion (the blanketing effect) is the best way to extinguish the fire.

82. **The correct answer is (D).** When resistance is placed against the drill bit, the tool will function satisfactorily and the hole will be made. As the bit approaches the underside of the wood, the resistance is lessened and the bit is forced through the wood, splintering the underside.

83. **The correct answer is (A).** Jumping into a net is not an activity with which many people have had experience. The person jumping might be unable to gauge distance accurately. Firefighters must be alert to position the net under the person.

84. **The correct answer is (B).** The metal spoon is a good conductor and will rapidly conduct heat away from the water. This reduces the amount of heat absorbed by the glass and prevents it from breaking.

85. **The correct answer is (B).** Friction in a pipe or a hose varies directly with the length. The longer the pipe or hose, the greater the friction. The immediate result of friction in a hose is to cut down the available pressure at the nozzle.

86. **The correct answer is (B).** This is a principle of pressure in fluids. Pressure applied to a confined fluid from within is transmitted in all directions without differentiation. Therefore, all points in the 100-foot pipe would have the same pressure.

87. **The correct answer is (B).** This is an example of a second-class lever. (The weight is between the fulcrum and the lift.) By moving the weight closer to the fulcrum, an increase in distance between the load and the lift point is achieved. This increases the mechanical advantage of the lever, and less effort is necessary to lift the load.

88. **The correct answer is (B).** This is stated in the third sentence. Alterations in old buildings frequently create voids or channels that allow fire to extend throughout a building. Removal of partitions and/or structural members have resulted in the collapse of buildings.

89. **The correct answer is (C).** This is the meaning of the fourth sentence.

90. **The correct answer is (B).** This is stated in the last sentence.

91. **The correct answer is (B).** See the third sentence.

92. **The correct answer is (C).** Many times, the firefighter is the first person at the scene of an accident. By being knowledgeable in first aid, he or she can give prompt and efficient care to an injured person. This is essential when attending to life-threatening emergencies.

93. **The correct answer is (D).** Proper treatment of shock is to avoid rough handling, control bleeding, assure breathing, and give nothing by mouth.

94. **The correct answer is (D).** Loosening the tourniquet can result in death by causing additional bleeding, by dislodging clots, and by causing tourniquet shock.

95. **The correct answer is (A).** An open fracture, as the name implies, is a fracture in which there is an open wound. Prompt and efficient care should be given to a closed fracture to prevent it from becoming an open or compound fracture.

96. **The correct answer is (B).** The primary objective of the supervisor is to get the job accomplished. Directing is instructing, informing, and ordering subordinates about what to do and sometimes how to do it to achieve the objectives of the department.

97. **The correct answer is (B).** The correct spelling is "atmosphere."

98. **The correct answer is (D).** The correct spelling is "aqueous."

99. **The correct answer is (A).** The word "after" is the best word to complete the sentence.

100. **The correct answer is (C).** The word "firefighters'" is the best word to complete the sentence. The apostrophe after the "s" in "firefighters'" indicates that the supplies belong to multiple firefighters.

# Practice Test 2

**ANSWER SHEET**

1. Ⓐ Ⓑ Ⓒ Ⓓ Ⓔ
2. Ⓐ Ⓑ Ⓒ Ⓓ Ⓔ
3. Ⓐ Ⓑ Ⓒ Ⓓ Ⓔ
4. Ⓐ Ⓑ Ⓒ Ⓓ Ⓔ
5. Ⓐ Ⓑ Ⓒ Ⓓ Ⓔ
6. Ⓐ Ⓑ Ⓒ Ⓓ Ⓔ
7. Ⓐ Ⓑ Ⓒ Ⓓ Ⓔ
8. Ⓐ Ⓑ Ⓒ Ⓓ Ⓔ
9. Ⓐ Ⓑ Ⓒ Ⓓ Ⓔ
10. Ⓐ Ⓑ Ⓒ Ⓓ Ⓔ
11. Ⓐ Ⓑ Ⓒ Ⓓ Ⓔ
12. Ⓐ Ⓑ Ⓒ Ⓓ Ⓔ
13. Ⓐ Ⓑ Ⓒ Ⓓ Ⓔ
14. Ⓐ Ⓑ Ⓒ Ⓓ Ⓔ
15. Ⓐ Ⓑ Ⓒ Ⓓ Ⓔ
16. Ⓐ Ⓑ Ⓒ Ⓓ Ⓔ
17. Ⓐ Ⓑ Ⓒ Ⓓ Ⓔ
18. Ⓐ Ⓑ Ⓒ Ⓓ Ⓔ
19. Ⓐ Ⓑ Ⓒ Ⓓ Ⓔ
20. Ⓐ Ⓑ Ⓒ Ⓓ Ⓔ
21. Ⓐ Ⓑ Ⓒ Ⓓ Ⓔ
22. Ⓐ Ⓑ Ⓒ Ⓓ Ⓔ
23. Ⓐ Ⓑ Ⓒ Ⓓ Ⓔ
24. Ⓐ Ⓑ Ⓒ Ⓓ Ⓔ
25. Ⓐ Ⓑ Ⓒ Ⓓ Ⓔ

26. Ⓐ Ⓑ Ⓒ Ⓓ Ⓔ
27. Ⓐ Ⓑ Ⓒ Ⓓ Ⓔ
28. Ⓐ Ⓑ Ⓒ Ⓓ Ⓔ
29. Ⓐ Ⓑ Ⓒ Ⓓ Ⓔ
30. Ⓐ Ⓑ Ⓒ Ⓓ Ⓔ
31. Ⓐ Ⓑ Ⓒ Ⓓ Ⓔ
32. Ⓐ Ⓑ Ⓒ Ⓓ Ⓔ
33. Ⓐ Ⓑ Ⓒ Ⓓ Ⓔ
34. Ⓐ Ⓑ Ⓒ Ⓓ Ⓔ
35. Ⓐ Ⓑ Ⓒ Ⓓ Ⓔ
36. Ⓐ Ⓑ Ⓒ Ⓓ Ⓔ
37. Ⓐ Ⓑ Ⓒ Ⓓ Ⓔ
38. Ⓐ Ⓑ Ⓒ Ⓓ Ⓔ
39. Ⓐ Ⓑ Ⓒ Ⓓ Ⓔ
40. Ⓐ Ⓑ Ⓒ Ⓓ Ⓔ
41. Ⓐ Ⓑ Ⓒ Ⓓ Ⓔ
42. Ⓐ Ⓑ Ⓒ Ⓓ Ⓔ
43. Ⓐ Ⓑ Ⓒ Ⓓ Ⓔ
44. Ⓐ Ⓑ Ⓒ Ⓓ Ⓔ
45. Ⓐ Ⓑ Ⓒ Ⓓ Ⓔ
46. Ⓐ Ⓑ Ⓒ Ⓓ Ⓔ
47. Ⓐ Ⓑ Ⓒ Ⓓ Ⓔ
48. Ⓐ Ⓑ Ⓒ Ⓓ Ⓔ
49. Ⓐ Ⓑ Ⓒ Ⓓ Ⓔ
50. Ⓐ Ⓑ Ⓒ Ⓓ Ⓔ

51. Ⓐ Ⓑ Ⓒ Ⓓ Ⓔ
52. Ⓐ Ⓑ Ⓒ Ⓓ Ⓔ
53. Ⓐ Ⓑ Ⓒ Ⓓ Ⓔ
54. Ⓐ Ⓑ Ⓒ Ⓓ Ⓔ
55. Ⓐ Ⓑ Ⓒ Ⓓ Ⓔ
56. Ⓐ Ⓑ Ⓒ Ⓓ Ⓔ
57. Ⓐ Ⓑ Ⓒ Ⓓ Ⓔ
58. Ⓐ Ⓑ Ⓒ Ⓓ Ⓔ
59. Ⓐ Ⓑ Ⓒ Ⓓ Ⓔ
60. Ⓐ Ⓑ Ⓒ Ⓓ Ⓔ
61. Ⓐ Ⓑ Ⓒ Ⓓ Ⓔ
62. Ⓐ Ⓑ Ⓒ Ⓓ Ⓔ
63. Ⓐ Ⓑ Ⓒ Ⓓ Ⓔ
64. Ⓐ Ⓑ Ⓒ Ⓓ Ⓔ
65. Ⓐ Ⓑ Ⓒ Ⓓ Ⓔ
66. Ⓐ Ⓑ Ⓒ Ⓓ Ⓔ
67. Ⓐ Ⓑ Ⓒ Ⓓ Ⓔ
68. Ⓐ Ⓑ Ⓒ Ⓓ Ⓔ
69. Ⓐ Ⓑ Ⓒ Ⓓ Ⓔ
70. Ⓐ Ⓑ Ⓒ Ⓓ Ⓔ
71. Ⓐ Ⓑ Ⓒ Ⓓ Ⓔ
72. Ⓐ Ⓑ Ⓒ Ⓓ Ⓔ
73. Ⓐ Ⓑ Ⓒ Ⓓ Ⓔ
74. Ⓐ Ⓑ Ⓒ Ⓓ Ⓔ
75. Ⓐ Ⓑ Ⓒ Ⓓ Ⓔ

76. Ⓐ Ⓑ Ⓒ Ⓓ Ⓔ
77. Ⓐ Ⓑ Ⓒ Ⓓ Ⓔ
78. Ⓐ Ⓑ Ⓒ Ⓓ Ⓔ
79. Ⓐ Ⓑ Ⓒ Ⓓ Ⓔ
80. Ⓐ Ⓑ Ⓒ Ⓓ Ⓔ
81. Ⓐ Ⓑ Ⓒ Ⓓ Ⓔ
82. Ⓐ Ⓑ Ⓒ Ⓓ Ⓔ
83. Ⓐ Ⓑ Ⓒ Ⓓ Ⓔ
84. Ⓐ Ⓑ Ⓒ Ⓓ Ⓔ
85. Ⓐ Ⓑ Ⓒ Ⓓ Ⓔ
86. Ⓐ Ⓑ Ⓒ Ⓓ Ⓔ
87. Ⓐ Ⓑ Ⓒ Ⓓ Ⓔ
88. Ⓐ Ⓑ Ⓒ Ⓓ Ⓔ
89. Ⓐ Ⓑ Ⓒ Ⓓ Ⓔ
90. Ⓐ Ⓑ Ⓒ Ⓓ Ⓔ
91. Ⓐ Ⓑ Ⓒ Ⓓ Ⓔ
92. Ⓐ Ⓑ Ⓒ Ⓓ Ⓔ
93. Ⓐ Ⓑ Ⓒ Ⓓ Ⓔ
94. Ⓐ Ⓑ Ⓒ Ⓓ Ⓔ
95. Ⓐ Ⓑ Ⓒ Ⓓ Ⓔ
96. Ⓐ Ⓑ Ⓒ Ⓓ Ⓔ
97. Ⓐ Ⓑ Ⓒ Ⓓ Ⓔ
98. Ⓐ Ⓑ Ⓒ Ⓓ Ⓔ
99. Ⓐ Ⓑ Ⓒ Ⓓ Ⓔ
100. Ⓐ Ⓑ Ⓒ Ⓓ Ⓔ

practice test

# PRACTICE TEST 2

## 100 Questions • 210 Minutes

**Directions:** The following passage describes an incident encountered by firefighters while engaging in their day-to-day activities. You will have 5 minutes to study this passage. Questions 1–15 are based on the details described in these paragraphs. Therefore, you should make every effort to memorize the highlights as they are described during the time allowed for memorization. Mark your answers on the answer sheet.

During the summer of 1981, a severe water shortage existed that was the result of an extended heat wave and drought. Hydrants opened by citizens in an effort to gain relief from the heat depleted water resources rapidly, and emergency measures were undertaken by the city to conserve water.

Fire department personnel conducted a survey that revealed that 600 hydrants were opened daily by citizens and an average of 1,100 gallons of water per minute were wasted at each open hydrant. Hydrant patrols were implemented using 75 engine companies on 2-hour shifts to close down open hydrants.

One day, while on hydrant patrol, Engine 299 stopped to shut down an open hydrant that was being used by area residents to gain relief from the summer heat. While the firefighters were in the act of closing the hydrant, they were subjected to severe verbal abuse by the crowd that quickly formed around them. A potentially explosive situation developed rapidly. The firefighters closed the hydrant and quickly left the scene to return to the firehouse.

They were in the firehouse for a very short time when an alarm was received to return to the same street for a reported fire in an occupied multiple-dwelling building. Engine 299 responded to the alarm. As they turned into the street, they were greeted with a barrage of rocks and bottles thrown from doorways and rooftops. The officer in command ordered the firefighters to take cover immediately and transmitted a radio message requesting police assistance. Police officers arrived and dispersed the crowd. The firefighters then checked the location of the reported fire and learned that there was no actual fire; it was a false alarm. They returned to the firehouse, and a report of the incident was forwarded to the fire commissioner.

The following day, the deputy chief requested that a meeting be arranged between fire department officials and community leaders in an effort to discuss and resolve the problem. At the meeting, fire department representatives informed community leaders of the importance of having adequate water pressure available to extinguish fires and attempted to convince residents that firefighters were there to help them. Residents also were informed that sprinkler caps for fire hydrants were available at the firehouse. The use of these caps would result in the saving of much water. It then would be possible for the residents to use the hydrants to cool themselves, and sufficient water pressure would be maintained to fight fires.

Community leaders must be convinced that responding to false alarms can result in the destruction of property and the deaths of citizens because firefighters are not able to respond to an actual fire if they are responding to a false alarm.

1. According to the passage, a severe water shortage existed in the city during the summer of 1981. This shortage was due to

   (A) broken water mains.

   (B) open hydrants.

   (C) an extended heat wave and drought.

   (D) misuse of water by industry.

2. The firefighters of Engine 299 stopped to shut down the open hydrant while

   (A) returning to quarters.

   (B) responding to an alarm.

   (C) on alarm box inspection duty.

   (D) on hydrant patrol duty.

3. From the information given in the passage, we can conclude that the main concern of the fire department was to

   (A) provide the area residents with sprinkler caps.

   (B) have water available for fire-fighting.

   (C) educate of the residents in the area.

   (D) show the cooperation between the police and fire departments.

4. If a lieutenant and five firefighters are on duty in each engine company, the total personnel hours spent by the fire department on hydrant patrol for one day was

   (A) 150 hours.

   (B) 350 hours.

   (C) 600 hours.

   (D) 900 hours.

5. The most important factor the fire officials wanted to convey to the community residents at the meeting was

   (A) to conserve water.

   (B) to use sprinkler caps whenever possible.

   (C) that false alarms can result in death.

   (D) not to abuse firefighters while they are on duty.

6. During extended periods of hot weather, hydrant patrol duty usually becomes a daily routine. Of the following, the action that should NOT be taken during these patrols is to

   (A) learn the location and the condition of the hydrants in the district.

   (B) use force on area residents when encountering opposition to the firefighters.

   (C) train chauffeurs to operate the fire apparatus.

   (D) shut down open hydrants to ensure adequate water pressure.

7. The officer in command ordered the firefighters to take shelter from the barrage of rocks and bottles being thrown. The lieutenant's actions were

   (A) improper; the lieutenant should have first ordered a search for the fire.

   (B) proper; however, the lieutenant should have tried to calm the crowd first.

   (C) improper; the lieutenant should have requested more fire department units to respond to the scene.

   (D) proper; the lieutenant's immediate responsibility was the safety and protection of the firefighters under attack.

8. The firefighters shut down the hydrant and hastily left the scene in order to

   (A) respond to an alarm for a fire in a multiple dwelling.

   (B) respond to a false alarm.

   (C) obtain police assistance.

   (D) avoid a physical confrontation with the crowd.

9. Fire department engine companies were deployed to shut down open hydrants. The total number of hours on patrol duty by fire department companies was

   (A) 150 hours.

   (B) 200 hours.

   (C) 250 hours.

   (D) 300 hours.

10. It would be incorrect to state that the police were called to the scene

   (A) to shut down the open hydrant.

   (B) to control the crowd.

   (C) for protection of the firefighters.

   (D) by the lieutenant in command of Engine 299.

11. Open hydrants were depleting the water supply rapidly. In a 2-hour period, an open hydrant was discharging approximately

   (A) 1,100 gallons of water.

   (B) 2,200 gallons of water.

   (C) 66,000 gallons of water.

   (D) 132,000 gallons of water.

12. According to the information contained in the passage, Engine 299 responded from headquarters to

(A) investigate a complaint of rubbish on a roof.

(B) a reported fire in an occupied private dwelling.

(C) a reported fire in a multiple dwelling.

(D) assist police at a false alarm.

13. What was the average number of hydrants shut down each day by each engine company?

(A) 4

(B) 6

(C) 8

(D) 10

14. The officer in command of Engine 299 forwarded a report of the rock- and bottle-throwing incident. This report was forwarded to the

(A) fire commissioner.

(B) deputy chief.

(C) community leaders.

(D) police department.

15. It would be reasonable to believe that the main purpose of the meeting between fire officials and the community leaders was to

(A) inform the residents of the importance of the firefighters.

(B) stress the importance of having adequate water pressure available for firefighting.

(C) remind the community that the firefighter is their friend.

(D) make the community aware of the sprinkler caps.

16. The expansion ratio for medium-expansion foams are from 50:1 to 300:1. If 3 gallons of expansion concentrate are used, how many gallons of foam will be produced?

(A) 150 to 500 gallons

(B) 150 to 900 gallons

(C) 100 to 600 gallons

(D) 300 to 900 gallons

(E) 200 to 950 gallons

17. A fire consumed portions of three floors of a building. On the first floor, 1,300 square feet were damaged; on the second floor, 12,000 square feet were damaged; on the third floor, 7,500 square feet were damaged. What is the total damaged area of the building?

(A) 13,300 square feet

(B) 19,500 square feet

(C) 20,800 square feet

(D) 23,500 square feet

(E) 25,300 square feet

18. If a water tanker can pump out water at 10,000 gallons per hour, how long will it take to pump out 1,000 gallons of water?

(A) 6 minutes

(B) 9 minutes

(C) 10 minutes

(D) 12 minutes

(E) 15 minutes

19. In an enclosed room, a house fire can create temperatures of 900°F. How many times hotter is this than an average room temperature of 75°F?

(A) 10

(B) 12

(C) 13

(D) 15

(E) 20

20. If a fully extended 24-foot ladder is needed to reach a second-story window, how long should a ladder be to reach a fifth-story window?

(A) 35 feet

(B) 40 feet

(C) 50 feet

(D) 60 feet

(E) 65 feet

21. In any city or suburb with a population of at least 20,000 people, there should be two fire hydrants for every square city block. If there are 75 square city blocks in one mile, how many fire hydrants should there be in two square miles?

(A) 150

(B) 200

(C) 225

(D) 250

(E) 300

**CHOOSE THE CORRECT SPELLING OF THE WORD TO COMPLETE THE SENTENCES IN QUESTIONS 22–23.**

22. There are _____ advantages to keeping equipment and protective clothing clean and free from defects.

(A) numerous

(B) numerus

(C) numirous

(D) numourus

(E) numirus

23. _____ rope often is used instead of natural rope in rescue operations.

(A) Sinthetic

(B) Synthettic

(C) Synthetick

(D) Sinthetik

(E) Synthetic

**CHOOSE THE APPROPRIATE WORD TO COMPLETE THE SENTENCES IN QUESTIONS 24–25.**

24. Firefighter Campbell _____ a captain at station 32.

    (A) am

    (B) is

    (C) are

    (D) were

    (E) being

25. Since the fire was over, Mr. Cook had _____ to collect what was left of his belongings.

    (A) began

    (B) begin

    (C) beginning

    (D) begins

    (E) begun

26. In New York, schools are required to hold at least 12 fire drills each school year, 8 of which must be conducted between September and December. Of the following, the best justification for holding most fire drills during the September 1 to December 1 period is that

    (A) most fires occur during that period.

    (B) pupils are trained in evacuation of their schools at the beginning of the school year.

    (C) the weather is milder during that period.

    (D) school attendance is higher at the beginning of the school year.

27. Following a severe smoky fire in the lobby of a hotel, a dead body was found in a room on the eighth floor. The coroner who examined the body found that the victim died from the inhalation of toxic gases and smoke. The most probable explanation of the occurrence is that

    (A) the coroner made an error in the findings.

    (B) the victim inhaled the gases and smoke in the lobby, fled to the eighth floor to escape the fire, and collapsed there.

    (C) the gases and smoke traveled through vertical openings to the upper stories where they were inhaled by the victim.

    (D) a second fire occurred simultaneously in the upper stories of the hotel.

28. The basic assumption of fire prevention educational programs is that people frequently

    (A) must be forced into obeying fire laws.

    (B) are unaware of the dangers involved in some of their actions.

    (C) do not care whether their actions are dangerous.

    (D) assume that fire insurance protects them against all fire loss.

**29.** The statement that is most accurate is that visibility in buildings on fire is

(A) not a serious problem mainly because fires give off sufficient light to enable firefighters to see clearly.

(B) a serious problem mainly because fires often knock out the electrical system of the building.

(C) not a serious problem mainly because most fires occur during daylight hours.

(D) a serious problem mainly because smoky conditions are often encountered.

**30.** While operating at a five-alarm fire, a firefighter is approached by an obviously intoxicated man who claims to be a former firefighter and who offers to help put out the fire. In this situation, the best of the following courses of action for the firefighter is to

(A) give the man some easy-to-perform task.

(B) refer the man to the officer in command of the fire.

(C) decline the offer of help and ask the man to remain outside the fire lines.

(D) ask to see the man's credentials and, if he is a former firefighter as he claims, put him to work stretching hoselines.

**31.** A civilian came into a firehouse to report that some adolescents had turned on a hydrant in the neighborhood. After getting the location of the hydrant, the firefighter on housewatch asked the civilian for his name and was told "Smith." The firefighter then thanked the civilian and proceeded to process the report. The firefighter's actions in this situation were

(A) proper.

(B) proper, except that the firefighter should not have wasted time asking for the civilian's name.

(C) proper, except that he should have found out whether the civilian's name really was Smith.

(D) proper, except that the firefighter should have obtained the civilian's first name and address as well as his family name.

**32.** Fire apparatus are not permitted to use their sirens when returning to headquarters. Of the following, the main justification for this restriction is that

(A) the chances of being involved in traffic accidents are reduced.

(B) there is no need for prompt return to quarters after a fire has been extinguished.

(C) apparatus return to quarters using less crowded streets than when responding to alarms.

(D) the officer in command is better able to exchange radio messages with the fire alarm dispatcher.

**33.** A fire escape platform is attached to a building flush with the windowsill leading onto it. The most important reason for this arrangement is that it reduces the chance that

**(A)** intruders will utilize the fire escapes for illegal purposes because they will be visible from inside the apartment.

**(B)** tenants will place garbage or other obstructions on the platform because these objects will be visible from inside the apartment.

**(C)** persons using the fire escape in an emergency will trip as they leave the apartment.

**(D)** sagging or loosening of the supports will occur without coming to the attention of persons occupying the apartment.

**34.** A civilian, on his way home from work one evening, hears an alarm ringing and sees water running out of a sprinkler discharge pipe on the side of a building. No smoke or other indications of fire are seen. There is a padlock on the entrance to the building, no fire alarm box is in sight, and a firehouse is located 1½ blocks away. In this situation, it would be most proper for the civilian to

**(A)** take no action because there is no evidence of a fire or other emergency.

**(B)** attempt to ascertain the name of the building's owner and notify him of the situation.

**(C)** send an alarm from the closest fire alarm box.

**(D)** go to the firehouse and inform the firefighter at the desk of the situation.

**35.** Running a hoseline between two rungs of a ladder into a building on fire generally is considered to be a

**(A)** good practice, mainly because resting the hose on the rungs of the ladder reduces the load that must be carried by the firefighters

**(B)** poor practice, mainly because the maneuverability of the ladder is reduced.

**(C)** good practice, mainly because the weight of the hoseline holds the ladder more securely against the side of the building.

**(D)** poor practice, mainly because the ladder can be damaged by the additional weight of the hoseline filled with water.

**36.** Suppose you are the firefighter on housewatch duty when a woman rushes into the firehouse and reports that a teenage gang is assaulting a man on the street several blocks away. In this situation, the best of the following courses of action for you to take first is to

**(A)** notify the police department of the reported incident.

**(B)** go to the scene of the disturbance and verify the woman's report.

**(C)** suggest to the woman that she report the matter to the police department.

**(D)** notify your fellow firefighters of the incident and all go to the aid of the man being assaulted.

37. A firefighter on the way to work is standing on a subway platform waiting for a train. Suddenly, a man standing nearby collapses to the floor, showing no sign of breathing. In this situation, the firefighter should

    (A) run to the nearest telephone and summon an ambulance.

    (B) designate a responsible person to look after the victim and then continue on to the firehouse.

    (C) administer first-aid measures to restore breathing.

    (D) take no action because the victim is obviously dead.

38. Fire companies often practice methods of firefighting and lifesaving in public places. The best justification for this practice is that it

    (A) provides facilities for practicing essential operations that are not readily available in fire department installations.

    (B) shows the taxpaying public that they are getting their money's worth.

    (C) attracts youngsters and interests them in firefighting as a career.

    (D) makes certain that all company equipment is in the best possible working condition.

39. The fire that generally would present the greatest danger to surrounding buildings from flying sparks or brands is a fire in a(n)

    (A) warehouse storing paper products.

    (B) factory manufacturing plastics.

    (C) outdoor lumberyard.

    (D) open-type parking garage.

40. A fire marshal, questioning firefighters at the scene of a suspicious fire, obtains some conflicting statements about details of the fire situation present upon their arrival. The most likely explanation of this conflict is that

    (A) the firefighters have not been properly trained to carry out their part in arson investigations.

    (B) the fire marshal did not give the firefighters time to collect and organize their impressions of the fire scene.

    (C) witnesses to an event see the situation from their own point of view and seldom agree on all details.

    (D) details are of little importance if there is general agreement on major matters.

41. A firefighter on duty at a theater who discovers standees obstructing aisles should immediately

    (A) report the situation to his or her superior officer.

    (B) order the standees to move out of the aisles.

    (C) ask the theater manager to correct the situation.

    (D) issue a summons to the usher assigned to that area.

42. At a fire on the fourth floor of an apartment house, the first engine company to arrive advanced a hoseline up the stairway to the third floor before charging the hose with water. The main reason that the firefighters delayed charging their line is that an empty line

    (A) is less likely to whip about and injure firefighters.

    (B) is easier to carry.

    (C) won't leak water.

    (D) is less subject to damage.

43. Suppose the owner of a burning tenement building complains that, although the fire is located on the first floor, firefighters are chopping holes in the roof. Of the following, the most appropriate reason you can give for their action is that the fire can be fought most effectively by permitting

    (A) smoke and hot gases to escape.

    (B) firefighters to attack the fire from above.

    (C) firefighters to gain access to the building through the holes.

    (D) immediate inspection of the roof for extension of the fire.

44. The fire department always endeavors to purchase the best apparatus and equipment and to maintain it in the best condition. The main justification for this policy is that

    (A) public confidence in the department is increased.

    (B) failure of equipment at a fire can have serious consequences.

    (C) replacement of worn-out parts often is difficult.

    (D) the dollar cost to the department is less in the long run.

45. The statement about smoke that is most accurate is that smoke is

    (A) irritating but not dangerous in itself.

    (B) irritating and dangerous only because it can reduce the oxygen content of the air breathed.

    (C) dangerous because it can reduce the oxygen content of the air breathed and often contains toxic gases.

    (D) dangerous because it supports combustion.

46. Suppose you are a firefighter making a routine inspection of a rubber goods factory. During the inspection, you discover some minor violations of the Fire Prevention Code. When you call these violations to the attention of the factory owner, he becomes annoyed and tells you that he is the personal friend of high officials in the fire department and city government. Under these circumstances, the best course for you to follow is to

    (A) summon a police officer to arrest the owner for attempting to intimidate a public official performing a duty.

    (B) make a very thorough inspection and serve summonses for every possible violation of the Fire Department Code.

    (C) ignore the owner's remarks and continue the inspection in your usual manner.

    (D) try to obtain from the owner the names and positions of his friends.

**47.** It has been suggested that property owners should be charged a fee each time the fire department is called to extinguish a fire on their property. Of the following, the best reason for rejecting this proposal is that

    **(A)** delays in calling the fire department might result.

    **(B)** many property owners don't occupy the property they own.

    **(C)** property owners might resent such a charge because they pay real estate taxes.

    **(D)** it might be difficult to determine on whose property a fire started.

**48.** Standpipe systems of many bridges are the dry pipe type. A dry pipe system has no water in the pipes when not in use. When water is required, it is necessary first to pump water into the system. The main reason for using a dry standpipe system is to prevent

    **(A)** corrosion of the pipes.

    **(B)** freezing of water in the pipes.

    **(C)** waste of water through leakage.

    **(D)** strain on the pumps.

**49.** Assume you are a firefighter on your way home after completing your tour of duty. Just as you are about to enter the subway, a man runs up to you and reports a fire in a house located five blocks away. You recognize the man as a mildly retarded but harmless person who frequently loiters around firehouses and at fires. Of the following, the best action for you to take is to

    **(A)** run to the house to see if there really is a fire.

    **(B)** call in an alarm from a nearby telephone.

    **(C)** ignore the report because of the man's mental condition.

    **(D)** call the police and ask that a radio patrol car investigate the report.

Rescue operations often take place in confined areas that do not allow for natural or forced ventilation. Due to the dangers that these types of situations present to a firefighter, it is extremely important to use a system of communication so that the firefighter in the confined space can signal to other firefighters if there is trouble. A method of signaling called the OATH method has been developed for this type of situation. The letter O indicates that the individual is okay, A indicates that the individual wants to advance, T indicates the individual wants to be taken up, and H indicates that the individual needs help. One tug on a safety line stands for O, two tugs for A, three tugs for T, and four tugs for H.

50. If a firefighter is working in a deep, narrow well and wants to be taken back up to the surface, how many times should he tug on the lifeline?

   **(A)** 1
   **(B)** 2
   **(C)** 3
   **(D)** 4
   **(E)** 5

**QUESTION 51–55 ARE BASED ON THE FOLLOWING PASSAGE:**

I. The sizes of living rooms shall meet the following requirements:

　　a. In each apartment, there shall be at least one living room containing at least 120 square feet of clear floor area; every other living room, except a kitchen, shall contain at least 70 square feet of clear floor area.

　　b. Every living room that contains less than 80 square feet of clear floor area or that is located in the cellar or basement shall be at least 9 feet high; every other living room shall be at least 8 feet high.

　　c. Apartments containing three or more rooms can have dining bays, which shall not exceed 55 square feet in floor surface area and shall not be deemed separate rooms or subject to the requirements for separate rooms. Every such dining bay shall be provided with at least one window containing an area at least one eighth of the floor surface area of such dining bay.

**51.** The minimum volume of a living room, other than a kitchen, that meets the minimum requirements listed in the passage is one that measures

**(A)** 70 cubic feet.

**(B)** 80 cubic feet.

**(C)** 630 cubic feet.

**(D)** 640 cubic feet.

**52.** A builder proposes to construct an apartment house containing an apartment consisting of a kitchen that measures 10 feet by 6 feet, a room that measures 12 feet by 12 feet, and a room that measures 11 feet by 7 feet. This apartment

**(A)** does not comply with the requirements listed in the passage.

**(B)** complies with the requirements listed in the passage provided that it is not located in the cellar or basement.

**(C)** complies with the requirements listed in the passage provided that the height of the smaller rooms is at least 9 feet.

**(D)** may or may not comply with the requirements listed in the passage depending on the clear floor area of the kitchen.

**53.** The definition of the term "living room" that is most in accord with its meaning in the passage is

**(A)** a sitting room or parlor.

**(B)** the largest room in an apartment.

**(C)** a room used for living purposes.

**(D)** any room in an apartment containing 120 square feet of clear floor area.

54. Assume that one room in a four-room apartment measures 20 feet by 10 feet and contains a dining bay of 8 feet by 6 feet. According to the passage, the dining bay must be provided with a window measuring at least

    (A) 6 square feet.

    (B) 7 square feet.

    (C) 25 square feet.

    (D) 55 square feet.

55. Kitchens, according to the preceding passage, are

    (A) not considered "living rooms."

    (B) considered "living rooms" and must therefore meet the height and area requirements of the passage.

    (C) considered "living rooms" but need meet only the height or area requirements of the passage, not both.

    (D) considered "living rooms" but need not meet area requirements.

**QUESTION 56–60 ARE BASED ON THE FOLLOWING PASSAGE:**

**EMPLOYEE-LEAVE REGULATIONS**

"As a full-time, permanent city employee under the Career and Salary Plan, firefighter Peter Smith earns an 'annual leave allowance.' This consists of a certain number of days off a year with pay and can be used for vacation, personal business, or observing religious holidays. As a newly appointed employee, during his first eight years of city service, he will earn an 'annual leave allowance' of twenty days off a year (an average of $1\frac{2}{3}$ days off a month). After he has finished eight full years of working for the city, he will begin earning an additional five days off a year. His 'annual leave allowance' will then be 25 days a year and will remain at this amount for seven full years. He will begin earning an additional two days off a year after he has completed a total of fifteen years of city employment. Therefore, in his sixteenth year of working for the city, Smith will be earning twenty-seven days off a year as his 'annual leave allowance' (an average of $2\frac{1}{4}$ days off a month).

A "sick leave allowance" of one day a month also is given to firefighter Smith, but it can be used only in the case of actual illness. When Smith returns to work after using 'sick leave allowance,' he must have a doctor's note if the absence is for a total of more than three days, but he also might be required to show a doctor's note for absences of one, two, or three days."

56. According to the preceding passage, Pete Smith's "annual leave allowance" consists of a certain number of days off each year that he

(A) does not get paid for.

(B) gets paid for at time and a half.

(C) can use for personal business.

(D) cannot use for observing religious holidays.

57. According to the preceding passage, after Pete Smith has been working for the city for nine years, his "annual leave allowance" will be

(A) 20 days a year.

(B) 25 days a year.

(C) 27 days a year.

(D) 37 days a year.

58. According to the preceding passage, Pete Smith will begin earning an average of $2\frac{1}{4}$ days off a month as his "annual leave allowance" after he has worked for the city for

(A) 7 full years.

(B) 8 full years.

(C) 15 full years.

(D) 17 full years.

59. According to the preceding passage, Pete Smith is given a "sick leave allowance" of

(A) 1 day every 2 months.

(B) 1 day per month.

(C) $1\frac{2}{3}$ days per month.

(D) $2\frac{1}{4}$ days a month.

60. According to the preceding passage, when Pete Smith uses "sick leave allowance," he might be required to show a doctor's note

(A) even if his absence is for only one day.

(B) only if his absence is for more than two days.

(C) only if his absence is for more than three days.

(D) only if his absence is for three days or more.

**QUESTIONS 61–64 ARE BASED ON THE FOLLOWING LIST OF WARNINGS CONCERNING ELECTRICITY:**

- Firefighters never should attempt to cut any wires. They should wait until trained utility workers are available to do any necessary cutting.

- When downed electrical wires are encountered, a danger zone should be established around the downed wire to ensure the safety of firefighters.

- All wires should be treated as though they are charged with a high voltage electrical current.

- Firefighters are in danger of not only shock and burns from electrical equipment but also eye injuries from electrical flashing. A firefighter should never look directly at flashing electrical lines.

- Specific attention must be paid to electrical dangers when raising or lowering hoselines, ladders, or equipment near overhead lines.

- Firefighters should be aware of any tingling sensation felt in the feet when working in an area where wires are down. This sensation might indicate that the ground is charged.

- Full protective clothing should always be worn in areas where electrical hazards might exist. Only approved insulated tools that are regularly tested should be used.

61. When a firefighter is a safe distance away from a downed wire, but the wire is sparking and flashing as the current jumps from the wire to a nearby car, the firefighter should

   (A) cut the wire near the power source.

   (B) assume the wire is not charged.

   (C) watch the flashing closely for flames.

   (D) ignore any tingling felt in the feet.

   (E) avoid looking directly at the flashing.

62. When in an area in which an electrical hazard might exist, firefighters should wear

   (A) street clothes.

   (B) full protective clothing.

   (C) only protective gloves.

   (D) only protective boots.

   (E) only protective gloves and boots.

63. Raising a ladder near an overhead power line

   (A) should never be done when there are possible electrical dangers.

   (B) is a job that should be conducted only by trained utility workers.

   (C) is not a serious concern if a safety zone has been established.

   (D) should be done with specific attention to electrical dangers.

   (E) is not a concern if a firefighter assumes the wires are not charged.

64. When working around a downed wire, a firefighter notices a tingling sensation in her feet. This sensation might indicate that

   (A) the ground is charged.

   (B) the wire is not charged.

   (C) the wire should be cut.

   (D) there is no electrical danger.

   (E) a danger zone is unnecessary.

### QUESTIONS 65–68 ARE BASED ON THE FOLLOWING PARAGRAPH:

Whenever a social group has become so efficiently organized that it has gained access to an adequate supply of food and has learned to distribute it among its members so well that wealth considerably exceeds immediate demands, it can be depended on to utilize its surplus energy in an attempt to enlarge the sphere in which it is active. The structure of ant colonies renders them particularly prone to this sort of expansionist policy. With very few exceptions, ants of any given colony are hostile to those of any other colony, even of the same species. This condition is bound to produce preliminary bickering among colonies that are closely associated.

65. According to the paragraph, a social group is wealthy when it

   (A) is efficiently organized.

   (B) controls large territories.

   (C) contains energetic members.

   (D) produces and distributes food reserves.

66. According to the paragraph, the structure of an ant colony is its

   (A) social organization.

   (B) nest arrangement.

   (C) territorial extent.

   (D) food-gathering activities.

67. It follows from the preceding paragraph that the least expansionist society would be one that has

   (A) great poverty generally.

   (B) more than sufficient wealth to meet its immediate demands.

   (C) great wealth generally.

   (D) wide inequality between its richest and poorest members.

68. According to the preceding paragraph, an ant generally is hostile EXCEPT to other

   (A) insects.

   (B) ants.

   (C) ants of the same species.

   (D) ants of the same colony.

**QUESTIONS 69–72 ARE BASED ON THE FOLLOWING TABLE:**

Carbon monoxide (CO) is a lethal gas that causes more deaths than any other product of combustion. Very small concentrations of carbon monoxide readily combine with chemicals in the blood and quickly cause damage to the brain and other body tissues by blocking the blood's ability to carry oxygen.

**Toxic Effects of Carbon Monoxide**

| Carbon Monoxide in Air (%) | Symptoms |
| --- | --- |
| 1.23 | Instant unconsciousness; death likely in 1 to 3 minute |
| .32 | Dizziness, headache, nausea after 5 to 10 minutes; unconsciousness after 30 minutes |
| .04 | Headache after 1 to 2 hours of exposure |
| .01 | No damage—no symptoms |

**Scenario A**

On a very cold day in January, Anthony starts up his car in the garage. Due to the cold temperatures outside, he decides to stay in the car and to leave the garage door closed until the car warms up. After about 10 minutes, Anthony has a headache and starts to feel dizzy and nauseous.

**69.** According to the information and table presented, the likely cause of Anthony's symptoms is

**(A)** the extremely cold temperature.

**(B)** the lack of carbon monoxide in his blood.

**(C)** the lack of oxygen in his blood.

**(D)** carbon monoxide levels of .04 percent.

**(E)** carbon monoxide levels of 1.23 percent.

**70.** According to the information and table, Anthony could become unconscious if he

**(A)** has no symptoms.

**(B)** has a headache 3 hours later.

**(C)** gets any more oxygen.

**(D)** stays in the car for another 20 minutes.

**(E)** is not exposed to more carbon monoxide.

**Scenario B**

Two firefighters enter a room that contains a smoldering fire. Against the advice of his partner, one of the firefighters takes off his breathing apparatus to clear some debris from the inside of his mask. He instantly falls to the floor unconscious.

71. The likely reason for the firefighter's unconsciousness is

(A) dizziness, headache, or nausea.

(B) carbon monoxide levels of .01 percent.

(C) carbon monoxide levels of .04 percent.

(D) carbon monoxide levels of .32 percent.

(E) carbon monoxide levels of 1.23 percent.

**Scenario C**

A fire started in a large apartment building and took several hours to extinguish. Evacuation of the building took over an hour to complete. Some of the last people to leave the building were complaining of headaches.

72. What was the likely concentration of carbon monoxide in the apartment building?

(A) 1.5 percent

(B) 1.23 percent

(C) .32 percent

(D) .04 percent

(E) .01 percent

**73.** The function of pinion gear (2) in the hand drill shown in the diagram is to

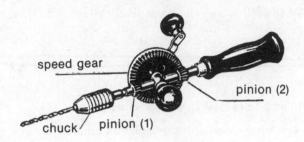

**(A)** increase the speed of the chuck.

**(B)** keep the speed gear from wobbling.

**(C)** double the turning force on the chuck.

**(D)** allow reverse rotation of the speed gear.

**74.** Assume that two identical insulated jugs are filled with equal quantities of water from a water tap. A block of ice is placed in one jug and the same quantity of ice, chopped into small cubes, is placed in the other jug. The statement that is most accurate is that the water in the jug containing the chopped ice, compared to the water in the other jug, will be chilled

**(A)** faster but to a substantially higher temperature.

**(B)** faster and to approximately the same temperature.

**(C)** slower but to a substantially higher temperature.

**(D)** slower and to approximately the same temperature.

**75.** The long pole and hook shown in the sketch is called a pike pole. Firefighters sometimes push the point and hook through plaster ceilings and then pull the ceiling down. Of the following, the most likely reason for this practice is to

**(A)** let heat and smoke escape from the room.

**(B)** trace defective electric wiring through the house.

**(C)** see if hidden fire is burning above the ceiling.

**(D)** remove combustible material that will provide fuel for the fire.

### QUESTION 76 IS BASED ON THE FOLLOWING PASSAGE:

Firefighters Jones and Smith were recently called to a traumatic fire involving a crashed passenger jet. Jones was assigned to a crew responsible for containing and putting out the fire. Smith was assigned to a crew responsible for rescuing victims from the plane.

Many of the victims Smith attempted to save were small children, and many did not survive. In the weeks following the plane crash, Jones noticed that Smith was very depressed and had a hard time focusing his attention on his duties at the station.

**76.** Considering the recent events, what is most likely the reason for Smith's behavior?

**(A)** Job stress

**(B)** Boredom

**(C)** Drugs

**(D)** Alcohol

**(E)** Laziness

**77.** As your company arrives at the scene of a fire in a large rooming house, a man is spied on the second-story fire escape, about to let himself down to the ground using a sheet he has torn into strips, knotted, and attached to some object in the building. Two middle-aged women and a young boy are observed descending to the street using the fire escape. Flames and smoke are issuing from the third-floor windows. Of the following, the best characterization of this man's behavior is that it is

**(A)** intelligent, because the fire has cut off egress by means of the stairway.

**(B)** not intelligent, because a firm anchor for the knotted sheet could be achieved by securing it to the fire escape.

**(C)** intelligent, because the fire might heat the fire escape to the point where it is red hot.

**(D)** not intelligent, because to descend the fire escape would be a quicker and safer means of escape.

**78.** Suppose the same quantity of water is placed in the cup and bowl pictured below and both are left on a table. The time required for the water to evaporate completely from the cup, as compared to the bowl, would be

**(A)** longer.

**(B)** shorter.

**(C)** equal.

**(D)** longer or shorter depending on the temperature and humidity in the room.

**79.** Some tools are known as all-purpose tools because they can be used for a great variety of purposes. Others are called special-purpose tools because they are suitable only for a particular purpose. In general, an all-purpose tool, as compared to a special tool for the same purpose, is

**(A)** cheaper.

**(B)** less efficient.

**(C)** safer to use.

**(D)** simpler to operate.

80. During a snowstorm, a passenger car with rear-wheel drive gets stuck in the snow. It is observed that the rear wheels are spinning in the snow, and the front wheels are not turning. The statement that best explains why the car is not moving is that

(A) moving parts of the motor are frozen or blocked by the ice and snow.

(B) the front wheels are not receiving power because of a defective or malfunctioning transmission.

(C) the rear wheels are not obtaining sufficient traction because of the snow.

(D) the distribution of the power to the front and rear wheels is not balanced.

81. When the inside face of a rubber suction cup is pressed against a smooth wall, it usually remains in place because of the

(A) force of molecular attraction between the rubber and the wall.

(B) pressure of the air on the rubber cup.

(C) suction caused by the elasticity of the rubber.

(D) static electricity generated by the friction between the rubber and the wall.

82. The self-contained breathing apparatus consists of a tank containing breathable air supplied to the user through flexible tubes. This type of breathing apparatus would be least effective in an atmosphere containing

(A) insufficient oxygen to sustain life.

(B) gases irritating to the skin.

(C) gases that cannot be filtered.

(D) a combination of toxic gases.

83. At a fire during very cold weather, a firefighter who was ordered to shut down a hoseline left it partially open so that a small amount of water continued to flow out of the nozzle. Leaving the nozzle partially open in this situation was a

(A) good practice, mainly because the hoseline can be put back into action more quickly.

(B) poor practice, mainly because the escaping water will form puddles, which will freeze on the street.

(C) good practice, mainly because the water in the hoseline will not freeze.

(D) poor practice, mainly because unnecessary water damage to property will result.

84. Some sprinkler systems are supplied with water from a single source. Others are supplied with water from two different sources. In general, firefighters consider the two-source system preferable to the single-source system because the two-source system

    (A) is cheaper to install.

    (B) is less likely to be put out of operation.

    (C) is capable of delivering water at higher pressure.

    (D) goes into operation faster.

85. Of the following statements about electric fuses, the one that is most valid is that they

    (A) should never be replaced by coins.

    (B) can be replaced by coins for a short time if there are no fuses available.

    (C) can be replaced by coins provided that the electric company is notified.

    (D) can be replaced by coins provided that care is taken to avoid overloading the circuit.

86. The function of the flat surface machined into the shaft in the following diagram is to

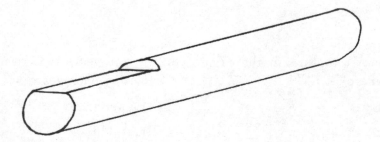

    (A) prevent slippage of a pulley positioned at this end on the shaft.

    (B) provide a nonskid surface to hold the shaft steady as it is machined.

    (C) prevent the shaft from rolling about when it is placed on a flat workbench.

    (D) reveal subsurface defects in the shaft.

practice test

**QUESTIONS 87–89 ARE BASED ON THE FOLLOWING DIAGRAM. GEAR A IS THE DRIVER. GEARS A AND D EACH HAVE TWICE AS MANY TEETH AS GEAR B, AND GEAR C HAS FOUR TIMES AS MANY TEETH AS GEAR B. THE DIAGRAM IS SCHEMATIC; THE TEETH GO ALL AROUND EACH GEAR.**

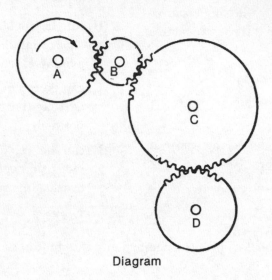

Diagram

87. Two gears that turn in the same direction are

 (A) A and B.

 (B) B and C.

 (C) C and D.

 (D) B and D.

88. The two gears that revolve at the same speed are gears

 (A) A and C.

 (B) A and D.

 (C) B and C.

 (D) B and D.

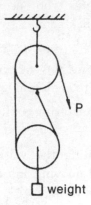

weight

**89.** In the preceding sketch of a block
and fall, if the end of the rope P is
pulled so that it moves one foot, the
distance the weight will be raised is

**(A)** $\frac{1}{2}$ foot.

**(B)** 1 foot.

**(C)** $1\frac{1}{2}$ feet.

**(D)** 2 feet.

**90.** The following sketches show four objects that have the same weight but different shapes. The object that is most difficult to tip over is numbered

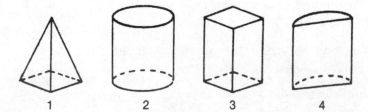

**(A)** 1

**(B)** 2

**(C)** 3

**(D)** 4

91. A firefighter, while taking some clothing to a dry cleaner in his neighborhood, noticed that inflammable cleaning fluid was stored in a way that created a fire hazard. The firefighter called this to the attention of the proprietor, explaining the danger involved. This method of handling the situation was

    (A) bad; the firefighter should not have interfered in a matter that was not his responsibility.

    (B) good; the proprietor probably will remove the hazard and be more careful in the future.

    (C) bad; the firefighter should have reported the situation to the fire inspector's office without saying anything to the proprietor.

    (D) good; because the firefighter was a customer, he should treat the proprietor more leniently than he would treat other violators.

92. During the course of fighting a fire, a firefighter encounters an unconscious middle-aged man who is bleeding profusely. It would be most proper for the firefighter to administer first aid by initially

    (A) attempting to bring the man back to consciousness.

    (B) calling for medical assistance and awaiting its arrival.

    (C) administering artificial respiration.

    (D) calling for medical assistance and seeking out and blocking the source of the bleeding.

93. A firefighter who encounters a victim of shock should give first aid to the victim by

    (A) covering the victim with a wet cloth to keep him or her cool.

    (B) administering artificial respiration.

    (C) loosening the victim's clothing and removing his or her shoes and stockings.

    (D) keeping the victim lying down and elevating extremities if there is no further injury.

94. Modern acceptable first-aid technique recommends that, whenever possible, bleeding caused by punctures should be controlled by the application of a

    (A) tourniquet.

    (B) pressure bandage.

    (C) coagulant.

    (D) figure-eight bandage.

95. A newly appointed firefighter becomes involved in an accident while on duty. Faced with this problem, it would be most appropriate for the firefighter's superior officer to first ascertain

    (A) if the firefighter is accident-prone.

    (B) the cause of the accident.

    (C) if the orientation program stresses employee safety sufficiently.

    (D) if the firefighter has ever been involved in an accident previously that was covered by workmen's compensation.

96. A veteran firefighter feels lightheaded soon after reporting for a tour of duty. It would be most proper for the firefighter to

    (A) ask to be excused from ascending heights during this tour of duty.

    (B) say nothing because a replacement is not available.

    (C) promptly report this condition to his or her superior officer.

    (D) discuss the matter with fellow firefighters to see if they have ever had the same problem.

97. A firefighter, after completing a special course of study on safety at work, is not convinced that a method taught during the course is the best way to avoid accidents. Upon returning to duty, the firefighter should

    (A) perform the task as it was taught.

    (B) perform the task in the manner he or she is convinced is the best way.

    (C) attempt to gain the support of fellow firefighters in his or her method.

    (D) request a transfer to other duty where it will not be necessary to perform this task.

98. A firefighter on inspection duty is offered a sum of money by a store owner to overlook a violation of the fire law. The best action to take would be to

    (A) order the store owner to clear the violation without delay and report the incident to his or her superior officer.

    (B) say nothing to the store owner but return to the firehouse to seek the advice of fellow firefighters.

    (C) inform the store owner that firefighters do not accept bribes.

    (D) refuse the money but offer to assist the store owner to clear the violation during his or her off-duty hours.

99. All newly appointed firefighters are put through an extensive training course before they are permitted to participate in the actual activity of fighting fires. The most probable reason for this is that

    (A) it would be unwise for inept firefighters to be exposed to the scrutiny of the public.

    (B) the skills necessary for effective firefighting are unique to the job.

    (C) the orientation training session is necessary to establish the officer–firefighter relationship.

    (D) newly appointed firefighters must be convinced of the dangers inherent in firefighting.

**100.** The statement "Modern technology has greatly influenced firefighting techniques" means most nearly that

**(A)** newly appointed firefighters are better equipped to fight fires than their predecessors because they are better educated.

**(B)** the need for investigating whether changes should be made in procedures becomes important when new firefighting equipment is acquired.

**(C)** established successful firefighting techniques should be retained whenever possible.

**(D)** newly acquired equipment always calls for changes in established procedures.

# ANSWERS AND EXPLANATIONS

| | | | | | | | | |
|---|---|---|---|---|---|---|---|---|
| 1. | C | 21. | E | 41. | C | 61. | E | 81. | B |
| 2. | D | 22. | A | 42. | B | 62. | B | 82. | B |
| 3. | B | 23. | E | 43. | A | 63. | D | 83. | C |
| 4. | D | 24. | B | 44. | B | 64. | A | 84. | B |
| 5. | C | 25. | E | 45. | C | 65. | D | 85. | A |
| 6. | B | 26. | B | 46. | C | 66. | A | 86. | A |
| 7. | D | 27. | C | 47. | A | 67. | A | 87. | D |
| 8. | D | 28. | B | 48. | B | 68. | D | 88. | B |
| 9. | A | 29. | D | 49. | B | 69. | C | 89. | A |
| 10. | A | 30. | C | 50. | C | 70. | D | 90. | A |
| 11. | D | 31. | D | 51. | C | 71. | E | 91. | B |
| 12. | C | 32. | A | 52. | C | 72. | D | 92. | D |
| 13. | C | 33. | C | 53. | C | 73. | B | 93. | D |
| 14. | A | 34. | D | 54. | A | 74. | B | 94. | B |
| 15. | B | 35. | B | 55. | D | 75. | C | 95. | B |
| 16. | B | 36. | A | 56. | C | 76. | A | 96. | C |
| 17. | C | 37. | C | 57. | B | 77. | D | 97. | A |
| 18. | A | 38. | A | 58. | C | 78. | A | 98. | A |
| 19. | B | 39. | C | 59. | B | 79. | B | 99. | B |
| 20. | D | 40. | C | 60. | A | 80. | C | 100. | D |

1.  **The correct answer is (C).** This is stated in the first sentence of the first paragraph.

2.  **The correct answer is (D).** This is stated in the first sentence of the third paragraph.

3.  **The correct answer is (B).** This is stated in the second sentence of the fifth paragraph.

4.  **The correct answer is (D).** 75 companies × 2 hours each = 150 hours × 6 persons per engine company = 900 personnel hours.

5.  **The correct answer is (C).** See the last paragraph.

6.  **The correct answer is (B).** If force is necessary to move people away from hydrants to shut the hydrants, the police department should be notified for assistance. The firefighter should not become involved in this action.

7.  **The correct answer is (D).** To effectively fight a fire, firefighters should be in good physical condition. One of the lieutenant's responsibilities is to keep the firefighters from becoming injured.

8. **The correct answer is (D).** "A potentially explosive situation developed rapidly." There was no reason for the firefighters to hang around waiting for trouble.

9. **The correct answer is (A).** 75 companies × 2 hours = 150 hours.

10. **The correct answer is (A).** The police were called by the lieutenant in command of Engine 299 to control the crowd and to protect firefighters, not to close the hydrant.

11. **The correct answer is (D).** 1,100 gallons per minute × 120 minutes (2 hours) = 132,000 gallons.

12. **The correct answer is (C).** See the beginning of the fourth paragraph.

13. **The correct answer is (C).** 600 hydrants ÷ 75 companies = 8 hydrants.

14. **The correct answer is (A).** This is stated in the last sentence of the fourth paragraph.

15. **The correct answer is (B).** This is stated in the second sentence of the fifth paragraph.

16. **The correct answer is (B).** For every 1 gallon of expansion concentrate, 50 to 300 gallons of foam are produced. Thus, if 3 gallons of concentrate are used, 150 (50 × 3) to 900 (300 × 3) gallons of foam will be produced.

17. **The correct answer is (C).** The total area of the building that was damaged is the sum of the square feet consumed on the three floors. Thus, 1,300 + 12,000 + 7,500 = 20,800 square feet.

18. **The correct answer is (A).** 1,000 gallons is $\frac{1}{10}$ of 10,000 gallons. If 10,000 gallons are pumped per hour, we want to know what $\frac{1}{10}$ of one hour is. Because there are 60 minutes in 1 hour, $60 \times \frac{1}{10} = 6$ minutes.

19. **The correct answer is (B).** 900 ÷ 75 = 12. To check your answer, 75 × 12 = 900.

20. **The correct answer is (D).** If it takes a 24-foot ladder to reach the second story, it takes 12 feet of ladder for each story (24 ÷ 2 = 12). Because there are five stories, 12 × 5 = 60.

21. **The correct answer is (E).** Because there are two hydrants for every block and 75 blocks for every square mile, there are 75 × 2 = 150 hydrants for every square mile. Because we are looking at 2 square miles, 150 × 2 = 300 hydrants for 2 square miles.

22. **The correct answer is (A).** The correct spelling is "numerous."

23. **The correct answer is (E).** The correct spelling is "Synthetic."

24. **The correct answer is (B).** The word "is" is the best word to complete the sentence.

25. **The correct answer is (E).** The word "begun" is the best word to complete the sentence. This is the correct past-tense form of the word "begin."

26. **The correct answer is (B).** In general, there are many new students in a school at the beginning of the school year who are unfamiliar with the exits. By having the majority of drills in the early part of the school year, the children become familiar with the evacuation plan and the location of the exits.

27. **The correct answer is (C).** Toxic gases and smoke from a fire are light and rise rapidly through vertical openings in a building. They "mushroom" on the upper floors. This is why ventilation at the roof is essential.

28. **The correct answer is (B).** People frequently create unsafe situations. Poor housekeeping or poor safety habits eventually could result in a destructive fire. Educating people about the existing dangers can save them from injury, death, and destruction of their property.

29. **The correct answer is (D).** A top priority in firefighting is ventilation of the building to remove smoke and toxic gases. Smoke removal heightens visibility, saves lives, reduces the chances of panic, and enables firefighters to perform more effectively.

30. **The correct answer is (C).** Only authorized, qualified firefighters are permitted to operate at a fire. Be firm. If the man insists upon "helping," have a police officer remove him from the scene.

31. **The correct answer is (D).** In general, it is a good policy to obtain as much information as you can from the complainant. It makes reporting easier and more complete. By having the full name and address of the complainant on record, contact could be made quite easily if necessary.

32. **The correct answer is (A).** When returning from an alarm, fire apparatus should adhere to traffic regulations just like any other vehicle on the road. Sirens can cause motorists and pedestrians to become confused and nervous, thereby increasing the chance of an accident.

33. **The correct answer is (C).** People using a fire escape in an emergency usually are excited. Any obstruction in their path can present a hazard. By having the platform level with the sill, people using the fire escape have better footing, reducing the chance of tripping.

34. **The correct answer is (D).** The alarm ringing and the discharge of water from the pipe are indications that one or more sprinkler heads have opened and there is water flow in the building. Because the location of the firehouse is known and nearby and the location of the alarm box is uncertain, the wise decision is to go to the firehouse and report the situation.

35. **The correct answer is (B).** If a hoseline is run between the rungs of a ladder, the ladder no longer is maneuverable until the hoseline is shut down and disconnected.

36. **The correct answer is (A).** Notifying the police department is the quickest and most proper action to take. Radio communications enable the police to be at the scene rapidly. They are trained to handle this type of emergency. You must remain available to respond to fires.

37. **The correct answer is (C).** The immediate administration of first aid by the firefighter to restore breathing gives the victim a chance of surviving. The firefighter also should direct a passerby to call for medical assistance, either by telephone or at the change booth. If a transit officer is nearby, notify the officer.

38. **The correct answer is (A).** By conducting a drill or practice at the site of a possible emergency, firefighting forces become familiar with the various aspects of a particular type of building or location. They get firsthand knowledge of the obstacles that might be encountered during an actual emergency.

39. **The correct answer is (C).** Lumberyards store a wide assortment of wood and wood products. Strong upward drafts can develop at a lumberyard fire and can carry sparks and large brands a considerable distance, especially if there is much wind.

40. **The correct answer is (C).** At a suspicious fire, the officer in charge should document pertinent facts as soon as possible. The passing of time generally results in a witness forgetting many of the details of what he or she saw originally. Documenting the facts will enable the firefighter to recall the incident.

41. **The correct answer is (C).** By asking the theater manager to correct the situation, the firefighter is making the manager aware of the existing violation. It is the manager's responsibility to correct the situation.

42. **The correct answer is (B).** An uncharged 50-foot length of $2\frac{1}{2}$-inch hose weighs approximately 40 pounds. Water weighs $8\frac{1}{3}$ pounds per gallon, and there are 13 gallons in the 50-foot length. Obviously, the uncharged line is much lighter and much easier to carry to the point where water is needed.

43. **The correct answer is (A).** Smoke and hot gases from the fire will rise and "mushroom" at the roof or upper floors. If the roof is not vented, the hot gases will ignite and spread fire to the roof area.

44. **The correct answer is (B).** The fire equipment must be maintained to respond immediately in an emergency. Delays caused by apparatus breakdowns could result in loss of life or excessive property damage. This is why extensive preventive maintenance is performed on high-quality department apparatus.

45. **The correct answer is (C).** Smoke is a product of combustion, so eliminate choice (D). The newspaper carries reports of death from smoke inhalation, so eliminate choice (A). Smoke, as the product of combustion, not only reduces the oxygen content of the air but contains particles, carbon monoxide, and whatever toxic gases might be produced by the substance burning.

46. **The correct answer is (C).** As long as the firefighter is performing his or her job properly, there is no reason to be intimidated by the owner.

47. **The correct answer is (A).** Charging a property owner a fee for calling the fire department is not a good policy. People might be unable or unwilling to afford the fee and might attempt to extinguish the fire themselves. Experience has shown that delays in calling the fire department have resulted in fires getting out of control, causing loss of life and extensive property damage.

48. **The correct answer is (B).** The use of dry pipe standpipe systems where the pipes are exposed to freezing temperatures avoids having pipes blocked by ice or pipes that burst from the pressure of frozen water.

49. **The correct answer is (B).** Calling in the alarm will bring firefighters to the reported location much faster than if the firefighter ran to the house to see if there really was a fire. Every report of fire must be taken seriously.

50. **The correct answer is (C).** The passage indicates that three tugs stands for T, which indicates that the individual wants to be taken up.

51. **The correct answer is (C).** A room must have at least 70 square feet of floor space, but if it has less than 80 square feet, it must be at least 9 feet high. 70 square feet × 9 feet = 630 cubic feet.

52. **The correct answer is (C).** This is stated in paragraph b. Every living room with less than 80 square feet of floor area shall be 9 feet high.

53. **The correct answer is (C).** When speaking of all rooms in apartments of three or more rooms as "living rooms," the passage must be referring to rooms used for living.

54. **The correct answer is (A).** The floor space of the dining bay is 6 × 8 = 48 square feet. One eighth of 48 = 6 square feet.

55. **The correct answer is (D).** The words "every other living room except a kitchen" imply that a kitchen is a living room but that it is not subject to the size requirements of other living rooms.

56. **The correct answer is (C).** See the second sentence.

57. **The correct answer is (B).** For 8 to 15 years of service, the annual leave is 25 days.

58. **The correct answer is (C).** The sixteenth year, in which leave is earned at the rate of 2¼ days per month, comes after 15 full years.

59. **The correct answer is (B).** See the first sentence of the second paragraph.

60. **The correct answer is (A).** See the last sentence.

61. **The correct answer is (E).** The fourth warning on the list states that firefighters should never look directly at flashing electrical lines.

62. **The correct answer is (B).** The last warning on the list states that full protective clothing always should be worn in areas in which electrical hazards might exist.

63. **The correct answer is (D).** The fifth warning on the list states that specific attention must be paid to electrical dangers when raising a ladder near an overhead power line.

64. **The correct answer is (A).** The sixth warning on the list states that any tingling sensation felt in the feet might indicate that the ground is charged.

65. **The correct answer is (D).** When supply exceeds current needs, the society is wealthy and can distribute its surplus.

66. **The correct answer is (A).** This information can be inferred from the first and second sentences.

67. **The correct answer is (A).** Poverty is the opposite of wealth. If there is poverty, there is no surplus and no expansionism.

68. **The correct answer is (D).** This is stated in the third sentence.

69. **The correct answer is (C).** The table lists headache, dizziness, and nausea as signs that someone has been exposed to an atmosphere with carbon monoxide levels of .32 percent for 5 to 10 minutes. The paragraph indicates that carbon monoxide causes damage by blocking the blood's ability to carry oxygen. Thus, choice (C) is the correct response.

70. **The correct answer is (D).** The table lists unconsciousness as a symptom of remaining in an atmosphere with carbon monoxide levels of .32 percent for more than 30 minutes. Because Anthony already exhibits the symptoms of headache, dizziness, and nausea after 10 minutes, he could become unconscious if he remains in the car for an additional 20 minutes.

71. **The correct answer is (E).** The table lists instant unconsciousness as a symptom of breathing air with carbon monoxide levels of 1.23 percent. Thus, choice (E) is the correct response.

72. **The correct answer is (D).** The table lists headache as a symptom of exposure to carbon monoxide levels of .04 percent for 1 to 2 hours. Because the last people to leave the building were in it for over an hour and were complaining of headaches, choice (D) is the correct response.

73. **The correct answer is (B).** The pinion gear is a small bevel or spur gear that meshes with the speed gear. The pinions "bind" the large gear to prevent it from wobbling.

74. **The correct answer is (B).** The small cubes of chopped ice have a greater surface area than the large block of ice. Therefore, more water will be exposed to the area of the chopped ice, and it will chill faster to the same temperature.

75. **The correct answer is (C).** In a fire, flame and hot gases can enter cracks or holes in a ceiling or wall and can burn undetected. If the ceilings are not opened and the fire extinguished, the fire will continue to burn and spread throughout the structure. The pike pole is used for this purpose.

76. **The correct answer is (A).** The passage indicates that the plane crash was very traumatic. Smith was assigned to rescue victims from the plane, many of whom did not survive. Judging from the information presented in the passage, it is likely that Smith's depression and lack of focus are due to job stress.

77. **The correct answer is (D).** There are people on the fire escape right now, so it obviously is not too hot. The fire escape would be quicker and safer than a sheet, which might pull loose, tear, or burn. During moments of panic, people sometimes use poor judgment.

78. **The correct answer is (A).** Evaporation occurs when the molecules near the surface of the liquid escape into the air. The water in the bowl has a greater surface area exposed to the air, causing it to evaporate faster than the water in the cup.

79. **The correct answer is (B).** The special-purpose tool generally is more efficient because it is designed for one specific purpose. The all-purpose tool can accomplish the intended job, but it usually requires more time and works less effectively. In firefighting, time and efficiency are essential.

80. **The correct answer is (C).** For the rear wheels (which are the driving wheels) to drive or push the car, friction or resistance must be present between the tires and the road surface. The ice and snow create a slippery road condition, causing the tires to lose traction. The front wheels do not turn because power from the engine is transmitted to the rear wheels only.

81. **The correct answer is (B).** When the suction cup is pressed against the wall, the air in the cup is forced out and a vacuum or negative pressure remains. Atmospheric air with a pressure of 14.7 psi (pounds per square inch) pushes against the rubber cup and holds it against the wall.

82. **The correct answer is (B).** Many toxic gases and acids are absorbed directly through the skin. To operate safely and effectively in this type of atmosphere, self-contained suits made of rubber or other nonporous material

should be used in conjunction with the self-contained breathing apparatus. The self-contained apparatus takes in no outside air, so filtration is not a problem.

83. **The correct answer is (C).** During very cold weather, water will freeze much more quickly when it is idle than when it is moving. By keeping the water flowing, there is a smaller chance that the water in the hoseline will freeze.

84. **The correct answer is (B).** Some sprinkler systems require two sources of water. In the event that one source is placed out of service, the other is available to keep the system operating. Two-source systems can be supplied by: (a) two water mains, each on a different street; (b) a street main and a gravity tank; (c) a street main and a pressure tank.

85. **The correct answer is (A).** The function of a fuse is to protect against overload or overheating of a circuit. A burned-out fuse indicates that a dangerous condition on a circuit has been avoided. By replacing the fuse with a coin that cannot burn out, you bypass the safety feature of the fuse and create a fire hazard.

86. **The correct answer is (A).** The flat surface of the shaft is called a keyway. A corresponding opening in a gear or pulley accommodates the keyway. This prevents slippage of the gear or pulley on the shaft.

87. **The correct answer is (D).** Adjacent gears turn in opposite directions. Gear A turns in a clockwise direction; therefore, gear B, which is adjacent to it, will turn in a counterclockwise direction. Continuing, gear C, which is adjacent to gear B, will turn in a clockwise direction; gear D, which is adjacent to gear C, will turn counterclockwise, the same as gear B.

88. **The correct answer is (B).** Gears of the same size connected into the same system will turn at the same speed.

89. **The correct answer is (A).** The sketch is of a fixed pulley and a moveable pulley. With this arrangement, the effort to move the weight will be one half. The effort distance (the length of the pull), however, will be twice the distance the weight travels. Therefore, a one-foot pull raised the weight one-half foot.

90. **The correct answer is (A).** With all the objects equal in weight, the pyramid is the most difficult to tip over. This is due to the fact that the bulk of the volume and weight is at the base.

91. **The correct answer is (B).** Surely the proprietor of the dry cleaning store does not want to create a dangerous situation for himself and for his property. He will be grateful for the information and will most likely change his manner of storing cleaning fluid. The firefighter's action falls into the category of friendly, helpful neighbor.

answers

92. **The correct answer is (D).** Profuse bleeding can be life threatening. While waiting for medical assistance, the firefighter should initially attempt to control the bleeding by applying direct pressure to the wound with a sterile dressing if one is available. If bleeding persists, the firefighter should reach into the wound and pinch the blood vessel.

93. **The correct answer is (D).** Proper treatment for a person in shock includes ensuring adequate breathing, controlling the bleeding, administering oxygen, elevating extremities, avoiding rough handling, preventing loss of body heat, keeping the victim lying down, and giving nothing by mouth.

94. **The correct answer is (B).** A pressure bandage will be sufficient. In general, when a person suffers a puncture wound, there is very little external bleeding. However, severe internal bleeding is possible.

95. **The correct answer is (B).** Before any corrective action can be taken, the cause of the accident must be known. By getting the facts, the officer is able to determine why the accident occurred. Knowing why the accident occurred can prevent a similar accident in the future.

96. **The correct answer is (C).** The superior officer, being aware of the firefighter's condition, can summon medical advice for the firefighter and, if necessary, can relieve him or her from duty. Getting lightheaded or dizzy in a hazardous situation can result in serious injury to the firefighter and to anyone he or she might be assisting.

97. **The correct answer is (A).** The information presented at the special course on safety apparently was compiled by using case studies, information gathered from accident reports, and firsthand experience. The firefighter could voice his or her opinion on the subject to the safety instructor or to his or her superior officers. However, the firefighter should perform the task as it was taught.

98. **The correct answer is (A).** A firefighter on inspection duty has a job to perform. That job is to note and report violations and to follow through to be certain violations are corrected. Accepting a bribe to overlook violations is a criminal act as well as a dereliction of duty.

99. **The correct answer is (B).** An extensive training program is necessary because of the nature of the occupation of firefighting. Fires and fire conditions vary with each situation. The tremendous hazards present when fighting fires are not found in any other occupation. The lives of the fire victims, as well as those of fellow firefighters, are at stake. For the firefighter to function effectively, he or she must be as well-trained as possible before being permitted to take part in actual firefighting.

100. **The correct answer is (D).** Changes in procedures are necessary due to the many technological advances that have been introduced into the field of firefighting. These technological advances enable the firefighter to function more efficiently and with greater effectiveness. Some advances that have dictated changes include improved self-contained breathing apparatus, handi-talkies, power saws, hydraulic cutting equipment, and sophisticated foams and fog nozzles.

# Practice Test 3

## ANSWER SHEET

1. Ⓐ Ⓑ Ⓒ Ⓓ Ⓔ
2. Ⓐ Ⓑ Ⓒ Ⓓ Ⓔ
3. Ⓐ Ⓑ Ⓒ Ⓓ Ⓔ
4. Ⓐ Ⓑ Ⓒ Ⓓ Ⓔ
5. Ⓐ Ⓑ Ⓒ Ⓓ Ⓔ
6. Ⓐ Ⓑ Ⓒ Ⓓ Ⓔ
7. Ⓐ Ⓑ Ⓒ Ⓓ Ⓔ
8. Ⓐ Ⓑ Ⓒ Ⓓ Ⓔ
9. Ⓐ Ⓑ Ⓒ Ⓓ Ⓔ
10. Ⓐ Ⓑ Ⓒ Ⓓ Ⓔ
11. Ⓐ Ⓑ Ⓒ Ⓓ Ⓔ
12. Ⓐ Ⓑ Ⓒ Ⓓ Ⓔ
13. Ⓐ Ⓑ Ⓒ Ⓓ Ⓔ
14. Ⓐ Ⓑ Ⓒ Ⓓ Ⓔ
15. Ⓐ Ⓑ Ⓒ Ⓓ Ⓔ
16. Ⓐ Ⓑ Ⓒ Ⓓ Ⓔ
17. Ⓐ Ⓑ Ⓒ Ⓓ Ⓔ
18. Ⓐ Ⓑ Ⓒ Ⓓ Ⓔ
19. Ⓐ Ⓑ Ⓒ Ⓓ Ⓔ
20. Ⓐ Ⓑ Ⓒ Ⓓ Ⓔ
21. Ⓐ Ⓑ Ⓒ Ⓓ Ⓔ
22. Ⓐ Ⓑ Ⓒ Ⓓ Ⓔ
23. Ⓐ Ⓑ Ⓒ Ⓓ Ⓔ
24. Ⓐ Ⓑ Ⓒ Ⓓ Ⓔ
25. Ⓐ Ⓑ Ⓒ Ⓓ Ⓔ

26. Ⓐ Ⓑ Ⓒ Ⓓ Ⓔ
27. Ⓐ Ⓑ Ⓒ Ⓓ Ⓔ
28. Ⓐ Ⓑ Ⓒ Ⓓ Ⓔ
29. Ⓐ Ⓑ Ⓒ Ⓓ Ⓔ
30. Ⓐ Ⓑ Ⓒ Ⓓ Ⓔ
31. Ⓐ Ⓑ Ⓒ Ⓓ Ⓔ
32. Ⓐ Ⓑ Ⓒ Ⓓ Ⓔ
33. Ⓐ Ⓑ Ⓒ Ⓓ Ⓔ
34. Ⓐ Ⓑ Ⓒ Ⓓ Ⓔ
35. Ⓐ Ⓑ Ⓒ Ⓓ Ⓔ
36. Ⓐ Ⓑ Ⓒ Ⓓ Ⓔ
37. Ⓐ Ⓑ Ⓒ Ⓓ Ⓔ
38. Ⓐ Ⓑ Ⓒ Ⓓ Ⓔ
39. Ⓐ Ⓑ Ⓒ Ⓓ Ⓔ
40. Ⓐ Ⓑ Ⓒ Ⓓ Ⓔ
41. Ⓐ Ⓑ Ⓒ Ⓓ Ⓔ
42. Ⓐ Ⓑ Ⓒ Ⓓ Ⓔ
43. Ⓐ Ⓑ Ⓒ Ⓓ Ⓔ
44. Ⓐ Ⓑ Ⓒ Ⓓ Ⓔ
45. Ⓐ Ⓑ Ⓒ Ⓓ Ⓔ
46. Ⓐ Ⓑ Ⓒ Ⓓ Ⓔ
47. Ⓐ Ⓑ Ⓒ Ⓓ Ⓔ
48. Ⓐ Ⓑ Ⓒ Ⓓ Ⓔ
49. Ⓐ Ⓑ Ⓒ Ⓓ Ⓔ
50. Ⓐ Ⓑ Ⓒ Ⓓ Ⓔ

51. Ⓐ Ⓑ Ⓒ Ⓓ Ⓔ
52. Ⓐ Ⓑ Ⓒ Ⓓ Ⓔ
53. Ⓐ Ⓑ Ⓒ Ⓓ Ⓔ
54. Ⓐ Ⓑ Ⓒ Ⓓ Ⓔ
55. Ⓐ Ⓑ Ⓒ Ⓓ Ⓔ
56. Ⓐ Ⓑ Ⓒ Ⓓ Ⓔ
57. Ⓐ Ⓑ Ⓒ Ⓓ Ⓔ
58. Ⓐ Ⓑ Ⓒ Ⓓ Ⓔ
59. Ⓐ Ⓑ Ⓒ Ⓓ Ⓔ
60. Ⓐ Ⓑ Ⓒ Ⓓ Ⓔ
61. Ⓐ Ⓑ Ⓒ Ⓓ Ⓔ
62. Ⓐ Ⓑ Ⓒ Ⓓ Ⓔ
63. Ⓐ Ⓑ Ⓒ Ⓓ Ⓔ
64. Ⓐ Ⓑ Ⓒ Ⓓ Ⓔ
65. Ⓐ Ⓑ Ⓒ Ⓓ Ⓔ
66. Ⓐ Ⓑ Ⓒ Ⓓ Ⓔ
67. Ⓐ Ⓑ Ⓒ Ⓓ Ⓔ
68. Ⓐ Ⓑ Ⓒ Ⓓ Ⓔ
69. Ⓐ Ⓑ Ⓒ Ⓓ Ⓔ
70. Ⓐ Ⓑ Ⓒ Ⓓ Ⓔ
71. Ⓐ Ⓑ Ⓒ Ⓓ Ⓔ
72. Ⓐ Ⓑ Ⓒ Ⓓ Ⓔ
73. Ⓐ Ⓑ Ⓒ Ⓓ Ⓔ
74. Ⓐ Ⓑ Ⓒ Ⓓ Ⓔ
75. Ⓐ Ⓑ Ⓒ Ⓓ Ⓔ

76. Ⓐ Ⓑ Ⓒ Ⓓ Ⓔ
77. Ⓐ Ⓑ Ⓒ Ⓓ Ⓔ
78. Ⓐ Ⓑ Ⓒ Ⓓ Ⓔ
79. Ⓐ Ⓑ Ⓒ Ⓓ Ⓔ
80. Ⓐ Ⓑ Ⓒ Ⓓ Ⓔ
81. Ⓐ Ⓑ Ⓒ Ⓓ Ⓔ
82. Ⓐ Ⓑ Ⓒ Ⓓ Ⓔ
83. Ⓐ Ⓑ Ⓒ Ⓓ Ⓔ
84. Ⓐ Ⓑ Ⓒ Ⓓ Ⓔ
85. Ⓐ Ⓑ Ⓒ Ⓓ Ⓔ
86. Ⓐ Ⓑ Ⓒ Ⓓ Ⓔ
87. Ⓐ Ⓑ Ⓒ Ⓓ Ⓔ
88. Ⓐ Ⓑ Ⓒ Ⓓ Ⓔ
89. Ⓐ Ⓑ Ⓒ Ⓓ Ⓔ
90. Ⓐ Ⓑ Ⓒ Ⓓ Ⓔ
91. Ⓐ Ⓑ Ⓒ Ⓓ Ⓔ
92. Ⓐ Ⓑ Ⓒ Ⓓ Ⓔ
93. Ⓐ Ⓑ Ⓒ Ⓓ Ⓔ
94. Ⓐ Ⓑ Ⓒ Ⓓ Ⓔ
95. Ⓐ Ⓑ Ⓒ Ⓓ Ⓔ
96. Ⓐ Ⓑ Ⓒ Ⓓ Ⓔ
97. Ⓐ Ⓑ Ⓒ Ⓓ Ⓔ
98. Ⓐ Ⓑ Ⓒ Ⓓ Ⓔ
99. Ⓐ Ⓑ Ⓒ Ⓓ Ⓔ
100. Ⓐ Ⓑ Ⓒ Ⓓ Ⓔ

practice test

## PRACTICE TEST 3

# 100 Questions • 210 Minutes

**Directions:** Each question has four possible answers. Choose the letter that best answers the question and mark your answer on the answer sheet.

1. While performing a routine inspection of a factory building, a firefighter is asked a question by the plant manager about a matter under the control of the health department and about which the firefighter has little knowledge. In this situation, the best course of action for the firefighter to take is to

   **(A)** answer the question to the best of his or her knowledge.

   **(B)** tell the manager that he or she is not permitted to answer the question because it does not relate to a fire department matter.

   **(C)** tell the manager that it will be referred to the health department.

   **(D)** suggest to the manager that he communicate with the health department about the matter.

2. A firefighter on duty who answers a departmental telephone should give his or her name and rank

   **(A)** at the start of the conversation, as a matter of routine.

   **(B)** only if asked for this information by the caller.

   **(C)** only if the caller is a superior officer.

   **(D)** only if the telephone message requires the firefighter to take some action.

3. At a five-alarm fire in the West End, several companies from Northside were temporarily assigned to occupy the quarters and take over the duties of companies engaged in fighting the fire. The main reason for relocating the Northside companies was to

   **(A)** protect the firehouses from robbery or vandalism that might occur if they were left vacant for a long period of time.

   **(B)** provide for speedy response to the fire if additional companies were required.

   **(C)** give the Northside companies an opportunity to become familiar with the problems of the West End area.

   **(D)** provide protection to the West End area in the event that other fires occurred.

4. Two firefighters, while on their way to report for duty early one morning, observe a fire in a building containing a supermarket on the street level and apartments on the upper stories. One firefighter runs into the building to spread the alarm to tenants. The other firefighter runs to a street alarm box two blocks away and sends an alarm. The latter firefighter should then

   **(A)** return to the building on fire and help evacuate the tenants.

   **(B)** remain at the fire alarm box to direct the first fire company that arrives to the location of the fire.

   **(C)** look for a telephone to call his own fire company and explain that he and his companion will be late in reporting for duty.

   **(D)** look for a telephone to call the health department and request that an inspector be sent to the supermarket to examine the food involved in the fire.

5. A man found an official fire department badge and gave it to his young son to use as a toy. The man's action was improper mainly because

   **(A)** it is disrespectful to the fire department to use the badge in this manner.

   **(B)** the boy might injure himself playing with the badge.

   **(C)** an effort first should have been made to locate the owner of the badge before giving it to the boy.

   **(D)** the badge should have been returned to the fire department.

6. In the case of a fire in a mailbox, the fire department recommends that an extinguishing agent that smothers the fire, such as carbon tetrachloride, be used. Of the following, the most likely reason for NOT recommending the use of water is that

   **(A)** water is not effective on fires in small, tightly enclosed spaces.

   **(B)** someone might have mailed chemicals that could explode in contact with water.

   **(C)** water might damage the mail untouched by fire so that it could not be delivered.

   **(D)** the smothering agent can be put on the fire faster than water can be.

7. Of the following, the main difficulty in obtaining accurate information about the causes of fires is that

   **(A)** firefighters are too busy putting out fires to have time to investigate their causes.

   **(B)** most people have little knowledge of fire hazards.

   **(C)** fires destroy much of the evidence that would indicate the causes of the fires.

   **(D)** fire departments are more interested in fire prevention than in investigating fires that already have occurred.

practice test

8. In an effort to discourage the sending of false alarms and to help apprehend people guilty of this practice, it is suggested that the handles of fire alarm boxes be covered with a dye that would stain the hand of a person sending an alarm and would not wash off for 24 hours. The dye would be visible only under an ultraviolet light. Of the following, the chief objection to such a device is that it would

   **(A)** require funds that can be better used for other purposes.

   **(B)** have no effect on false alarms transmitted by telephone.

   **(C)** discourage some persons from sending alarms for real fires.

   **(D)** punish the innocent as well as the guilty.

9. Automatic fire-extinguishing sprinkler systems sometimes are not effective on fires accompanied by explosions chiefly because

   **(A)** these fires do not generate enough heat to start sprinkler operation.

   **(B)** the pipes supplying the sprinklers usually are damaged by the explosion.

   **(C)** fires in explosive materials usually cannot be extinguished by water.

   **(D)** sprinkler heads usually are clogged by dust created by the explosion.

10. When a fire occurs in the vicinity of a subway system, there is the possibility that water from the firefighters' hose streams will flood underground portions of the subway lines through sidewalk gratings. Of the following methods of reducing this danger, the one that generally would be suitable is for the officer in command to order subordinates to

   **(A)** use fewer hoselines and smaller quantities of water than they would ordinarily.

   **(B)** attack the fire from positions that are distant from the sidewalk gratings.

   **(C)** cover the sidewalk gratings with canvas tarpaulins.

   **(D)** advise the subway dispatcher to reroute the subway trains.

11. When responding to alarms, fire department apparatus generally follow routes established in advance. The least valid justification for this practice is that

   **(A)** motorists living in the area become familiar with these routes and tend to avoid them.

   **(B)** the likelihood of collision between two pieces of fire department apparatus is reduced.

   **(C)** the fastest response generally is obtained.

   **(D)** road construction, road blocks, detours, and similar conditions can be avoided.

12. From a distance, an off-duty firefighter sees a group of teenage boys set fire to a newspaper and then toss the flaming pages into the open window of a building being torn down. In this situation, the first action that should be taken by the firefighter is to

(A) send a fire alarm from the closest street alarm box.

(B) chase the boys and attempt to catch one of them.

(C) investigate whether a fire has been started.

(D) call the police from the closest police alarm box or telephone.

13. When responding to an alarm, officers are not to talk to the driver except to give orders or directions. Of the following, the best justification for this rule is that it

(A) gives the officer an opportunity to make preliminary plans for handling the fire problem.

(B) enables the driver to concentrate on driving the apparatus.

(C) maintains the proper relationship between ranks while on duty.

(D) permits the officer to observe the driver's skill, or lack thereof, in driving the apparatus.

14. The approved method of reporting a fire by telephone in most major cities is to dial

(A) the central headquarters of the fire department.

(B) 911.

(C) the local fire station.

(D) the telephone operator.

15. If there should be two fire hydrants for every square city block and there are 75 square city blocks in a mile, what percentage of a mile would be covered by 120 fire hydrants?

(A) 60 percent

(B) 70 percent

(C) 80 percent

(D) 90 percent

(E) 100 percent

16. A fire has started 3.2 miles from the firehouse. How long will it take the firefighters to arrive at the fire if they travel at an average speed of 30 mph?

(A) 6.6 minutes

(B) 7.1 minutes

(C) 8.3 minutes

(D) 8.5 minutes

(E) 9.2 minutes

17. $13 \times 13 + (9 - 7) =$

(A) 178

(B) 176

(C) 160

(D) 162

(E) 171

18. If water pressure (P) = 500, water temperature (W) = 45, and spray angle = 3.2, what is the correct answer to the equation $P \div A - 2(W) = ?$

(A) 23.4

(B) 42

(C) 56.32

(D) 66.25

(E) 74.3

**19.** In Hewson County, Missouri, last year, there were calls for 21 house fires, 17 industrial fires, 2 industrial accidents, and 46 car accidents. Fire station A received 34 of these calls last year. What percentage of the total number of calls did station A receive?

  **(A)** 23 percent

  **(B)** 35 percent

  **(C)** 40 percent

  **(D)** 42 percent

  **(E)** 46 percent

**CHOOSE THE CORRECT SPELLING OF THE WORD TO COMPLETE THE SENTENCES IN QUESTIONS 20–21.**

**20.** Extensive training is necessary for a firefighter to become _____ at using the equipment involved in fire-fighting.

  **(A)** profishant

  **(B)** proficient

  **(C)** prophicent

  **(D)** profiscent

  **(E)** proficent

**21.** Firefighters might have difficulty locating _____ victims in a smoke-filled building.

  **(A)** unconshous

  **(B)** unconscus

  **(C)** unconscious

  **(D)** unconshious

  **(E)** unconscous

**CHOOSE THE APPROPRIATE WORD TO COMPLETE THE SENTENCES IN QUESTIONS 22–24.**

22. Unfortunately, many of the clues leading to the cause of the fire were _____ beneath the rubble.

    **(A)** hidden

    **(B)** hid

    **(C)** hide

    **(D)** hidded

    **(E)** hides

23. Firefighters Johnson and Fernandez helped many people escape from a large industrial fire. They were later rewarded for _____ heroic efforts.

    **(A)** they

    **(B)** them

    **(C)** they're

    **(D)** their

    **(E)** there

24. It often is better for at least two firefighters to attempt to _____ a ladder.

    **(A)** raise

    **(B)** rise

    **(C)** rose

    **(D)** risen

    **(E)** raised

25. Suppose you are a firefighter on housewatch duty when a civilian enters the firehouse. He introduces himself as a British firefighter visiting the country to study American firefighting methods. He asks you for permission to ride on the fire apparatus when it responds to alarms so he can observe operations firsthand. You know it is against departmental policy to permit civilians to ride apparatus without written permission from headquarters. In this situation, you should

    **(A)** refuse the request but suggest that he follow the apparatus in his own car when it responds to an alarm.

    **(B)** call headquarters and request permission to permit the visitor to ride the apparatus.

    **(C)** refuse the request and suggest that he apply to headquarters for permission.

    **(D)** refuse the request and suggest that he return the next time the fire department holds an open house.

**26.** When operating at a pier fire, firefighters usually avoid driving their apparatus onto the pier itself. The main reason for this precaution is to reduce the possibility that the apparatus will be

(A) delayed in returning to quarters.

(B) driven off the end of the pier.

(C) destroyed by a fire that spreads rapidly.

(D) in the way of the firefighters.

**27.** Pumpers purchased by the fire department are equipped with enclosed cabs. In the past, fire department apparatus were open with no cab or roof. The main advantage of the enclosed cab is that it provides

(A) additional storage space for equipment.

(B) a place of shelter for firefighters operating in an area of radioactivity.

(C) protection for firefighters from weather conditions and injury.

(D) emergency first-aid and ambulance facilities.

**28.** Heavy blizzards greatly increase the problems and work of the fire department. When such a situation occurs, the fire commissioner could reasonably be expected to

(A) order members of the fire department to perform extra duty.

(B) limit parking on city streets.

(C) station firefighters at fire alarm boxes to prevent the sending of false alarms.

(D) prohibit the use of kerosene heaters.

**29.** Regulations of the fire department require that, when placing a hose on a fire wagon, care should be taken to avoid bending the hose at places where it had been bent previously. The most important reason for this requirement is that repeated bending of a hose at the same places will cause

(A) kinks in the hose at those places.

(B) weakening of the hose at those places.

(C) discoloration of the hose at those places.

(D) dirt to accumulate and clog the hose at those places.

**30.** While fighting a fire in an apartment when the occupants are not at home, a firefighter finds a sum of money in a closet. Under these circumstances, the firefighter should turn over the money to

(A) a responsible neighbor.

(B) the desk sergeant of the nearest police station.

(C) the superintendent of the apartment house.

(D) his or her superior officer.

**31.** "When it is necessary to remove a cornice, every effort should be made to pull it back on the roof." The most important reason for the direction in this quotation is that pulling back the cornice on the roof rather than dropping it to the street below

(A) requires less time.

(B) is safer for the people on the street.

(C) makes it possible to reuse the cornice.

(D) is less dangerous to firefighters working on the roof.

**32.** After a fire has been extinguished, one firefighter often remains at the scene after the others have left. Of the following, the main reason for this practice is that this firefighter can

**(A)** prevent looters from stealing valuables.

**(B)** watch for any rekindling of the fire.

**(C)** search the area for lost valuables.

**(D)** examine the premises for evidence of arson.

**33.** Firefighters usually attempt to get as close as possible to the seat of a fire so they can direct their hose streams with accuracy. Intense heat, however, sometimes keeps them at a distance. An unsatisfactory method of overcoming this problem is to

**(A)** have firefighters use a solid object, such as a wall, as a shield.

**(B)** keep firefighters cool by wetting them with small streams of water.

**(C)** use large high-pressure streams and operate at a great distance from the fire.

**(D)** use a water spray to break down the heat waves coming from the fire.

**34.** Suppose you are an off-duty firefighter driving your car in the midtown area. As you cross an intersection, you hear sirens and, looking back, see fire apparatus approaching. In this situation, the best action for you to take is to

**(A)** attempt to clear a path for the fire apparatus by driving rapidly and sounding your horn.

**(B)** drive to the next intersection and direct traffic until the apparatus has passed.

**(C)** permit the apparatus to pass and then follow it closely to the fire, sounding your horn as you drive.

**(D)** pull to the curb, permit the apparatus to pass, and then continue on your way.

**35.** Whenever a public performance given in a theater involves the use of scenery or machinery, a firefighter is assigned to be present. The main reason for this assignment is that

**(A)** theaters are located in high-property-value districts.

**(B)** the use of scenery and machinery increases the fire hazard.

**(C)** theatrical districts have heavy traffic, making for slow response of apparatus.

**(D)** emergency exits may be blocked by the scenery or machinery.

practice test

**QUESTIONS 36–39 ARE BASED ON THE FOLLOWING PARAGRAPH:**

A plastic does not consist of a single substance; it is a blended combination of several substances. In addition to the resin, it may contain various fillers, plasticizers, lubricants, and coloring material. Depending on the type and quantity of substances added to the binder, the properties, including combustibility, might be altered considerably. The flammability of plastics depends on their composition and, as with other materials, on their physical size and condition. Thin sections, sharp edges, or powdered plastics will ignite and burn more readily than the same amount of identical material in heavy sections with smooth surfaces.

36. According to the paragraph, all plastics contain a

   (A) resin.

   (B) resin and a filler.

   (C) resin, filler, and plasticizer.

   (D) resin, filler, plasticizer, lubricant, and coloring material.

37. The conclusion best supported by the paragraph is that the flammability of plastics

   (A) generally is high.

   (B) generally is moderate.

   (C) generally is low.

   (D) varies considerably.

38. According to the paragraph, plastics can best be described as

   (A) a trade name.

   (B) the name of a specific product.

   (C) the name of a group of products that have some similar and some dissimilar properties.

   (D) the name of any substance that can be shaped or molded during the production process.

39. In a manufacturing process, large thick sheets of a particular plastic are cut, buffed, and formed into small tools. The statement most in accord with the information in the preceding paragraph is that

   (A) the dust particles of the plastics are more flammable than the tools or the sheets.

   (B) the plastic tools are more flammable than the dust particles or the sheets.

   (C) the sheets of the plastic are more flammable than the dust particles or the tools.

   (D) there is insufficient information to determine the relative flammability of sheets, tools, and dust particles.

**QUESTIONS 40–42 ARE BASED ON THE FOLLOWING PARAGRAPH:**

To guard against overheating of electrical conductors in buildings, an overcurrent protective device is provided for each circuit. This device is designed to open the circuit and to cut off the flow of current whenever the current exceeds a predetermined limit. The fuse, which is a common form of overcurrent protection, consists of a fusible metal element that, when heated by the current to a certain temperature, melts and opens the circuit.

**40.** According to the paragraph, a circuit that is NOT carrying an electric current is a(n)

(A) open circuit.

(B) closed circuit.

(C) circuit protected by a fuse.

(D) circuit protected by an overcurrent protective device other than a fuse.

**41.** As used in the paragraph, the best example of a conductor is a(n)

(A) metal table that comes in contact with a source of electricity.

(B) storage battery generating electricity.

(C) electrical wire carrying an electrical current.

(D) dynamo converting mechanical energy into electrical energy.

**42.** According to the paragraph, the maximum number of circuits that can be handled by a fuse box containing six fuses is

(A) 3

(B) 6

(C) 12

(D) Cannot be determined from the information given in the paragraph

## QUESTIONS 43–45 ARE BASED ON THE FOLLOWING PARAGRAPH:

Unlined linen hose essentially is a fabric tube made of closely woven linen yarn. Due to the natural characteristics of linen, very shortly after water is introduced, the wet threads swell, closing the minute spaces between them and making the tube practically watertight. This type of hose tends to deteriorate rapidly if not thoroughly dried after use or if installed where it will be exposed to dampness or weather conditions. It is not ordinarily built to withstand frequent service or use in which the fabric will be subjected to chafing from rough or sharp surfaces.

43. Seepage of water through an unlined linen hose is observed when the water is first turned on. From the preceding paragraph, we can conclude that the seepage

    (A) indicates that the hose is defective.

    (B) does not indicate that the hose is defective provided that the seepage is proportionate to the water pressure.

    (C) does not indicate that the hose is defective provided that the seepage is greatly reduced when the hose becomes thoroughly wet.

    (D) does not indicate that the hose is defective provided that the seepage takes place only at the surface of the hose.

44. Unlined linen hose is most suitable for use

    (A) as a garden hose.

    (B) on fire department apparatus.

    (C) as emergency fire equipment in buildings.

    (D) in fire department training schools.

45. The use of unlined linen hose would be least appropriate in a(n)

    (A) outdoor lumberyard.

    (B) nonfireproof office building.

    (C) department store.

    (D) cosmetic manufacturing plant.

**QUESTIONS 46–47 ARE BASED ON THE FOLLOWING PARAGRAPH:**

One of the most common emergency situations that a firefighter might face is a motor vehicle accident. Victims in these situations are commonly trapped in the vehicle, making it difficult for a rescuer to administer proper first aid. It is common, however, for injuries involved in a motor vehicle accident to be very serious and even life threatening. It therefore is important to quickly gain access to the vehicle so that at least one rescuer can be placed in the vehicle to begin stabilizing the victim. After a rescuer gains access to the vehicle, he or she should conduct an initial survey to assess any life-threatening injuries. While conducting the initial survey, the rescuer should keep the victim's airway open, perform CPR if necessary, and treat any uncontrolled bleeding. These procedures are listed in terms of importance and always should be conducted in that order.

**46.** It might be difficult for a rescuer to administer first aid to the victim of a motor vehicle accident because the victim often is

(A) stubborn.

(B) trapped.

(C) breathing.

(D) young.

(E) stabilized.

**47.** The first thing a rescuer should assess after access is gained to the vehicle is whether

(A) there is uncontrolled bleeding.

(B) CPR is necessary.

(C) the victim is trapped.

(D) any bones are broken.

(E) the victim's airway is open.

*practice test*

**QUESTIONS 48–50 ARE BASED ON THE FOLLOWING LIST OF PROCEDURES:**

Fire Chief Williams has prepared the following list of procedures, which are not in sequence, for rescuing victims from burning buildings:

1. From the time of entrance, an internal building search always should be conducted on hands and knees. Visibility in a burning building often is poor, making it dangerous to walk upright.

2. Firefighters involved in rescue operations always should enter a burning building in groups of at least two.

3. Victims who have been removed from a building always should be placed in the custody of an individual who can ensure that the victim will not attempt to reenter the building.

4. Full protective clothing and a protective breathing apparatus always should be donned and double-checked for malfunctions prior to any rescue operation.

5. After a room has been thoroughly searched, the firefighter always should leave a sign to other firefighters that the room has been searched. Hang tags should be left on door handles to indicate that the room has been searched.

6. An internal search always should begin with an outside wall. Windows typically are located on the outside wall and can be opened for ventilation.

48. The logical order for the procedures listed is

   (A) 1, 5, 3, 6, 2, 4
   (B) 6, 5, 4, 2, 3, 1
   (C) 6, 5, 1, 3, 4, 2
   (D) 4, 2, 1, 6, 5, 3
   (E) 4, 6, 2, 1, 5, 3

49. According to the procedures listed, a firefighter who walks inside of a burning building is in violation of which procedural rule?

   (A) 2
   (B) 4
   (C) 5
   (D) 1
   (E) 6

50. The procedures listed imply that tags should be left on the door handles of searched rooms so that

   (A) victims can find their way out of the building.
   (B) other firefighters do not search the room again.
   (C) victims do not attempt to reenter the building.
   (D) windows can be located for ventilation.
   (E) protective equipment can be double-checked.

**51.** A firefighter on inspection duty should behave in a manner that is likely to gain the support of the public that his or her agency serves. The firefighter can accomplish this by engaging in all but one of the following practices. The practice that will NOT accomplish this is

**(A)** being courteous and neat in appearance.

**(B)** being prepared to give explanations of the Fire Prevention Code.

**(C)** overlooking minor infractions of the Fire Prevention Code because correcting them would be expensive.

**(D)** performing inspection duties in a business-like manner.

**52.** As a firefighter, you receive a call during the night while off duty from a neighbor who says his house is on fire. The most appropriate action for you to take first is to

**(A)** call for the assistance of the fire department if it has not already been called.

**(B)** go to the neighbor's home to evaluate the extent of the fire.

**(C)** call other off-duty firefighters who live close by to help fight the fire.

**(D)** advise the neighbor that you will be over as soon as you are dressed.

**QUESTIONS 53–57 ARE BASED ON THE FOLLOWING PARAGRAPH:**

When several victims are involved in a motor vehicle accident, it is essential that firefighters quickly assess the extent of each victim's injuries so the people can be treated and removed from their vehicles in a logical sequence. Victims who are not injured or who are able to exit the vehicle on their own should be quickly removed from the vehicle to make room for firefighters to attend to the most seriously injured individuals. After the most seriously injured individuals have been stabilized, firefighters should give priority to victims trapped in their vehicles.

Last week, firefighters Warren and Mauro were the first to arrive at the scene of a multicar accident that occurred in a rural area. Realizing that the next emergency vehicle would not be able to reach the location for another 20 minutes, the firefighters had to quickly assess which victims should be treated and/or removed from their vehicles.

Three vehicles were involved in the accident. Each vehicle had a driver and one passenger. The first vehicle was severely damaged with the driver trapped inside. The driver of the car was unconscious and did not appear to be breathing. The passenger did not appear to have any life-threatening injuries.

The second car was not damaged so severely. The driver of the car appeared to have a broken leg that was trapped by the steering column but had no other serious injuries. The passenger of the car appeared to be quite shaken from the accident but had no serious injuries.

The third car was the least damaged of the three cars involved in the accident. The driver, however, was not wearing a seat belt and hit the windshield with her head. She had some minor cuts on her forehead and was extremely disoriented. The passenger in her car had a life-threatening compound fracture of his leg that was bleeding profusely.

53. According to the first paragraph, which two victims should be quickly removed from the cars to make room for rescue operations?

    (A) Driver 1, passenger 1
    (B) Passenger 1, passenger 2
    (C) Driver 2, driver 3
    (D) Driver 3, passenger 3
    (E) Driver 1, passenger 3

54. If firefighter Mauro is treating driver 1 and no other victims have been treated, who should firefighter Warren attend to first?

    (A) Passenger 1
    (B) Passenger 3
    (C) Driver 2
    (D) Driver 3
    (E) Passenger 2

55. After the most seriously injured individuals have been stabilized, who should firefighters Warren and Mauro attend to?

    (A) Driver 1, driver 2
    (B) Passenger 1, passenger 2
    (C) Driver 1, passenger 1
    (D) Driver 2, passenger 2
    (E) Driver 3, passenger 1

56. A clear sign that driver 3 might have suffered a major head injury is that

    (A) she had cuts on her forehead.
    (B) her car had the least damage.
    (C) she did not appear to be breathing.
    (D) she was extremely disoriented.
    (E) she was trapped in the car.

57. Passenger 1 should be quickly removed from the car because he/she

    (A) was seriously injured.
    (B) was bleeding profusely.
    (C) was not injured.
    (D) had a broken leg.
    (E) was trapped in the car.

58. The best explanation for why smoke usually rises from a fire is that

    (A) cooler, heavier air displaces lighter, warm air.
    (B) heat energy from the fire propels smoke upward.
    (C) suction from the upper air pulls the smoke upward.
    (D) burning matter is chemically changed into heat energy.

59. The following diagram shows various types of ramps leading to a loading platform. The ramp that would permit the load to be moved up to the platform with the least amount of force is

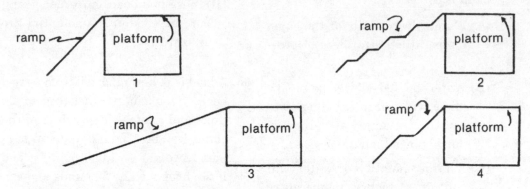

(A) 1

(B) 2

(C) 3

(D) 4

60. The practice of racing a car engine to warm it up in cold weather generally is

(A) good, mainly because repeated stalling of the engine and drain on the battery is avoided.

(B) bad, mainly because too much gas is used to get the engine heated.

(C) good, mainly because the engine becomes operational in the shortest period of time.

(D) bad, mainly because proper lubrication is not established rapidly enough.

61. Ice on sidewalks often can be melted by sprinkling salt on it. The melting of the ice results from

(A) a chemical reaction between the salt and the ice that produces heat.

(B) attraction of sun rays by the salt to the ice.

(C) lowering of the freezing point of water by the salt.

(D) the heat of friction caused by persons walking on the salt.

**62.** Only one of the following statements relating to the temperature at which water boils is correct. Which statement is correct?

**(A)** Water always boils at the same temperature regardless of pressure.

**(B)** Water heated slowly by a low flame will boil at a higher temperature than water heated quickly by a high flame.

**(C)** A large quantity of water will boil at a higher temperature than a small quantity.

**(D)** Water heated at sea level will boil at a higher temperature than water heated on the top of a mountain.

**63.** A substance that is a good conductor of heat is most likely to be a poor

**(A)** conductor of electricity.

**(B)** insulator of heat.

**(C)** vibrator of sound.

**(D)** reflector of light.

**64.** At a meeting concerned with fire prevention, this was said: "The fact that fire loss has been maintained near its previous levels is encouraging evidence that our increasing efforts over the years in the field of public education in fire protection have not been unavailing." Of the following, the most essential assumption that must be made if this statement is accepted is that

**(A)** further public education in fire protection is desirable.

**(B)** fire loss has been computed on the basis of real value rather than insured value.

**(C)** reference is made here to losses due to fires caused by carelessness rather than sabotage.

**(D)** there has been, in recent years, an increase in the potential fire hazard.

**65.** A load is to be supported from a steel beam by a chain consisting of 20 links and a hook. If each link of the chain weighs 1 pound and can support a weight of 1,000 pounds, and if the hook weighs 5 pounds and can support a weight of 5,000 pounds, the maximum load that can be supported from the hook is most nearly

**(A)** 25,000 pounds.

**(B)** 5,000 pounds.

**(C)** 1,000 pounds.

**(D)** 975 pounds.

**66.** Ice formation in water pipes often causes bursting of the pipes because

**(A)** the additional weight of ice overloads the pipes.

**(B)** water cannot pass the ice block and builds up great pressure on the pipes.

**(C)** the cold causes contraction of the pipes and causes them to pull apart.

**(D)** water expands upon freezing and builds up great pressure on the pipes.

67. The suggestion has been made that groups of firefighters without apparatus of any kind should be kept in reserve at a few centrally located points throughout the city. Of the following, the most valid justification for this proposal is that

(A) when second or third alarms are sent, the need often is for more firefighters rather than more apparatus.

(B) the fire department is understaffed.

(C) the fire districts in a city should be revised periodically to meet population trends.

(D) discipline is as important as apparatus in extinguishing fires quickly.

68. In the diagram below, crossing the V-belt as shown by the dotted lines will result in

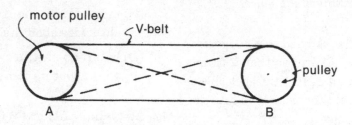

(A) pulley A reversing direction.

(B) no change in the direction of either pulley.

(C) pulley B reversing direction.

(D) stoppage of the motor.

**QUESTIONS 69–71 ARE BASED ON THE FOLLOWING DIAGRAM. ASSUME THAT THE TEETH OF THE GEARS ARE CONTINUOUS ALL THE WAY AROUND EACH GEAR.**

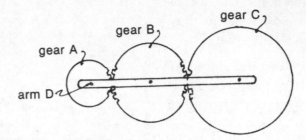

69. Fastening gear A to arm D at another point in addition to its shaft will result in

    **(A)** gear B rotating on its shaft in a direction opposite to gear A.

    **(B)** gear C rotating on its shaft in a direction opposite to gear A.

    **(C)** arm D rotating around the shaft of gear C.

    **(D)** the locking of all gears.

70. If gear C is fastened to a supporting frame (not shown) so that it cannot rotate, and if gear A turns clockwise on its shaft, then gear B will turn

    **(A)** counterclockwise and arm D will turn clockwise around the shaft of gear C.

    **(B)** counterclockwise and arm D will turn counterclockwise around the shaft of gear C.

    **(C)** clockwise and arm D will turn clockwise around the shaft of gear C.

    **(D)** clockwise and arm D will turn counterclockwise around the shaft of gear C.

71. If gear B is fastened to a supporting frame (not shown) so that it cannot rotate, and if arm D rotates clockwise around the shaft of gear B, then for each complete revolution arm D makes, gear A will make

    **(A)** more than one turn clockwise about its own shaft.

    **(B)** less than one turn clockwise about its own shaft.

    **(C)** more than one turn counterclockwise about its own shaft.

    **(D)** less than one turn counterclockwise about its own shaft.

72. As a ship sails away from shore, it appears to go below the horizon. The best explanation for this observation is that it results from the

    **(A)** rise and fall of tides.

    **(B)** curvature of the earth's surface.

    **(C)** refraction of light.

    **(D)** effect of gravity on moving bodies.

**73.** Of the following, the most important reason for lubricating moving parts of machinery is to

(A) reduce friction.

(B) prevent rust formation.

(C) increase inertia.

(D) reduce the accumulation of dust and dirt on the parts.

**74.** A canvas tarpaulin measures 6 ft × 9 ft. The largest circular area that can be covered completely by this tarpaulin is a circle with a diameter of

(A) 9 feet.

(B) 8 feet.

(C) 7 feet.

(D) 6 feet.

**75.** A firefighter caught a civilian attempting to reenter a burning building despite several warnings to stay outside of the fire lines. The civilian insisted frantically that he needed to save some very valuable documents from the fire. The firefighter then called a police officer to remove the civilian. The firefighter's action was

(A) wrong; it is bad public relations to order people about.

(B) right; the firefighter is charged with the responsibility of protecting lives.

(C) wrong; the firefighter should have explained to the civilian why he should not enter the building.

(D) right; civilians must be excluded from the fire zone.

**QUESTIONS 76–78 ARE BASED ON THE FOLLOWING DIAGRAM. IN THE DIAGRAM, PULLEY A AND PULLEY B ARE BOTH FIRMLY ATTACHED TO THE SHAFT SO THAT BOTH PULLEYS AND THE SHAFT CAN TURN ONLY AS A SINGLE UNIT. THE RADIUS OF PULLEY A IS 4 INCHES, AND THE RADIUS OF PULLEY B IS 1 INCH.**

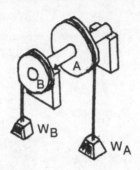

**76.** If weight WA weighs 20 pounds, the system will be in balance if weight WB weighs

(A) 5 pounds.

(B) 10 pounds.

(C) 20 pounds.

(D) 80 pounds.

**77.** If the rope on pulley A is pulled downward so that it unwinds, the rope on pulley B will

(A) wind slower.

(B) wind faster.

(C) unwind slower.

(D) unwind faster.

**78.** When pulley A makes one complete revolution, pulley B will make

(A) one quarter of a revolution.

(B) one revolution.

(C) four revolutions.

(D) a number of revolutions that cannot be determined from the information given.

**79.** For combustion to take place, the atmosphere must contain at least 12 percent oxygen. The maximum percent of oxygen in an atmosphere that can support combustion is

(A) 21 percent.

(B) 25 percent.

(C) 50 percent.

(D) 100 percent.

**80.** Firefighters, when not responding to alarms, are assigned to perform housekeeping chores on the equipment in the firehouse. The primary reason for these assignments is to

(A) keep the firefighters busy when not responding to alarms.

(B) keep the equipment and quarters in the best possible condition.

(C) save money by not hiring maintenance workers.

(D) make firehouses competitive with each other.

81. If a firefighter in the act of extinguishing a fire is forced to run through an area of dense smoke, it would be best for the firefighter to

    **(A)** take deep, slow breaths of air.

    **(B)** breathe deeply before starting and then slowly while passing through the area.

    **(C)** take a shallow breath before starting and a deep breath while passing through the area.

    **(D)** refrain from breathing while passing through the area.

82. Empty hoses are dragged to the site of a fire before the water supply is turned on. The reason for this is that

    **(A)** full hoses are liable to catch on protruding objects.

    **(B)** full hoses are likely to be damaged while being dragged.

    **(C)** the source of the water might be too far removed from the fire.

    **(D)** empty hoses can be dragged more easily and quickly.

83. An intense fire develops in a room in which carbon dioxide cylinders are stored. The principal hazard in this situation is that

    **(A)** the $CO_2$ might catch fire.

    **(B)** toxic fumes might be released.

    **(C)** the cylinders might explode.

    **(D)** released $CO_2$ might intensify the fire.

84. At a fire involving the roof of a five-story building, the firefighters trained their hose stream on the fire from a vacant lot across the street, aiming the stream at a point about 15 feet above the roof. In this situation, water in the stream would be traveling at the greatest speed

    **(A)** as it leaves the hose nozzle.

    **(B)** at a point midway between the ground and the roof.

    **(C)** at the maximum height of the stream.

    **(D)** as it drops on the roof.

85. Of the following circumstances concerning a fire, which one indicates most strongly the possibility of arson?

    **(A)** There was heavy charting of wood around the point of origin of the fire.

    **(B)** Three fires apparently broke out simultaneously in different parts of the building.

    **(C)** The heat was so intense that glass in the building became molten and fused.

    **(D)** The smoke was very heavy when the firefighters arrived.

86. A firefighter searching an area after an explosion comes upon a man in great pain with a leg injury bleeding severely. Under these circumstances, the first action the firefighter should take is to attempt to

    **(A)** immobilize the man's leg.

    **(B)** cover the man with a blanket or coat.

    **(C)** stop the bleeding.

    **(D)** make the man comfortable.

**87.** Spontaneous combustion might be the reason for a pile of oily rags catching fire. In general, spontaneous combustion is the direct result of

**(A)** application of flame.

**(B)** falling sparks.

**(C)** intense sunlight.

**(D)** chemical action.

**88.** Most fire departments advocate that firefighters become involved in community activities such as coaching or sponsoring a Little League team. This course of action is likely to

**(A)** produce expert Little League coaching.

**(B)** lead to better understanding of the activities of the fire department.

**(C)** encourage community members to assist in fighting fires when needed.

**(D)** enlist community support of the firefighters' quest for higher salaries.

**89.** There is a common bond among all firefighters functioning throughout the United States and even in some other parts of the world. There are many nationwide organizations of firefighters, and it is not uncommon for them to drop in at the local fire department when they are visiting other cities. The most likely reason for this is the

**(A)** similarity of techniques, equipment, and problems from one fire department to another.

**(B)** need for firefighters to be strongly organized to achieve equitable pay.

**(C)** extreme dangers involved in fighting fires.

**(D)** usual competitiveness of police and fire departments.

**90.** A firefighter stationed in a firehouse is approached by an irate citizen who declares that firefighters are paid well but spend most of their time lounging around the firehouse doing nothing. It would be most appropriate for the firefighter to respond by saying which of the following?

**(A)** "You had better forward your complaint to the mayor."

**(B)** "If you did my job for one week, you would change your opinion."

**(C)** "How hard do you work for a living?"

**(D)** "The job of a firefighter is to always be ready to respond to an alarm."

**91.** Firefighters often visit schools to give talks about fire prevention to children. The type of firefighter best suited for this duty is

**(A)** one whose demeanor depicts authority but who has little experience in public speaking.

**(B)** a newly appointed firefighter who is a recent college graduate who majored in speech.

**(C)** a veteran firefighter who projects a strong image and who has addressed audiences in the past.

**(D)** the firefighter who has achieved the highest rating on the eligible list.

**92.** During an emergency situation, a firefighter might be the recipient of an order from his or her superior that is delivered in an abrupt manner. The recipient of this type of an order should assume that

(A) he or she is in disfavor with the superior officer.

(B) the superior officer feels that he or she will not respond properly to any other type of order.

(C) he or she is about to be brought up on charges.

(D) emergency situations call for this type of order.

**93.** Studies of accidents have indicated that more accidents occur at the end of a tour of duty than at its beginning. The most likely reason for this is

(A) carelessness on the part of the victims.

(B) improper training in accident prevention.

(C) lack of interest in safety.

(D) mental and physical fatigue on the part of firefighters.

**94.** Firefighters can best avoid accidents while doing heavy lifting by

(A) keeping the knees bent and the back straight.

(B) keeping both knees and the back straight.

(C) using the arms only.

(D) using the method best suited to the individual.

**95.** The least likely reason for including an accident-prevention program in the training program for newly appointed firefighters is the

(A) nature of the work being performed.

(B) number of reported accidents.

(C) number of levels of command.

(D) time lost as a result of accidents.

**96.** A training officer assigned to teach safety to firefighters can best gain their cooperation by

(A) exercising the authority consistent with his or her rank.

(B) waiting until a serious accident occurs.

(C) making a distinct effort to become "one of the boys" during the training session.

(D) indicating to the trainees how an accident can be costly both to the individual and to the department.

**97.** A veteran firefighter observes a newly appointed firefighter performing a maintenance task in an unsafe manner. It would be most appropriate for the veteran firefighter to

(A) tell the superior officer of the indiscretion at once.

(B) illustrate the proper manner for performing this task to the rookie firefighter.

(C) not do anything because this is not an assigned responsibility.

(D) tell another rookie firefighter to demonstrate the proper method because it is likely that there is a common bond between them.

**98.** A well-executed accident report is extremely valuable principally because it

    **(A)** isolates the cause of the accident so that future similar accidents can be avoided.

    **(B)** forms the basis of statistical studies.

    **(C)** complies with existing regulations.

    **(D)** informs everybody that an accident has taken place.

**99.** It is essential that accident reports be submitted as soon after the accident has occurred as is practical. The most important reason for this course of action is that

    **(A)** witnesses tend to change their stories as time passes.

    **(B)** accident reports should be considered trivial and therefore should be gotten out of the way.

    **(C)** the details of the report will still be fresh in the mind of the reporter.

    **(D)** all reports should be submitted promptly even if they are incomplete.

**100.** Following an accident involving a firefighter on the job, it is normal procedure for a fire department to spend considerable effort in an attempt to identify the cause of the accident. The most important reason for this procedure is

    **(A)** that no accident report is complete without the cause of the accident.

    **(B)** for statistical purposes.

    **(C)** so that the guilty firefighter can be disciplined.

    **(D)** so that future similar accidents can be avoided.

## ANSWERS AND EXPLANATIONS

| | | | | | | | | | |
|---|---|---|---|---|---|---|---|---|---|
| 1. | D | 21. | C | 41. | C | 61. | C | 81. | D |
| 2. | A | 22. | A | 42. | D | 62. | D | 82. | D |
| 3. | D | 23. | D | 43. | C | 63. | B | 83. | C |
| 4. | B | 24. | A | 44. | C | 64. | A | 84. | A |
| 5. | D | 25. | C | 45. | A | 65. | D | 85. | B |
| 6. | C | 26. | C | 46. | B | 66. | D | 86. | C |
| 7. | C | 27. | C | 47. | E | 67. | A | 87. | D |
| 8. | C | 28. | A | 48. | D | 68. | C | 88. | B |
| 9. | B | 29. | B | 49. | D | 69. | D | 89. | A |
| 10. | C | 30. | D | 50. | B | 70. | B | 90. | D |
| 11. | A | 31. | B | 51. | C | 71. | A | 91. | C |
| 12. | C | 32. | B | 52. | A | 72. | B | 92. | D |
| 13. | B | 33. | B | 53. | B | 73. | A | 93. | D |
| 14. | B | 34. | D | 54. | B | 74. | D | 94. | A |
| 15. | C | 35. | B | 55. | A | 75. | B | 95. | C |
| 16. | A | 36. | A | 56. | D | 76. | D | 96. | D |
| 17. | E | 37. | D | 57. | C | 77. | A | 97. | B |
| 18. | D | 38. | C | 58. | A | 78. | B | 98. | A |
| 19. | C | 39. | A | 59. | C | 79. | D | 99. | C |
| 20. | B | 40. | A | 60. | D | 80. | B | 100. | D |

1. **The correct answer is (D).** The most effective course of action is to suggest that the manager contact the health department so he is assured of receiving correct information.

2. **The correct answer is (A).** This immediately identifies the person receiving the call and is likely to save time and prevent misunderstandings.

3. **The correct answer is (D).** No area should remain without proper protection to cover possible fires. If a company is busy fighting one fire, coverage should be provided in the event that a second fire occurs in the same area.

4. **The correct answer is (B).** The responding fire company does not know the actual location of the fire. It will respond to the signal received from the box. When the fire apparatus arrives at the location of the box, the firefighter will be able to direct it to the actual location of the fire.

5. **The correct answer is (D).** The badge might fall into the wrong hands and be used improperly. The badge belongs to the fire department. Any found property should be returned to the rightful owner, especially if the owner can be identified.

6. **The correct answer is (C).** Consideration should be given to the fact that as much mail as possible should be salvaged from a mailbox fire. Very often, the contents of the mail are extremely important and valuable.

7. **The correct answer is (C).** One of the difficulties in determining the cause of a fire is that a fire is likely to destroy the evidence of what caused it.

8. **The correct answer is (C).** This device would discourage many people from using fire alarm boxes due to the fear of possible incrimination.

9. **The correct answer is (B).** Water pipes leading to the sprinklers often are blown apart by an explosion.

10. **The correct answer is (C).** This course of action would prevent large amounts of water from entering the subway system.

11. **The correct answer is (A).** Note that the choice that is NOT true is the correct answer. Motorists do not avoid routes of travel because fire apparatus might use them.

12. **The correct answer is (C).** The first action to be taken is to determine whether a fire has been started. The sooner a fire is discovered, the easier it will be to extinguish. This generally is true for all fires.

13. **The correct answer is (B).** Safe driving calls for total concentration. Firefighters should refrain from distracting the chauffeur while he or she is driving the apparatus.

14. **The correct answer is (B).** Most major cities have established a 911 emergency number that accepts all calls and sends assistance for police, fire, and medical emergencies.

15. **The correct answer is (C).** There are 150 hydrants for every square mile. To figure out what percentage of 150 fire hydrants 120 hydrants is, take $120 \div 150 = .80$. To convert this to a percent, take $100\% \times .80 = 80\%$.

16. **The correct answer is (A).** In 1 hour, the firefighters would have traveled 30 miles by traveling 30 mph. It will take them a fraction of that time to travel 3.2 miles. Thus, take $3.2 \div 30 = .11$. Because there are 60 minutes in an hour, $60 \times .11 = 6.6$ minutes.

17. **The correct answer is (E).** First, do what is in the parentheses, $9 - 7 = 2$. $13 \times 13 = 169$. $169 + 2 = 171$.

18. **The correct answer is (D).** To figure out the equation, simply replace the letters with the corresponding numbers. $500 \div 3.2 - 2(45)$. First multiply 2 by what is in the parentheses. Then divide 500 by 3.2. Now we have $156.25 - 90 = 66.25$.

19. **The correct answer is (C).** The total number of calls was $21 + 17 + 2 + 46 = 86$. Because station A received 34 calls, this is equal to $34 \div 86 = 40\%$ of the calls.

20. **The correct answer is (B).** The correct spelling is "proficient."

21. **The correct answer is (C).** The correct spelling is "unconscious."

22. **The correct answer is (A).** The word "hidden" is the best word to complete the sentence.

23. **The correct answer is (D).** The word "their" is the best word to complete the sentence. This is the proper possessive form of the word "they."

24. **The correct answer is (A).** The word "raise" is the best word to complete the sentence.

25. **The correct answer is (C).** The fire department accepts liability when it permits a nonmember to ride on its equipment. The decisions to accept this liability should come from headquarters.

26. **The correct answer is (C).** If the pier is damaged during the fighting of the fire, the apparatus might fall from the pier into the water.

27. **The correct answer is (C).** These closed cabs provide greater protection for firefighters from inclement weather and hazardous conditions.

28. **The correct answer is (A).** Fires that occur during severe weather conditions are more difficult to extinguish. Therefore, increased manpower might be needed to combat a fire in a blizzard.

29. **The correct answer is (B).** Bends weaken the hose and therefore should be kept to a minimum.

30. **The correct answer is (D).** The decision as to what to do with the money should come from a higher authority in the fire department.

31. **The correct answer is (B).** A section of a falling cornice might injure a pedestrian or a member of the fire department working on the street level.

32. **The correct answer is (B).** There always is a possibility that a fire might rekindle from the debris and ashes.

33. **The correct answer is (B).** Wetting the firefighters with small streams of water will accomplish nothing but making the firefighters wet.

34. **The correct answer is (D).** There is no reason for you to do anything but get out of the way of the oncoming fire apparatus. Driving rules require all vehicles to pull to the curb to allow fire apparatus to pass.

35. **The correct answer is (B).** Because of the use of scenery, which might be composed largely of wood and other highly combustible materials, and machinery, which might involve electrical wiring or other devices creating a potential fire hazard, a firefighter is assigned to be present to ensure that these materials are handled correctly and that, in the event of a fire, the problem can be remedied as quickly as possible.

36. **The correct answer is (A).** The second sentence indicates this fact.

37. **The correct answer is (D).** This fact is indicated in the third and fourth sentences of the paragraph.

38. **The correct answer is (C).** This is stated in the first, second, and third sentences.

39. **The correct answer is (A).** This can be derived from the last sentence of the paragraph.

40. **The correct answer is (A).** An open circuit does not carry electric current. When a fuse melts due to overheating, an open circuit condition results.

41. **The correct answer is (C).** An electrical wire carrying electrical current is a common example of a conductor.

42. **The correct answer is (D).** There is no information about the ratio of circuits to fuses in this paragraph.

43. **The correct answer is (C).** The second sentence indicates that the introduction of water will cause the threads to swell, thus reducing the space between threads and making the hose practically watertight.

44. **The correct answer is (C).** Unlined linen hose cannot withstand very hard wear; thus, it is best used for emergencies.

45. **The correct answer is (A).** Unlined linen hose could withstand neither the punishment of being dragged over rough lumber nor the exposure to weather conditions.

46. **The correct answer is (B).** The second sentence indicates that victims often are trapped, making it difficult for a rescuer to administer proper first aid.

47. **The correct answer is (E).** The passage indicates that keeping the victim's airway open, performing CPR if necessary, and treating any uncontrolled bleeding are listed in terms of importance. Therefore, a rescuer should first assess whether the victim's airway is open.

48. **The correct answer is (D).** Full protective clothing must be put on before doing anything else. The next step is to enter the building in groups of at least two. After you have entered the building, the search should be conducted on hands and knees. The search then should begin on an outside wall. After a room has been searched, the door should be tagged. Then, victims who have been removed should be placed in someone's custody.

49. **The correct answer is (D).** Walking inside a burning building is a violation of procedure 1. An internal building search always should be conducted on hands and knees.

50. **The correct answer is (B).** Procedure 5 indicates that tagging the door of a room is a sign to other firefighters that the room has already been searched. This prevents other firefighters from searching the room again.

51. **The correct answer is (C).** It is not the firefighter's responsibility to decide which laws should be enforced and which should not, regardless of expense. Nonprofessional behavior will not gain respect.

52. **The correct answer is (A).** The sooner the fire department arrives at the scene of a fire, the more likely it is that damage to life and property will be kept to a minimum. A single firefighter would not be able to combat the fire alone without the assistance of other firefighters and the use of firefighting apparatus.

53. **The correct answer is (B).** The first paragraph states that victims who are not injured or who are able to leave the vehicle on their own should be quickly removed from the vehicle to make room for the rescuers. Passenger 1 and passenger 2 did not have any serious injuries and therefore should have been quickly removed from the cars to make room for the rescuers.

54. **The correct answer is (B).** The first paragraph states that the most seriously injured individuals should be stabilized first. If firefighter Mauro is treating driver 1, passenger 3 is the most seriously injured individual left and should have been treated next.

55. **The correct answer is (A).** The first paragraph states that, once the most seriously injured individuals have been stabilized, priority should be given to victims trapped in their vehicles. Driver 1 and driver 2 both were trapped in their cars and should have been attended to next.

56. **The correct answer is (D).** The cuts on her forehead do not necessarily indicate that she suffered a head injury; therefore, choice (A) is incorrect. Her extreme disorientation, however, might indicate that she has suffered a head injury.

57. **The correct answer is (C).** The first paragraph states that individuals who are not injured should be quickly removed from their cars.

58. **The correct answer is (A).** The cooler, heavier air, because it is dense, pushes the hot air upward.

59. **The correct answer is (C).** The incline of ramp 3 is not nearly so steep as that of ramp 1. Ramps 2 and 4 have irregularities in their inclines that would make it even more difficult to move a heavy load to the top of the platform.

60. **The correct answer is (D).** Oil pressure should be established before the engine is raced.

61. **The correct answer is (C).** Salt affects ice in this manner.

62. **The correct answer is (D).** Water boils at 212°F at sea level, a lower temperature at lower air pressure. Cooking and baking instructions often vary for certain altitudes above sea level.

63. **The correct answer is (B).** It follows that a substance that carries heat well cannot be used as an insulator of heat.

64. **The correct answer is (A).** If some public education has helped maintain fire losses at an even level (costs only go up, so maintaining losses shows the effect of education), then increasing public education should reduce the level of future losses.

65. **The correct answer is (D).** A chain is only as strong as its weakest link. All links in this chain are of equal strength. Each can support 1,000 pounds. The weight of the hook reduces this amount slightly.

66. **The correct answer is (D).** Water expands as it freezes. If the expanding water is in an enclosed area, as in a pipe, the expansion must cause the pipe to burst.

67. **The correct answer is (A).** We first met this principle, the need for extra manpower, in the paragraph for questions 53–56.

68. **The correct answer is (C).** When it is desired that one pulley be turned in the opposite direction from its connecting pulley, drive belts must be crossed as shown in the diagram.

69. **The correct answer is (D).** With the center of gear A fastened at one point on the shaft, the gear is able to rotate at that point. If gear A is fastened at an additional point on the shaft, the gear becomes rigid and unable to rotate. It therefore cannot turn gear B, and all gears will lock.

70. **The correct answer is (B).** By anchoring gear C, gear C then becomes the pivoting point. The remaining gears will then rotate around the pivot, which is gear C.

71. **The correct answer is (A).** Gear A, being smaller than gears B and C, must revolve at a greater rate of speed to travel the same distance. Therefore, gear A will make more than one rotation on its own shaft for every revolution that arm D makes.

72. **The correct answer is (B).** Because of the curvature of the earth's surface, a person cannot see beyond the horizon.

73. **The correct answer is (A).** Lubrication with oil, grease, or any other suitable substance will reduce friction between moveable parts of machinery. It is not possible to operate most machinery well without lubrication. In addition, the machinery would wear out quickly because of the friction.

74. **The correct answer is (D).** Because the tarpaulin measures 6 feet by 9 feet, it is not possible to completely cover any circular area that exceeds 6 feet in diameter.

75. **The correct answer is (B).** The firefighter already had given warnings and presumably some explanation to the civilian. The job of the firefighter is to save lives by fighting fires. This firefighter acted correctly in calling a police officer to remove the civilian from danger so that the firefighter could return to fighting the fire.

76. **The correct answer is (D).** There is a 1:4 ratio between pulleys B and A. To balance the system, the weight on the smaller pulley must be 4 times the weight on the larger pulley. The weight on the large pulley is 20 pounds; therefore, 4 times 20 pounds equals 80 pounds, which is choice (D).

77. **The correct answer is (A).** Pulley A has a circumference that is 4 times greater than pulley B. With each revolution, pulley B will be traveling one fourth the distance traveled by pulley A. This results in one fourth the amount of rope being wound on the pulley. Therefore, pulley B would wind at a slower rate.

78. **The correct answer is (B).** Both pulleys are attached firmly to the shaft; therefore, both will make the same number of revolutions.

79. **The correct answer is (D).** Oxygen, though noncombustible, will support combustion. An atmosphere of 100-percent oxygen will cause burning material to accelerate with explosive rapidity.

80. **The correct answer is (B).** Housekeeping chores are part of the job of a firefighter. The motivated firefighter will take pride in the condition of the equipment used in the performance of his or her duties.

81. **The correct answer is (D).** By not breathing, the firefighter will expose his or her lungs to a minimum amount of irritation from the smoke.

82. **The correct answer is (D).** Full hoses weigh considerably more than empty hoses because of the weight of the water. Therefore, it is much easier to carry or drag empty hoses to the site of a fire.

83. **The correct answer is (C).** The immediate danger involving the cylinders of $CO_2$ is that the contents of the cylinders might expand and explode from overheating. Not only will the explosion cause immediate danger to the premises, but personnel also might be injured from the fragments of metal flying from the exploding cylinders.

84. **The correct answer is (A).** The water pressure will diminish gradually as the water gets farther from the nozzle of the hose.

85. **The correct answer is (B).** It would be most unlikely for three fires to break out simultaneously in the same building from accidental causes. Deliberate setting of the fires must be suspected.

86. **The correct answer is (C).** Whenever severe bleeding occurs, it should be given attention immediately because such bleeding can quickly result in the loss of life.

87. **The correct answer is (D).** Spontaneous combustion is combustion (burning) that occurs from no external cause. Flame and falling sparks are obvious external causes. The heat of intense sunlight might act as a catalyst, speeding up chemical activity in a pile of rags or another hazardous pile of debris, but the actual cause of spontaneous combustion is a chemical reaction within the heap.

88. **The correct answer is (B).** When firefighters become involved in community activities, it is likely that the residents will become more aware of the purpose and objectives of the fire department.

89. **The correct answer is (A).** Firefighters have the same responsibilities throughout the world—to extinguish fires, to save lives, and to minimize property damage. The equipment and techniques used are basically the same throughout the world, although larger cities might have more sophisticated firefighting systems.

90. **The correct answer is (D).** There always will be citizens under the impression that firefighters do not work hard because they often can be seen at the firehouse instead of out fighting fires. It is very important that these citizens be made to understand that, after housekeeping chores are done, the firefighters' prime responsibility is to make themselves ready to answer an alarm.

91. **The correct answer is (C).** Firefighting is a serious business; therefore, a proper image of the fire department must be created in the minds of children. The firefighter selected must be skilled in public speaking and must be able to put forth a strong and serious image.

92. **The correct answer is (D).** Emergency situations call for immediate action. Very often, courtesies are brushed over in the transmission of orders. The firefighter should realize that the abrupt manner in which he or she is addressed does not reflect on him or her personally.

93. **The correct answer is (D).** The tired worker is likely to become involved in accidents due to lack of alertness to dangerous situations.

94. **The correct answer is (A).** The weight of the object is evenly distributed with the knees bent and the back straight. No part of the body receives the bulk of the weight.

95. **The correct answer is (C).** The number of levels of command has no bearing on the necessity for such a program.

96. **The correct answer is (D).** The best approach to convince firefighters to take a course in safety and accident prevention seriously is to convince them of how costly an accident can be to themselves as well as to the department.

97. **The correct answer is (B).** The newly appointed firefighter should be corrected at once before an accident occurs and before this method of working becomes a habit.

98. **The correct answer is (A).** Isolating the cause of an accident will alert firefighters to be aware of such situations in the future. This should reduce the chances of a similar accident occurring.

99. **The correct answer is (C).** Time plays tricks on an individual's memory. Therefore, the sooner the report is submitted, the more accurate it is likely to be.

100. **The correct answer is (D).** After an accident occurs, little can be done but suffer its consequences. Similar accidents can be avoided in the future, however, if the cause of such an accident is identified.

answers

# Practice New York City Exam

The examination that follows is an actual firefighter examination given in New York City, and it is reprinted here by permission of the city. If you will be taking a New York City exam for firefighters, it will NOT be the same as this one. The format may be similar, but the questions will be different. If you plan to take a firefighter exam anywhere outside New York City, you cannot assume that your exam will look like this one at all. Practicing with this examination, however, still should provide excellent preparation for whatever your own town or city might have in store.

All of the instructions and cautions are included here along with the exam itself so you can see how seriously this exam must be taken. Keep in mind how important this exam is as you study and prepare yourself for the test.

## PRACTICE NEW YORK CITY EXAM ANSWER SHEET

| | | | |
|---|---|---|---|
| 1. Ⓐ Ⓑ Ⓒ Ⓓ Ⓔ | 26. Ⓐ Ⓑ Ⓒ Ⓓ Ⓔ | 51. Ⓐ Ⓑ Ⓒ Ⓓ Ⓔ | 76. Ⓐ Ⓑ Ⓒ Ⓓ Ⓔ |
| 2. Ⓐ Ⓑ Ⓒ Ⓓ Ⓔ | 27. Ⓐ Ⓑ Ⓒ Ⓓ Ⓔ | 52. Ⓐ Ⓑ Ⓒ Ⓓ Ⓔ | 77. Ⓐ Ⓑ Ⓒ Ⓓ Ⓔ |
| 3. Ⓐ Ⓑ Ⓒ Ⓓ Ⓔ | 28. Ⓐ Ⓑ Ⓒ Ⓓ Ⓔ | 53. Ⓐ Ⓑ Ⓒ Ⓓ Ⓔ | 78. Ⓐ Ⓑ Ⓒ Ⓓ Ⓔ |
| 4. Ⓐ Ⓑ Ⓒ Ⓓ Ⓔ | 29. Ⓐ Ⓑ Ⓒ Ⓓ Ⓔ | 54. Ⓐ Ⓑ Ⓒ Ⓓ Ⓔ | 79. Ⓐ Ⓑ Ⓒ Ⓓ Ⓔ |
| 5. Ⓐ Ⓑ Ⓒ Ⓓ Ⓔ | 30. Ⓐ Ⓑ Ⓒ Ⓓ Ⓔ | 55. Ⓐ Ⓑ Ⓒ Ⓓ Ⓔ | 80. Ⓐ Ⓑ Ⓒ Ⓓ Ⓔ |
| 6. Ⓐ Ⓑ Ⓒ Ⓓ Ⓔ | 31. Ⓐ Ⓑ Ⓒ Ⓓ Ⓔ | 56. Ⓐ Ⓑ Ⓒ Ⓓ Ⓔ | 81. Ⓐ Ⓑ Ⓒ Ⓓ Ⓔ |
| 7. Ⓐ Ⓑ Ⓒ Ⓓ Ⓔ | 32. Ⓐ Ⓑ Ⓒ Ⓓ Ⓔ | 57. Ⓐ Ⓑ Ⓒ Ⓓ Ⓔ | 82. Ⓐ Ⓑ Ⓒ Ⓓ Ⓔ |
| 8. Ⓐ Ⓑ Ⓒ Ⓓ Ⓔ | 33. Ⓐ Ⓑ Ⓒ Ⓓ Ⓔ | 58. Ⓐ Ⓑ Ⓒ Ⓓ Ⓔ | 83. Ⓐ Ⓑ Ⓒ Ⓓ Ⓔ |
| 9. Ⓐ Ⓑ Ⓒ Ⓓ Ⓔ | 34. Ⓐ Ⓑ Ⓒ Ⓓ Ⓔ | 59. Ⓐ Ⓑ Ⓒ Ⓓ Ⓔ | 84. Ⓐ Ⓑ Ⓒ Ⓓ Ⓔ |
| 10. Ⓐ Ⓑ Ⓒ Ⓓ Ⓔ | 35. Ⓐ Ⓑ Ⓒ Ⓓ Ⓔ | 60. Ⓐ Ⓑ Ⓒ Ⓓ Ⓔ | 85. Ⓐ Ⓑ Ⓒ Ⓓ Ⓔ |
| 11. Ⓐ Ⓑ Ⓒ Ⓓ Ⓔ | 36. Ⓐ Ⓑ Ⓒ Ⓓ Ⓔ | 61. Ⓐ Ⓑ Ⓒ Ⓓ Ⓔ | 86. Ⓐ Ⓑ Ⓒ Ⓓ Ⓔ |
| 12. Ⓐ Ⓑ Ⓒ Ⓓ Ⓔ | 37. Ⓐ Ⓑ Ⓒ Ⓓ Ⓔ | 62. Ⓐ Ⓑ Ⓒ Ⓓ Ⓔ | 87. Ⓐ Ⓑ Ⓒ Ⓓ Ⓔ |
| 13. Ⓐ Ⓑ Ⓒ Ⓓ Ⓔ | 38. Ⓐ Ⓑ Ⓒ Ⓓ Ⓔ | 63. Ⓐ Ⓑ Ⓒ Ⓓ Ⓔ | 88. Ⓐ Ⓑ Ⓒ Ⓓ Ⓔ |
| 14. Ⓐ Ⓑ Ⓒ Ⓓ Ⓔ | 39. Ⓐ Ⓑ Ⓒ Ⓓ Ⓔ | 64. Ⓐ Ⓑ Ⓒ Ⓓ Ⓔ | 89. Ⓐ Ⓑ Ⓒ Ⓓ Ⓔ |
| 15. Ⓐ Ⓑ Ⓒ Ⓓ Ⓔ | 40. Ⓐ Ⓑ Ⓒ Ⓓ Ⓔ | 65. Ⓐ Ⓑ Ⓒ Ⓓ Ⓔ | 90. Ⓐ Ⓑ Ⓒ Ⓓ Ⓔ |
| 16. Ⓐ Ⓑ Ⓒ Ⓓ Ⓔ | 41. Ⓐ Ⓑ Ⓒ Ⓓ Ⓔ | 66. Ⓐ Ⓑ Ⓒ Ⓓ Ⓔ | 91. Ⓐ Ⓑ Ⓒ Ⓓ Ⓔ |
| 17. Ⓐ Ⓑ Ⓒ Ⓓ Ⓔ | 42. Ⓐ Ⓑ Ⓒ Ⓓ Ⓔ | 67. Ⓐ Ⓑ Ⓒ Ⓓ Ⓔ | 92. Ⓐ Ⓑ Ⓒ Ⓓ Ⓔ |
| 18. Ⓐ Ⓑ Ⓒ Ⓓ Ⓔ | 43. Ⓐ Ⓑ Ⓒ Ⓓ Ⓔ | 68. Ⓐ Ⓑ Ⓒ Ⓓ Ⓔ | 93. Ⓐ Ⓑ Ⓒ Ⓓ Ⓔ |
| 19. Ⓐ Ⓑ Ⓒ Ⓓ Ⓔ | 44. Ⓐ Ⓑ Ⓒ Ⓓ Ⓔ | 69. Ⓐ Ⓑ Ⓒ Ⓓ Ⓔ | 94. Ⓐ Ⓑ Ⓒ Ⓓ Ⓔ |
| 20. Ⓐ Ⓑ Ⓒ Ⓓ Ⓔ | 45. Ⓐ Ⓑ Ⓒ Ⓓ Ⓔ | 70. Ⓐ Ⓑ Ⓒ Ⓓ Ⓔ | 95. Ⓐ Ⓑ Ⓒ Ⓓ Ⓔ |
| 21. Ⓐ Ⓑ Ⓒ Ⓓ Ⓔ | 46. Ⓐ Ⓑ Ⓒ Ⓓ Ⓔ | 71. Ⓐ Ⓑ Ⓒ Ⓓ Ⓔ | 96. Ⓐ Ⓑ Ⓒ Ⓓ Ⓔ |
| 22. Ⓐ Ⓑ Ⓒ Ⓓ Ⓔ | 47. Ⓐ Ⓑ Ⓒ Ⓓ Ⓔ | 72. Ⓐ Ⓑ Ⓒ Ⓓ Ⓔ | 97. Ⓐ Ⓑ Ⓒ Ⓓ Ⓔ |
| 23. Ⓐ Ⓑ Ⓒ Ⓓ Ⓔ | 48. Ⓐ Ⓑ Ⓒ Ⓓ Ⓔ | 73. Ⓐ Ⓑ Ⓒ Ⓓ Ⓔ | 98. Ⓐ Ⓑ Ⓒ Ⓓ Ⓔ |
| 24. Ⓐ Ⓑ Ⓒ Ⓓ Ⓔ | 49. Ⓐ Ⓑ Ⓒ Ⓓ Ⓔ | 74. Ⓐ Ⓑ Ⓒ Ⓓ Ⓔ | 99. Ⓐ Ⓑ Ⓒ Ⓓ Ⓔ |
| 25. Ⓐ Ⓑ Ⓒ Ⓓ Ⓔ | 50. Ⓐ Ⓑ Ⓒ Ⓓ Ⓔ | 75. Ⓐ Ⓑ Ⓒ Ⓓ Ⓔ | 100. Ⓐ Ⓑ Ⓒ Ⓓ Ⓔ |

**City of New York**

**Department of Personnel**

**EXAMINATION NO. 7022**

**FIREFIGHTER**

Social Security No. _____

Seat No. _____

Room No. _____

School _____

Written Test: Weight 50, pass mark to be determined.

GENERAL INSTRUCTIONS FOR CANDIDATES—READ NOW

GENERAL INFORMATION: Read the General Information and the Test Instructions now and fill out the information requested at the top of this booklet. The following are explanations of the information requested.

The Social Security No. is your correct Social Security Number. If your Social Security Number is not correct on your admission card, fill out the pink correction form (DP-148) so the correction can be made after the test. Your Seat No. is the number shown on the sheet of scrap paper on your desk. Your Room No. and School are written on the blackboard.

THE FORMS: You must now fill out your fingerprint card and answer sheet as explained below. Use pencil only.

Fingerprint Card: Print in pencil all information requested including your Social Security Number. Your fingerprints will be taken approximately 15 minutes after the Seventh Signal.

Answer Sheet: Print in pencil all the information requested on the Answer Sheet, including your Social Security No. and the Exam. No., in the boxes at the top of the columns, ONE NUMBER IN A BOX. In each column, darken (with pencil) the oval containing the number in the box at the top of the column; only ONE OVAL in each COLUMN should be darkened. Then, using your pencil, print the Exam, Title, School or Building, Room No., Seat No., and Today's Date in the appropriate spaces.

After filling in this information at the top of the Answer Sheet, please read the Confidential Questionnaire at the bottom of the Answer Sheet and answer the questions accurately.

After answering the Confidential Questionnaire, please complete the Survey Questionnaire.

You are not permitted to use a calculator on this test.

TEST INSTRUCTIONS

EIGHT SIGNALS WILL BE USED DURING THIS TEST. The following is an explanation of these signals.

<u>FIRST SIGNAL:</u> The First Memory Booklet is distributed. Write your Room, Seat, and Social Security Numbers on the cover. DO <u>NOT</u> OPEN THE BOOKLET UNTIL THE SECOND SIGNAL.

<u>SECOND SIGNAL:</u> OPEN THE FIRST MEMORY BOOKLET. You will be given 5 minutes to remember as many details of the scene as you can. You may <u>not</u> write or make any notes while studying this material. The first eight questions will be based on this material.

<u>THIRD SIGNAL:</u> CLOSE THE FIRST MEMORY BOOKLET. The monitor will collect this booklet and distribute the Second Memory Booklet. Write your Room, Seat, and Social Security Numbers on top of the Second Memory Booklet. DO <u>NOT</u> OPEN THE SECOND MEMORY BOOKLET OR WRITE ANYTHING ELSE UNTIL THE NEXT SIGNAL.

<u>FOURTH SIGNAL:</u> OPEN THE SECOND MEMORY BOOKLET AND ANSWER QUESTIONS 1–8. You will be given 5 minutes to answer the questions. DO NOT TURN TO ANY OTHER PART OF THE BOOKLET UNTIL THE NEXT SIGNAL. After you have answered questions 1–8 on your Answer Sheet, you may, for future reference, circle your answers in the Second Memory Booklet.

<u>FIFTH SIGNAL:</u> TURN TO THE DIAGRAM IN THE SECOND MEMORY BOOKLET. You will be given 5 minutes to try to remember as many details about the diagram as you can. You may not write or make any notes during this time. <u>DO NOT TURN TO ANY OTHER PAGE DURING THIS TIME.</u>

<u>SIXTH SIGNAL:</u> CLOSE THE SECOND MEMORY BOOKLET. Monitors will collect this booklet and distribute the Question Booklet. Do <u>not</u> open the Question Booklet until the next signal. <u>DO NOT WRITE ANYTHING DURING THIS PERIOD.</u>

<u>SEVENTH SIGNAL:</u> OPEN THE QUESTION BOOKLET AND ANSWER QUESTIONS 9–16 DEALING WITH THE SECOND MEMORY SECTION. Then continue answering the rest of the questions in the Question Booklet. This test consists of 100 questions. After you finish the memory questions, check to make sure the Question Booklet has all the questions from 9–100 and is <u>not</u> defective. If you believe that your booklet is defective, inform the monitor. You will have <u>4</u> hours from this signal

to complete all the questions. YOU WILL BE FINGERPRINTED DURING THIS PERIOD.

EIGHTH SIGNAL: END OF TEST. STOP WORKING. If you finish before this signal and want to leave, raise your hand.

ANYONE DISOBEYING ANY OF THESE INSTRUCTIONS MAY BE

DISQUALIFIED—THAT IS, YOU MAY RECEIVE A SCORE OF ZERO FOR

THE ENTIRE TEST!

### DIRECTIONS FOR ANSWERING QUESTIONS

Answer all the questions on the Answer Sheet before the Eighth Signal is given. ONLY YOUR ANSWER SHEET WILL BE MARKED, although you also should circle your answers in the Question Booklet before the end of the test for future reference. Use a soft pencil (No. 2) to mark your answers. If you want to change an answer, erase it and then mark your new answer. For each question, pick the best answer. Then, on your Answer Sheet in the row with the same number as the question, blacken the oval with the same letter as your answer. Do not make any stray pencil dots, dashes, or marks any place on the Answer Sheet. Do NOT fold, roll, or tear the Answer Sheet.

Here is a sample of how to mark your answers:

SAMPLE 0: The sum of 5 and 3 is

(A) 11 (B) 9 (C) 8 (D) 2

Because the answer is 8, your Answer Sheet should be marked like this:

SAMPLE 0: (A) ◯  (B) ◯  (C) ●  (D) ◯

WARNING: You are not allowed to copy answers from anyone or to use books or notes. It is against the law to take the test for somebody else or to let somebody else take the test for you. There is to be NO SMOKING anywhere in the building.

LEAVING: After the test starts, candidates may not leave the building until after they are fingerprinted. No one may leave the building until after 10:45 a.m. No one may come in after 10:45 a.m. During the test, you may not leave the room unless accompanied by a monitor. If you want to drop out of the test and not have your answers marked, write "I withdraw" on your Answer Sheet and sign your name.

You may take your Question Booklet and two Memory Booklets with you when you finish. The room monitor will return your Memory Booklets to you when you hand in your Answer Sheet.

DEPARTMENT OF PERSONNEL

Social Security No. _____

Room No. _____

Seat No. _____

School _____

First Memory Booklet                    **FIREFIGHTER**

**EXAMINATION NO. 7022**

<u>DO NOT OPEN THIS BOOKLET UNTIL THE SECOND SIGNAL IS GIVEN!</u>

Write your Social Security Number, Room Number, Seat Number, and School in the appropriate spaces at the top of this page.

You must follow the instructions found in the <u>CANDIDATES' INSTRUCTION BOOKLET.</u>

<u>ANYONE DISOBEYING ANY OF THE INSTRUCTIONS FOUND IN THE CANDIDATES' INSTRUCTION BOOKLET MAY BE DISQUALIFIED—RECEIVE A ZERO ON THE ENTIRE TEST.</u>

This booklet contains one scene. Try to remember as many details in the scene as you can.

DO <u>NOT</u> WRITE OR MAKE <u>ANY</u> NOTES WHILE STUDYING THE SCENE.

<u>DO NOT OPEN THIS BOOKLET UNTIL THE SECOND SIGNAL IS GIVEN!</u>

<u>Note:</u> You will have five (5) minutes to study the scene shown.

The New York City Department of Personnel makes no commitment, and no inference is to be drawn, regarding the content, style, or format of any future examination for the position of firefighter.

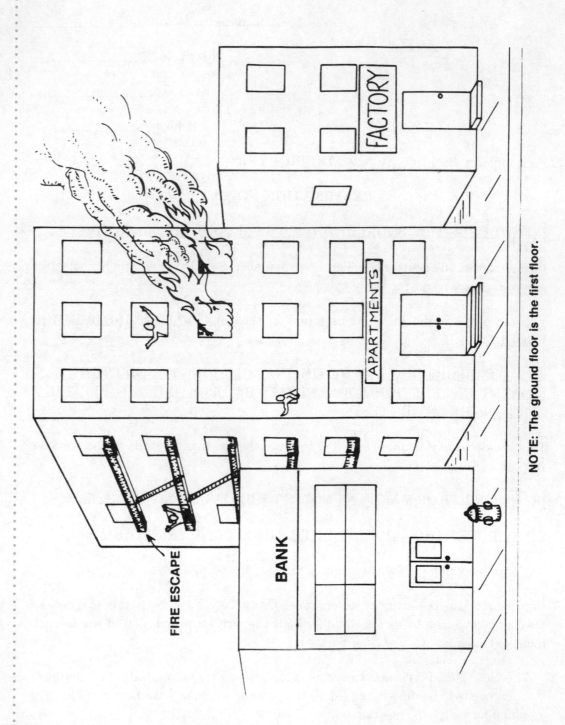

NOTE: The ground floor is the first floor.

CITY OF NEW YORK
DEPARTMENT OF PERSONNEL

Social Security No. _____

Room No. _____

Seat No. _____

School _____

Second Memory Booklet        **FIREFIGHTER**

**EXAMINATION No. 7022**

<u>DO NOT OPEN THIS BOOKLET UNTIL THE FOURTH SIGNAL IS GIVEN!</u>

Write your Social Security Number, Room Number, Seat Number, and School in the appropriate spaces at the top of this page.

You <u>must</u> follow the instructions found in the <u>CANDIDATES' INSTRUCTION BOOKLET.</u>

<u>ANYONE DISOBEYING ANY OF THE INSTRUCTIONS FOUND IN THE CANDIDATES' INSTRUCTION BOOKLET MAY BE DISQUALIFIED—RECEIVE A ZERO ON THE ENTIRE TEST.</u>

This booklet contains questions 1–8, which are based on the scene shown in the First Memory Booklet. Answer questions 1–8 after the fourth signal.

At the <u>FIFTH SIGNAL,</u> you should turn to the Second Memory Diagram. You will have five (5) minutes to memorize the floor plan presented in the diagram.

DO NOT OPEN THIS BOOKLET UNTIL THE FOURTH SIGNAL IS GIVEN!

The New York City Department of Personnel makes no commitment, and no inference is to be drawn, regarding the content, style, or format of any future examination for the position of firefighter.

# PRACTICE NEW YORK CITY EXAM
## 100 Questions • 210 Minutes

**Directions:** Answer questions 1–8 based on the scene in the first memory booklet.

1. The fire is located on the
   (A) first floor.
   (B) fifth floor.
   (C) fourth floor.
   (D) top floor.

2. The smoke and flames are blowing
   (A) up and to the left.
   (B) down and to the left.
   (C) up and to the right.
   (D) down and to the right.

3. There is a person on a fire escape on the
   (A) second floor.
   (B) third floor.
   (C) fourth floor.
   (D) fifth floor.

4. People are visible in windows at the front of the building on fire on the
   (A) second and third floors.
   (B) third and fifth floors.
   (C) fourth and sixth floors.
   (D) fifth and sixth floors.

5. The person who is closest to the flames is in a
   (A) front window on the third floor.
   (B) front window on the fifth floor.
   (C) side window on the fifth floor.
   (D) side window on the third floor.

6. A firefighter is told to go to the roof of the building on fire. It would be correct to state that the firefighter can cross directly to the roof from
   (A) the roof of the bank.
   (B) the roof of the factory.
   (C) either the bank or the factory.
   (D) neither the bank nor the factory.

7. On which side of the building on fire are fire escapes visible?
   (A) Left
   (B) Front
   (C) Right
   (D) Rear

8. The hydrant on the sidewalk is
   (A) in front of the bank.
   (B) between the bank and the apartments.
   (C) in front of the apartments.
   (D) between the apartments and the factory.

## DO <u>NOT</u> TURN PAGE UNTIL NEXT SIGNAL.

**Directions:** Look at this floor plan of an apartment. It is on the third floor of the building. The floor plan also indicates the public hallway.

Doors are shown as ⌐

Windows are shown as ▬▭▬

Doorways are shown as ▬ ▬

Stairs are shown as ▦▦▦

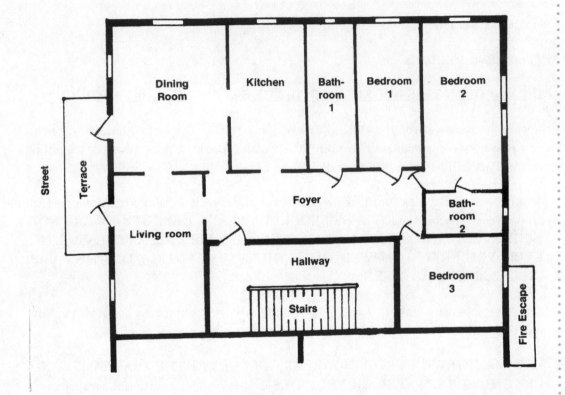

practice test

CITY OF NEW YORK
DEPARTMENT OF PERSONNEL

Social Security No. _____

Room No. _____

Seat No. _____

School _____

Question Booklet          **FIREFIGHTER**

**EXAMINATION NO. 7022**

Written Test: Weight 50

Time allowed: 4 Hours

<u>DO NOT OPEN THIS BOOKLET UNTIL THE SEVENTH SIGNAL IS GIVEN!</u>

Record your answers on the Official Answer Sheet before the last signal. If you wish, you also may record your answers in the Question Booklet before the last signal is given. This will be your only record of answers.

This Question Booklet contains procedures and definitions that are to be used to answer certain questions. THESE PROCEDURES AND DEFINITIONS ARE NOT NECESSARILY THOSE OF THE NEW YORK CITY FIRE DEPARTMENT. HOWEVER, YOU ARE TO ANSWER THESE QUESTIONS SOLELY ON THE BASIS OF THE MATERIAL GIVEN.

After the seventh signal is given, open this Question Booklet and begin work. You will have 4 hours to complete this test.

YOU MAY WRITE IN THIS BOOKLET (INCLUDING THE DIAGRAMS) AND USE THE SCRAP PAPER ON YOUR DESK. If you need additional scrap paper, ask the monitor.

Remember, only your Official Answer Sheet will be rated, so be sure to mark <u>all</u> your answers on the Official Answer Sheet before the eighth signal. <u>No additional time</u> will be given for marking your answers after the test has ended.

<u>DO NOT OPEN THIS BOOKLET UNTIL THE SEVENTH SIGNAL IS GIVEN!</u>

The New York City Department of Personnel makes no commitment, and no inference is to be drawn, regarding the content, style, or format of any future examination for the position of firefighter.

practice test

**QUESTIONS 9–16 ARE BASED ON THE FLOOR PLAN ON PAGE 331.**

9. Which room is farthest from the fire escape?

    **(A)** Bedroom 2.

    **(B)** Bedroom 3.

    **(C)** The kitchen.

    **(D)** The dining room.

10. Which one of the following rooms has only one door or doorway?

    **(A)** The living room.

    **(B)** Bedroom 1.

    **(C)** The kitchen.

    **(D)** The dining room.

11. Which room can firefighters reach directly from the fire escape?

    **(A)** The dining room.

    **(B)** The living room.

    **(C)** Bedroom 1.

    **(D)** Bedroom 3.

12. Which room does NOT have a door or doorway leading directly to the foyer?

    **(A)** Bathroom 1.

    **(B)** Bathroom 2.

    **(C)** Bedroom 1.

    **(D)** The dining room.

13. A firefighter leaving bathroom 2 would be in

    **(A)** bedroom 1.

    **(B)** bedroom 2.

    **(C)** bedroom 3.

    **(D)** the foyer.

14. Firefighters on the terrace would be able to enter directly into which rooms?

    **(A)** Bedroom 1 and bathroom 1.

    **(B)** Bedroom 2 and bathroom 2.

    **(C)** The dining room and the kitchen.

    **(D)** The dining room and the living room.

15. Which rooms have at least one window on two sides of the building?

    **(A)** Bedroom 2 and the dining room.

    **(B)** Bedroom 2 and bedroom 3.

    **(C)** The dining room and the living room.

    **(D)** The dining room, bedroom 2, and bedroom 3.

16. Firefighters can enter the kitchen directly from the foyer and

    **(A)** bedroom 1.

    **(B)** the living room.

    **(C)** bathroom 1.

    **(D)** the dining room.

NOTE: Please check your Answer Sheet to make sure you have written in and blackened your Social Security Number correctly. If you have not yet written in and blackened your Social Security Number, please do so immediately. If you have incorrectly written in or blackened your Social Security Number, please correct the mistake immediately. Failure to correctly blacken your Social Security Number may result in your Answer Sheet NOT being rated.

17. When the door to a fire area is opened, all firefighters _____ positioned on one side of the entrance.

   (A) have been
   (B) should have
   (C) want to
   (D) should be
   (E) will have

18. Soon after the fire yesterday, it _____ necessary to turn off the sprinkler system to prevent further damage.

   (A) will be
   (B) would be
   (C) is
   (D) has been
   (E) was

19. Firefighter Curtis _____ up for promotion next week.

   (A) was
   (B) has
   (C) are
   (D) is
   (E) will

20. It is important that the first firefighters arriving at a fire scene act _____.

   (A) quickest
   (B) quicker
   (C) quickly
   (D) quick
   (E) quickliest

21. Firefighter McDaniel, _____ courage should be commended, saved the little boy's life.

   (A) whom
   (B) whose
   (C) that
   (D) who
   (E) which

**QUESTIONS 22–24 ARE BASED ON THE FOLLOWING PASSAGE:**

When there is a fire in a subway train, it may be necessary for firefighters to evacuate people from the trains by way of the tunnels. In every tunnel, emergency exit areas have stairways that can be used to evacuate people to the street from the track area. All emergency exits can be recognized by an exit sign near a group of five white lights.

There is a Blue Light Area located every 600 feet in the tunnel. These areas contain a power removal box, a telephone, and a fire extinguisher. Removal of power from the third rail is the first step firefighters must take when evacuating people through tunnels. When a firefighter uses the power removal box to turn off electrical power during evacuation procedures, the firefighter must immediately telephone the trainmaster and explain the reason for the power removal. Communication between the firefighter and the trainmaster is essential. If the trainmaster does not receive a phone call within 4 minutes after power removal, the power will be restored to the third rail.

22. When evacuating passengers through the subway tunnel, firefighters must first

   **(A)** telephone the trainmaster for assistance.

   **(B)** remove electrical power from the third rail.

   **(C)** locate the emergency exit in the tunnel.

   **(D)** go to the group of five white lights.

23. Immediately after using the power removal box to turn off the electrical power, a firefighter should

   **(A)** wait 4 minutes before calling the trainmaster.

   **(B)** begin evacuating passengers through the tunnel.

   **(C)** call the trainmaster and explain why the power was turned off.

   **(D)** touch the third rail to see if the electrical power has been turned off.

24. A group of five white lights in a subway tunnel indicates that

   **(A)** a telephone is available.

   **(B)** the electrical power is off in the third rail.

   **(C)** a fire extinguisher is available.

   **(D)** an emergency exit is located there.

**QUESTIONS 25–26 ARE BASED ON THE FOLLOWING PASSAGE:**

A new firefighter learns the following facts about his company's response area: All the factories are located between 9th Avenue and 12th Avenue from 42nd Street to 51st Street; all the apartment buildings are located between 7th Avenue and 9th Avenue from 47th Street to 51st Street; all the private houses are located between 5th Avenue and 9th Avenue from 42nd Street to 47th Street; and all the stores are located between 5th Avenue and 7th Avenue from 47th Street to 51st Street.

The firefighter also learns that the apartment buildings are all between four and six stories; the private houses are all between one and three stories; the factories are all between three and five stories; and the stores are all either one or two stories.

25. An alarm is received for a fire located on 8th Avenue between 46th Street and 47th Street. The firefighter should assume that the fire is in a

  (A) private house between 1 and 3 stories.

  (B) private house between 4 and 6 stories.

  (C) factory between 3 and 5 stories.

  (D) factory between 4 and 6 stories.

26. The company responds to a fire on 47th Street between 6th Avenue and 7th Avenue. The firefighter should assume that he will be responding to a fire in a(n)

  (A) store of either 1 or 2 stories.

  (B) factory between 3 and 5 stories.

  (C) apartment building between 4 and 6 stories.

  (D) private house between 4 and 6 stories.

**QUESTION 27 IS BASED ON THE FOLLOWING PASSAGE:**

During a recent day tour with an engine company, firefighter Sims was assigned to the control position on the hose. The company responded to the following alarms during this tour:

Alarm 1:    At 9:30 a.m., the company responded to a fire on the first floor of an apartment building. At the fire scene, firefighter Sims pulled the hose from the fire engine and assisted the driver in attaching the hose to the hydrant.

Alarm 2:    At 11 a.m., the company responded to a fire on the third floor of a vacant building. Firefighter Sims pulled the hose from the fire engine and went to the building on fire.

Alarm 3:    At 1 p.m., the company responded to a fire in a first-floor laundromat. Firefighter Sims pulled the hose from the fire engine and assisted the driver in attaching the hose to the hydrant.

Alarm 4:    At 3 p.m., the company responded to a fire on the fourth floor of an apartment building. Firefighter Sims pulled the hose from the fire engine and went to the building on fire.

Alarm 5:    At 5:45 p.m., the company responded to a fire on the second floor of a private house. Firefighter Sims pulled the hose from the fire engine and assisted the driver in attaching the hose to the hydrant.

27. The firefighter assigned to the control position assists the driver in attaching the hose to a hydrant when the fire is

   **(A)** in an apartment building.

   **(B)** above the second floor.

   **(C)** in a vacant building.

   **(D)** below the third floor.

**QUESTIONS 28–30 ARE BASED ON THE FOLLOWING FLOOR PLAN:**

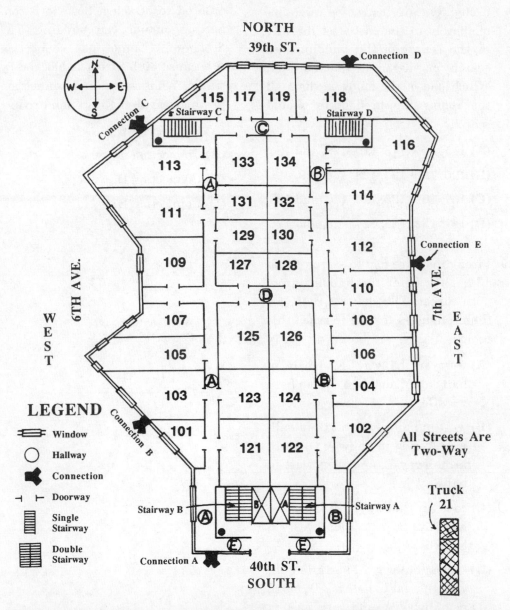

The floor plan represents a typical high-rise office building in midtown Manhattan. Numbers shown indicate room numbers. The pipe connections for the water supply system are outside the building at street level. Firefighters attach hoses to these connections to send water into the pipes in the building.

**QUESTIONS 28–30 REFER TO A FIRE ON THE FIRST FLOOR IN ROOM 111:**

**28.** After fighting the fire in Room 111, firefighters are instructed to go immediately to the east-west hallway in the center of the building and search for victims in that hallway. Which one of the following lists all the rooms the firefighters should search?

(A) 115, 117, 118, 119, 133, and 134

(B) 125, 126, 127, 128, and 129

(C) 107, 109, 125, 126, 127, and 128

(D) 121, 122, 123, 124, 125, and 126

**29.** Firefighters are told to search Room 134. They enter the building from 40th Street. The shortest route for the firefighters to take to reach this room is

(A) west in hallway E, north in hallway A, and then east in hallway C.

(B) west in hallway E, north in hallway A, east in hallway D, north in hallway B, and then west in hallway C.

(C) east in hallway E, north in hallway B, and then west in hallway C.

(D) east in hallway E, north in hallway B, west in hallway D, north in hallway A, and then east in hallway C.

**30.** Firefighters in Truck 21 have been ordered to attach a hose to a connection outside the building. The firefighters cannot use connection A because 40th Street is blocked by traffic. What is the first connection to which the firefighters can drive?

(A) Connection B

(B) Connection C

(C) Connection D

(D) Connection E

**QUESTIONS 31–33 ARE BASED ON THE FOLLOWING PASSAGE:**

Firefighters often know the appearance and construction features of apartments by recognizing general features on the outside of the building. The following are some general features of different types of buildings in New York City.

1.  Old Law Tenements
    Height—5 to 7 stories
    Width—25 feet
    Fire Escapes—There will be a rear fire escape if there are two apartments per floor. There will be front and rear fire escapes if there are four apartments per floor.

2.  Row Frames
    Height—2 to 5 stories
    Width—20 feet to 30 feet
    Fire Escapes—There will be a rear fire escape if the building is higher than 2 stories.

3.  Brownstones
    Height—3 to 5 stories
    Width—20 feet to 25 feet
    Fire Escapes—If the Brownstone has been changed from a private home to a multiple dwelling, there will be a rear fire escape. Unchanged Brownstones have no fire escapes.

**31.** Upon arrival at a fire, a firefighter observes that the building is three stories high and 25 feet wide. There are fire escapes only in the rear of the building. The firefighter should conclude that the building is either a

(A) row frame or an unchanged brownstone.

(B) row frame or an old law tenement with two apartments per floor.

(C) changed brownstone or an old law tenement with four apartments per floor.

(D) row frame or a changed brownstone.

**32.** At another fire, the building is five stories high and 25 feet wide. There is a front fire escape. The firefighters should conclude that this building has

(A) a rear fire escape because the building is a row frame higher than two stories.

(B) a rear fire escape because the building is an old law tenement with four apartments per floor.

(C) no rear fire escape because the building is a brownstone that has been changed into a multiple dwelling.

(D) no rear fire escape because the building has a front fire escape.

**33.** At another fire, the building is four stories high and 30 feet wide. The building has no front fire escape. The firefighter should conclude that the building is a(n)

**(A)** row frame that has no rear fire escape.

**(B)** old law tenement that has four apartments per floor.

**(C)** row frame that has a rear fire escape.

**(D)** brownstone that has been changed from a private home to a multiple dwelling.

### QUESTIONS 34–36 ARE BASED ON THE FOLLOWING PASSAGE:

Firefighters use two-way radios to alert other firefighters to dangerous conditions and to the need for help. Messages should begin with "MAY DAY" or "URGENT." "MAY DAY" messages have priority over "URGENT" messages. The following is a list of specific emergencies and the messages that should be sent.

"MAY DAY" Messages:

1. When a collapse is probable in the area where the firefighters are working: "MAY DAY—MAY DAY, collapse probable, GET OUT."

2. When a collapse has occurred in the area where the firefighters are working: "MAY DAY—MAY DAY, collapse occurred." The firefighter also should give the location of the collapse. If there are trapped victims, the number and condition of the trapped victims also should be given.

3. When a firefighter appears to be a heart attack victim: "MAY DAY—MAY DAY, CARDIAC." The location of the victim also should be given.

4. When anyone has a serious, life-threatening injury: "MAY DAY—MAY DAY." The firefighter also should describe the injury and should give the condition and the location of the victim.

"URGENT" Messages:

1. When anyone has a less serious injury that requires medical attention (for example, a broken arm): "URGENT—URGENT." The firefighter also should give the type of injury and the location of the victim.

2. When the firefighters should leave the building and fight the fire from the outside: "URGENT—URGENT, back out." The firefighter also should indicate the area to be evacuated.

3. "URGENT" messages also should be sent when firefighters' lives are endangered due to a drastic loss of water pressure in the hose.

34. Firefighters are ordered to extinguish a fire on the third floor of an apartment building. As the firefighters are operating the hose on the third floor, the stairway collapses and cuts the hose. What message should the firefighters send?

(A) "URGENT—URGENT, back out."

(B) "URGENT—URGENT, we have a loss of water on the third floor."

(C) "MAY DAY—MAY DAY, collapse occurred on third floor stairway."

(D) "MAY DAY—MAY DAY, collapse probable, GET OUT."

35. Two firefighters on the second floor of a vacant building are discussing the possibility of the floor's collapse. One of the firefighters clutches his chest and falls down. What message should the other firefighter send?

(A) "MAY DAY—MAY DAY, firefighter collapse on the second floor."

(B) "MAY DAY—MAY DAY, CARDIAC on the second floor."

(C) "URGENT—URGENT, firefighter unconscious on the second floor."

(D) "URGENT—URGENT, collapse probable on the second floor."

36. A firefighter has just decided that a collapse of the third floor is probable when he falls and breaks his wrist. What is the first message he should send?

(A) "URGENT—URGENT, broken wrist on the third floor."

(B) "MAY DAY—MAY DAY, broken wrist on the third floor."

(C) "MAY DAY—MAY DAY, collapse probable, GET OUT."

(D) "URGENT—URGENT, back out, third floor."

_practice test_

**QUESTIONS 37–38 ARE BASED ON THE FOLLOWING DIAGRAM:**

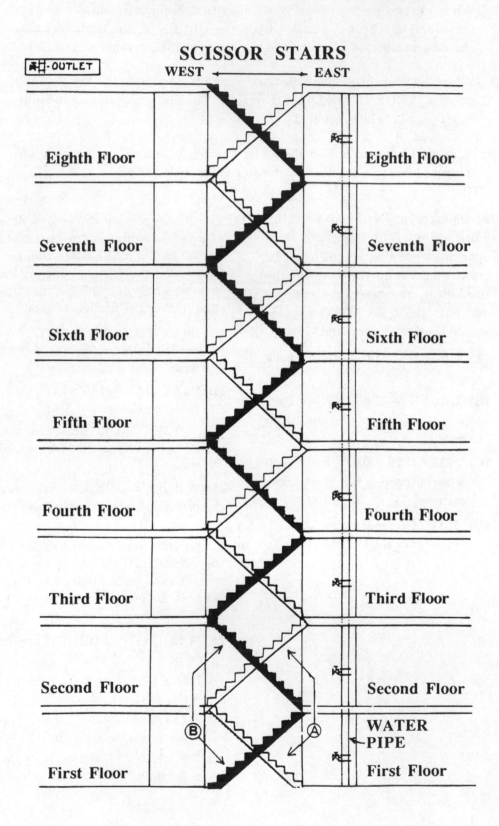

An eight-story apartment building has scissor stairs beginning on the first floor and going to the roof. Scissor stairs are two separate stairways (stairway A and stairway B) that crisscross each other and lead to opposite sides of the building on each floor. Once a person has entered either stairway, the only way to cross over to the other stairway on any floor is to leave the stairway and use the hallway on that floor. A person entering stairway A, which starts on the east side of the building on the first floor, would end up on the west side of the building on the second floor and back on the east side on the third floor. Similarly, a person entering stairway B, which starts on the west side of the building on the first floor, would end up on the east side of the building on the second floor and back on the west side on the third floor.

The apartment building has one water pipe for fighting fires. The pipe runs in a straight line near the stairway on the east side of the building from the first floor to the roof. There are water outlets for this pipe on each floor.

**QUESTIONS 37–38 INVOLVE A FIRE IN AN APARTMENT ON THE WEST SIDE OF THE SIXTH FLOOR.**

37. Firefighters are ordered to connect a hose to the nearest outlet below the fire. Upon reaching this outlet, they find that it is not usable. The next available outlet is on the

(A) fifth floor near stairway B.

(B) third floor near stairway A.

(C) fourth floor near stairway B.

(D) fourth floor near stairway A.

38. A firefighter working on the west side of the seventh floor is ordered to search for victims on the west side of the eighth floor. The door leading to the stairway on the west side of the seventh floor is jammed shut. To reach the victims, the firefighter should take

(A) stairway A to the eighth floor and then go across the hallway to the west side of the floor.

(B) stairway B to the eighth floor and then go across the hallway to the west side of the floor.

(C) the hallway to the east side of the seventh floor and go up stairway A.

(D) the hallway to the east side of the seventh floor and go up stairway B.

39. Firefighters refer to the four sides of a building on fire as "exposures." The front of the fire building is referred to as Exposure 1. Exposures 2, 3, and 4 follow in clockwise order. Firefighters are working at a building with a front entrance that faces south. A firefighter in the center of the roof is ordered to go to Exposure 3. To reach Exposure 3, the direction in which he must walk is

(A) east.

(B) west.

(C) south.

(D) north.

**QUESTIONS 40-44 ARE BASED ON THE FOLLOWING PASSAGE:**

Ventilation is an important process involved in the extinguishing of a fire. A building on fire usually is ventilated by creating a hole in the roof or a wall of the structure. This allows heated air, smoke, and gases produced by a fire to be replaced with cool, fresh air. There are several advantages to the use of ventilation in firefighting including advantages in rescue operations, fire extinguishing, and property conservation.

Ventilation can make the rescue process faster and more efficient by removing gases, smoke, and heated air that can be harmful to people who might be trapped in the building. By removing harmful substances produced by the fire and replacing them with fresh air, trapped individuals will have safer air to breathe and may have a better chance of survival. The removal of dark smoke and gases also can improve the visibility within the burning structure so that firefighters can locate victims faster.

The process of extinguishing a fire also can be improved by the proper use of ventilation. By creating a hole in the roof directly above a fire, gases, smoke, and heat will be effectively removed from the building. This will keep the fire from spreading to other parts of the building. At the same time, a properly placed hole will make conditions within the structure safer for firefighters by reducing heat within the building, removing dangerous gases, and improving visibility by removing smoke from the structure. This can help firefighters to quickly locate and extinguish the fire.

Property that is not damaged directly by the fire can be damaged by the heat and smoke produced by the fire as well as the water the firefighters use to extinguish the fire. This damage can be reduced, however, by the effective use of ventilation. By helping firefighters locate the source of a fire faster, ventilation can help confine the fire to a smaller area within the structure. This will reduce the amount of damaging heat and smoke while also reducing the amount of water necessary to extinguish the fire. With the fire confined to a small area, firefighters also can try to salvage some property by removing it from the structure even before the fire is completely extinguished.

40. Ventilation enables heated air, smoke, and gases to be replaced with

    (A) new water.

    (B) victims.

    (C) fire.

    (D) fresh air.

    (E) firefighters.

41. Property that is not damaged directly by a fire can be damaged by heat and smoke as well as the _____ that the firefighters use.

    (A) chemicals

    (B) hoses

    (C) water

    (D) ventilation

    (E) axes

42. Improved visibility created by proper ventilation will improve the rescue process by helping firefighters locate the

    (A) water main.

    (B) smoke detector.

    (C) hose.

    (D) ceiling or wall.

    (E) victims.

43. Creating a hole in the roof directly above the fire will keep the _____ from spreading to other parts of the building.

    (A) water

    (B) victims

    (C) firefighters

    (D) fire

    (E) ventilation

44. The preceding passage mentions three advantages to the use of ventilation in firefighting. These are advantages in

    (A) fire extinguishing, rescue operations, and fire control.

    (B) fire extinguishing, rescue operations, and property conservation.

    (C) rescue operations, property conservation, and wind control.

    (D) rescue operations, property conservation, and fire conservation.

    (E) property conservation, fire control, and rescue operations.

**QUESTIONS 45–50 ARE BASED ON THE FOLLOWING DIAGRAM:**

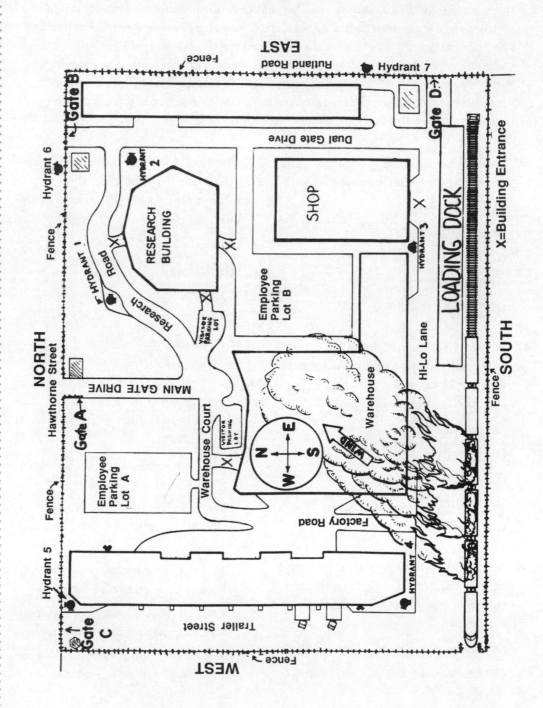

At 3 o'clock in the morning, a fire alarm is received for the area shown in the diagram. A train loaded with highly flammable material is on fire. The entire area is surrounded by a 10-foot high fence. At the time of the fire, Gate A is open but Gates B, C, and D are locked.

45. The first engine company arrives at the fire scene. The security guard at Gate A informs the firefighters of the location of the fire. Firefighter Jensen knows the area. He should inform the lieutenant that the way to drive to a hydrant that is as close to the fire as possible without passing through the smoke and flames is by going

    (A) south on Main Gate Drive, east on Research Road, south on Dual Gate Drive, and west on Hi-Lo Lane to hydrant 3.

    (B) south on Main Gate Drive, west on Warehouse Court, south on Factory Road, and west on Hi-Lo Lane to hydrant 4.

    (C) south on Main Gate Drive and east on Research Road to hydrant 1.

    (D) east on Hawthorne Street and south on Rutland Road to hydrant 7.

46. Firefighters at Employee Parking Lot A are ordered to drive their truck to the fence outside Gate D. The shortest route the firefighters could take from Warehouse Court is

    (A) south on Factory Road, west on Hi-Lo Lane, and north on Trailer Street.

    (B) east on Research Road and south on Dual Gate Drive.

    (C) north on Main Gate Drive, east on Hawthorne Street, and south on Rutland Road.

    (D) north on Main Gate Drive, west on Hawthorne Street, south on Trailer Street, and west on Hi-Lo Lane.

47. The first ladder company arrives at the fire scene. As they are driving north on Rutland Road, firefighters see the fire through Gate D. They cut the locks and enter Gate D. The lieutenant orders a firefighter to go on foot from Gate D to the Research Building and to search it for occupants. The entrance to the Research Building that is closest to the firefighter is

    (A) connected to the Visitor Parking Lot.

    (B) located on Research Road.

    (C) connected to Parking Lot B.

    (D) located on Dual Gate Drive.

48. The second engine company to arrive is ordered to attach a hose to a hydrant located outside the fenced area and then to await further orders. The hydrant outside of the fenced area that is closest to the flames is

    (A) hydrant 6.

    (B) hydrant 3.

    (C) hydrant 4.

    (D) hydrant 7.

**49.** The second ladder company to arrive at the fire scene is met at Gate C by a security guard who gives them the keys to open all the gates. They drive south on Trailer Street to the corner of Hi-Lo Lane and Trailer Street. The company is then ordered to drive to the corner of Research Road and Dual Gate Drive. The shortest route for the company to take without being exposed to the smoke and flames is

**(A)** east on Hi-Lo Lane, north on Factory Road, and east on Warehouse Court to Research Road.

**(B)** east on Hi-Lo Lane and north on Dual Gate Drive.

**(C)** north on Trailer Street, east on Hawthorne Street, and south on Dual Gate Drive.

**(D)** north on Trailer Street, east on Hawthorne Street, south on Main Gate Drive, and east on Research Road.

**50.** The heat from the fire in the railroad cars ignites the warehouse on the other side of Hi-Lo Lane. The officer of the first ladder company orders two firefighters on the west end of the loading dock to break the windows on the north side of the warehouse. Of the following, the shortest way for the firefighters to reach the northwest corner of the warehouse without passing through the smoke and flames is to go

**(A)** east on Hi-Lo Lane, north on Dual Gate Drive, and then west on Research Road to the entrance on Warehouse Court.

**(B)** west on Hi-Lo Lane, north on Factory Road, and then east on Warehouse Court to the Visitor Parking Lot on Warehouse Court.

**(C)** east on Hi-Lo Lane, north on Rutland Road, west on Hawthorne Street, and then south on Main Gate Drive to the Visitor Parking Lot on Warehouse Court.

**(D)** east on Hi-Lo Lane, north on Dual Gate Drive, west on Hawthorne Street, and then south on Main Gate Drive to the entrance on Warehouse Court.

practice test

Firefighters might have to use tools to force open an entrance door. Before the firefighters use the tools, they should turn the doorknob to see if the door is unlocked. If the door is locked, one firefighter should use an ax and another firefighter should use the halligan tool. This tool is used to pry open doors and windows. Firefighters must take the following steps in the order given to force open the door:

1. Place the prying end of the halligan tool approximately 6 inches above or below the lock. If there are two locks, the halligan tool should be placed between them.

2. Tilt the halligan tool slightly downward so that a single point on the prying end is at the door's edge.

3. Strike the halligan tool with the ax until the first point is driven in between the door and the door frame.

4. Continue striking with the ax until the door and frame are spread apart and the lock is broken.

5. Apply pressure toward the door and the door will spring open.

51. Firefighters respond to a fire on the second floor of a three-story apartment building. Two firefighters, one equipped with an ax and the other with a halligan tool, climb the stairs to the apartment on fire. They see that there are two locks on the apartment door. The firefighters should now

(A) place the prying end of the halligan tool about 6 inches above or below the locks.

(B) turn the doorknob to determine whether the door is locked.

(C) tilt the halligan tool toward the floor before striking.

(D) place the prying end of the halligan tool between the locks.

**QUESTIONS 52–53 ARE BASED ON THE FOLLOWING PASSAGE:**

The firefighter assigned to the roof position at a fire in a brownstone building should perform the following steps in the order given:

1. Go to the roof using one of the following options:
   a. First choice—The aerial ladder.
   b. Second choice—An attached building of the same height as the fire building.
   c. Third choice—A rear fire escape.
   d. Fourth choice—A 35-foot portable ladder.

2. Upon arrival on the roof, look around to determine whether any people are trapped who cannot be seen from the street.
   a. If a trapped person is observed, notify the officer and the driver that a lifesaving rope rescue is required. While waiting for assistance to conduct this rescue, assure the victim that help is on the way and proceed to Step 3.
   b. If no trapped persons are visible, proceed directly to Step 3.

3. Remove the cover from the opening in the roof.
   a. If there is no smoke or very little smoke coming from the opening, report to the officer for further orders.
   b. If heavy smoke comes from the opening, proceed to Step 4.

4. Remove the glass from the skylight.

**52.** Firefighters arriving at a fire in a brownstone are using the aerial ladder to make an immediate rescue. The firefighter assigned to the roof position should go to the roof of the building on fire via

**(A)** a 35-foot portable ladder.

**(B)** a rear fire escape.

**(C)** an attached building of the same height.

**(D)** the inside stairway of the fire building.

**53.** The firefighter assigned to the roof position at a fire in a brownstone arrives at the roof and finds that no persons are trapped. He then removes the roof cover from the opening in the roof. Which one of the following steps should be performed next?

**(A)** He should remove the glass from the skylight if heavy smoke is coming from the opening.

**(B)** He should remove the glass from the skylight if no smoke is coming from the roof opening.

**(C)** He should go to the top floor to assist in the search for trapped persons if heavy smoke is coming from the roof opening.

**(D)** He should report to the officer if heavy smoke is coming from the roof opening.

**QUESTIONS 54–55 ARE BASED ON THE FOLLOWING PASSAGE:**

Firefighters often are required to remove people who are trapped in elevators. At this type of emergency, firefighters perform the following steps in the order given:

1. Upon entering the building, determine the location of the elevator involved.

2. Reassure the trapped occupants that the fire department is on the scene and that firefighters are attempting to free them.

3. Determine whether there are any injured people in the elevator.

4. Determine whether all the doors from the hallways into the elevator shaft are closed.

5. If all the doors are closed, call for an elevator mechanic.

6. Wait until a trained elevator mechanic arrives before attempting to remove any trapped persons from the elevator unless they can be removed through the door to the hallway. However, firefighters must remove the trapped persons by any safe method if any one of the following conditions exists:

   a. There is a fire in the building.

   b. Someone in the elevator is injured.

   c. The people trapped in the elevator are in a state of panic.

54. Firefighters arrive at an elevator emergency in an office building. When they arrive, a maintenance man directs them to an elevator stuck between the fourth and fifth floors. He informs the firefighters that there is a young man in the elevator who apparently is calm and unhurt. Which one of the following steps should the firefighters perform next?

**(A)** Determine whether the young man is injured.

**(B)** Reassure the young man that the fire department is on the scene and that firefighters are attempting to free him.

**(C)** Check to make sure all the doors to the elevator and hallways are closed.

**(D)** Call for an elevator mechanic and await his arrival.

55. Firefighters are called to an elevator emergency at a factory building. The freight elevator has stopped suddenly between floors. The sudden stop caused heavy boxes to fall on the elevator operator, breaking his arm. Upon arrival, the firefighters determine the location of the elevator. They tell the trapped operator that they are on the scene, are aware of his injury, and are attempting to free him. They determine that all the hallway doors leading into the elevator shaft are closed. The firefighters' next step should be to

   (A) call for an ambulance and wait until it arrives.

   (B) remove the trapped person through the door to the hallway.

   (C) call for an elevator mechanic.

   (D) remove the trapped person by any safe method.

**QUESTION 56 IS BASED ON THE FOLLOWING PASSAGE:**

The preferred order of actions for firefighters to take when removing a victim from an apartment on fire is as follows:

1. First choice—Remove the victim to the street level through the public hallway.

2. Second choice—Remove the victim to the street level by the fire escape.

3. Third choice—Remove the victim to the street level using either a portable ladder or an aerial ladder.

4. Fourth choice—Lower the victim to the street level with a lifesaving rope.

56. Firefighters answering an alarm are not able to use the entrance to a building on fire to reach a victim on the third floor because there is a fire in the public hallway. The victim is standing at the front window of an apartment on fire that has a fire escape. A firefighter places a portable ladder against the building, climbs the ladder, and enters the window where the victim is standing. The firefighter is carrying a lifesaving rope and a two-way radio. The radio enables her to communicate with the firefighter who operates the aerial ladder. The firefighter should then remove this victim from the apartment by using the

(A) fire escape.

(B) aerial ladder.

(C) portable ladder.

(D) lifesaving rope.

practice test

**QUESTIONS 57–61 ARE BASED ON THE FOLLOWING FLOOR PLAN:**

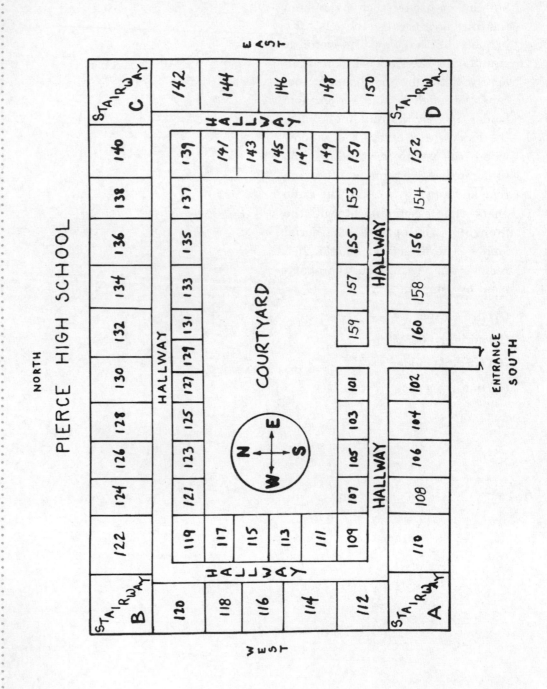

practice test

Because many children might need to be rescued in the event of a school fire, New York City firefighters must become familiar with the floor layouts of public schools. Firefighters can develop this familiarity by conducting training drills at the schools.

A ladder company and an engine company recently conducted a drill at Pierce High School. The firefighters determined that the room layout is the same on all floors.

Several days after the drill, the ladder and engine companies report to a fire at Pierce High School in classroom 304, which is on the third floor. The fire has spread into the hallway in front of Room 304, blocking the hallway.

**57.** A firefighter is instructed to search for victims in the southwest area of the third floor. He wants to search as many rooms as possible and start his search as close to the fire as possible without passing through the fire. From the street, the firefighter should use his ladder to enter

**(A)** Room 302.

**(B)** Room 306.

**(C)** Room 312.

**(D)** Room 352.

**58.** A firefighter goes to the third floor by way of the southwest building stairway. In Room 317, he finds a child who has been overcome by smoke. Upon returning to the hallway, he finds that the stairway he came up is now blocked by fire hoses. The closest stairway the firefighter can use to bring the child to the street level is

**(A)** Stairway A.

**(B)** Stairway B.

**(C)** Stairway C.

**(D)** Stairway D.

**59.** Fire has spread from Room 304 to the room directly across the hall. As a result of heavy smoke, firefighters are ordered to break the windows of this room from the closest room on the floor above. Which room should the firefighters go to?

**(A)** Room 403

**(B)** Room 305

**(C)** Room 313

**(D)** Room 413

**60.** The fire is in Rooms 303 and 305. Firefighters are told to go to the rooms in the north corridor facing the courtyard that are directly opposite 303 and 305. Which rooms should the firefighters go to?

**(A)** Rooms 323 and 325

**(B)** Rooms 326 and 328

**(C)** Rooms 333 and 335

**(D)** Rooms 355 and 357

**61.** Another fire breaks out in Room 336, blocking the entire hallway. Fire-fighters have brought a hose up the northeast stairway to fight this fire. Another hose must be brought up another stairway so that firefighters can approach the fire from the same direction. What is the closest stairway the firefighters could use?

**(A)** Stairway A

**(B)** Stairway B

**(C)** Stairway C

**(D)** Stairway D

**62.** When attempting to rescue a person trapped at a window, firefighters frequently use an aerial ladder with a rotating base. The rotating base enables the ladder to move up and down and from side to side. In a rescue situation, the ladder should be placed so that both sides of the ladder rest fully on the windowsill after it has been raised into position. Which one of the following diagrams shows the proper placement of the rotating base in relation to the window where the person is trapped?

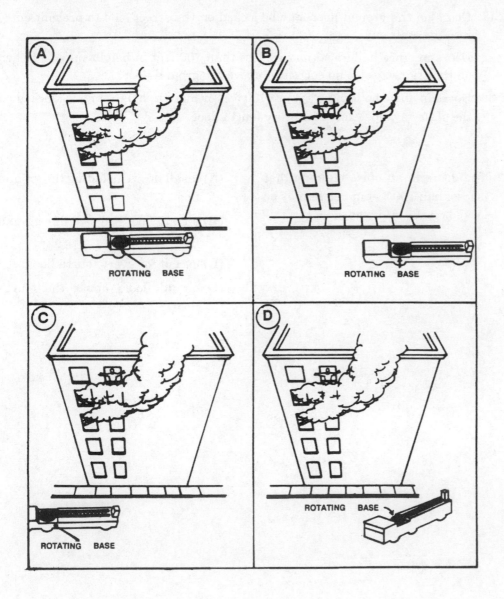

## QUESTION 63 IS BASED ON THE FOLLOWING PASSAGE:

Firefighters must regularly inspect office buildings to determine whether fire-prevention laws have been obeyed. Some of these fire-prevention laws are as follows:

DOORS: Doors should be locked as follows:

1. Doors on the ground floor may be locked on the street side to prevent entry into the stairway.
2. Doors in office buildings that are less than 100 feet in height may be locked on the stairway side on each floor above the ground floor.
3. Doors in office buildings that are 100 feet or more in height may be locked on the stairway side except for every fourth floor.

63. The doors in an office building that is less than 100 feet in height may be locked on the stairway side

(A) on all floors including the ground floor.

(B) on all floors above the ground floor.

(C) except for every fourth floor.

(D) on all floors above the fourth floor.

**QUESTION 64 IS BASED ON THE FOLLOWING PASSAGE:**

SIGNS: Signs concerning stairways should be posted in the following manner:

1.  A sign shall be posted near the elevator on each floor, stating "IN CASE OF FIRE, USE STAIRS UNLESS OTHERWISE INSTRUCTED." The sign shall contain a diagram showing the location of the stairs and the letter identification of the stairs.

2.  Each stairway shall be identified by an alphabetical letter on a sign posted on the hallway side of the stair door.

3.  Signs indicating the floor number shall be attached to the stairway side of each door.

4.  Signs indicating whether reentry can be made into the building (and the floors where reentry can be made) shall be posted on the stairway side of each door.

64. Which one of the following correctly lists the information that should be posted on the stairway side of a door?

    **(A)** A sign should indicate the floor number, whether reentry can be made into the building, and the floors where reentry can be made.

    **(B)** A sign should indicate the alphabetical letter of the stairway, whether reentry can be made into the building, and the floors where reentry can be made.

    **(C)** A sign should indicate the alphabetical letter of the stairway and the floor number.

    **(D)** A sign should indicate the alphabetical letter of the stairway, the floor number, whether reentry can be made into the building, and the floors where reentry can be made.

**QUESTION 65 IS BASED ON THE FOLLOWING PASSAGE:**

Every firefighter must know the proper first-aid procedures for treating injured people when there is a subway fire. In general, anyone suffering from smoke inhalation or heat exhaustion should be removed to fresh air and given oxygen immediately. Heart attack victims should be kept calm and should receive oxygen and medical attention immediately. Persons suffering from broken bones should not be moved until a splint is applied to the injury. In a situation in which there is a smoky fire in the subway, however, and the passengers needing immediate first aid are in danger from the fire, firefighters must first evacuate the passengers and perform first aid later, regardless of the injury.

65. The proper first-aid procedure for a man who apparently has suffered a heart attack on the station platform is to

(A) have the man take the next train to the nearest hospital.

(B) remove the man to the street and administer oxygen.

(C) turn off the electrical power and evacuate the man through the tunnel.

(D) keep the man calm and administer oxygen.

**QUESTIONS 66–67 IS BASED ON THE FOLLOWING PASSAGE:**

A firefighter is responsible for a variety of duties other than fighting fires. One such duty is housewatch.

A firefighter's primary responsibility during housewatch is to properly receive alarm information. This enables firefighters to respond to alarms for fires and emergencies. The alarms are received at the firehouse by one of the following methods: computer teleprinter messages, fire department telephone, or verbal alarm. The computer teleprinter and the telephone are used to alert the fire companies. These two types of alarms are transmitted by a dispatcher from a central communication office to the firehouse closest to the fire. The verbal alarm occurs when someone comes to the firehouse or stops the fire truck on the street to report a fire. Once an alarm has been received, the firefighter on housewatch duty alerts the rest of the firefighters to respond to the alarm.

Other housewatch responsibilities include keeping the appearance of the housewatch area neat and orderly, keeping the front of the firehouse clear of all vehicles and obstructions, and receiving telephone calls and visitors with complaints about fire hazards. The firefighter on housewatch duty also keeps an accurate and complete record of all administrative matters in a journal.

66. The methods a dispatcher uses to transmit alarms to the firehouse are the

(A) computer teleprinter, fire department telephone, and verbal alarm.

(B) verbal alarm and computer teleprinter.

(C) fire department telephone and verbal alarm.

(D) computer teleprinter and fire department telephone.

67. The primary responsibility of a firefighter on housewatch duty is to

(A) properly assign firefighters to specific duties.

(B) properly receive alarm information.

(C) keep the housewatch area neat and orderly.

(D) write all important information in the company journal.

One duty of a firefighter on housewatch is to ensure that the computer tele-printer is working properly. A company officer should be notified immediately of any equipment problems. The firefighter on housewatch should check the amount of paper in the teleprinter and refill it when necessary. The firefighter also should check the selector panel on the computer. The selector panel has a series of buttons that are used by the firefighter to let the dispatcher know that an alarm has been received and that the fire company is responding. These buttons have lights. To make sure the computer is functioning properly, the firefighter should press the button marked "test" and then release the button. If the computer lights go on and then go off after the "test" button has been released, the computer is working properly. In addition, the light next to the "test" button should always be blinking.

**68.** To make sure the selector panel of the computer is working properly, the firefighter on housewatch duty presses the button marked "test" and then releases the button. The firefighter should conclude that the computer is working properly if the

**(A)** computer lights stay on.

**(B)** computer lights keep blinking.

**(C)** computer lights go on and then off.

**(D)** "test" light stays on.

**69.** A firefighter on housewatch duty notices that the teleprinter is almost out of paper. In this situation, the firefighter should

**(A)** test the computer panel by pushing the "test" button.

**(B)** notify the officer to replace the paper.

**(C)** place a new supply of paper in the teleprinter.

**(D)** notify the dispatcher that the paper is being changed.

**QUESTIONS 70–71 ARE BASED ON THE FOLLOWING PASSAGE:**

When Sara was coming home from work one day, she noticed a strange man leaving through the back door of her neighbor's house. She saw him from a distance, but it appeared that the man was white, in his mid 20s, 5' 11", 180 lbs., had long brown hair, and was wearing blue jeans, a green sweatshirt, and a red baseball cap. About 20 minutes later, she noticed smoke coming from her neighbor's house and immediately called the fire department. When the firefighters arrived, she told them about the man she saw leaving the house. Later that evening, she was told that they suspected the fire was caused by arson. She was asked to indicate if the man she had seen leaving her neighbor's house was among the suspects. The following suspects were in the lineup

1. Male, white, late 20s, 6', 150 lbs., long brown hair, wearing blue jeans, a blue T-shirt, and a red baseball cap.

2. Male, white, mid 20s, 5' 11", 200 lbs., long blond hair, wearing blue jeans, a green sweatshirt, and a red baseball cap.

3. Male, white, mid 20s, 5' 11", 190 lbs., long brown hair, wearing blue jeans, a green sweatshirt, and a red baseball cap.

4. Male, white, mid 20s, 6' 2", 190 lbs., long brown hair, wearing blue jeans, a blue sweatshirt, and a red baseball cap.

5. Male, white, mid 30s, 6', 180 lbs., long brown hair, wearing blue jeans, a blue sweatshirt, and a red baseball cap.

**70.** Which one of the suspects in the lineup is most likely the man Sara saw leaving her neighbor's house?

**(A)** Suspect 1

**(B)** Suspect 2

**(C)** Suspect 3

**(D)** Suspect 4

**(E)** Suspect 5

**71.** Which suspect can be ruled out as the man leaving the house because of his hair and weight?

**(A)** Suspect 1

**(B)** Suspect 2

**(C)** Suspect 3

**(D)** Suspect 4

**(E)** Suspect 5

**QUESTION 72 IS BASED ON THE FOLLOWING PASSAGE:**

Firefighters from the first arriving ladder company work in teams while fighting fires in private homes. The inside team enters the building through the first-floor entrance and then searches the first floor for victims. The outside team uses ladders to enter upper-level windows for a quick search of the bedrooms on the second floor and above. The assignments for the members of the outside team are as follows:

Roof person—This member places a ladder at the front porch and enters the second-floor windows from the roof of the porch.

Outside vent person and driver—These members work together and place a portable ladder at a window on the opposite side of the house from which the roof person is working. If the aerial ladder can be used, however, the outside vent person and driver climb the aerial ladder in the front of the house, and the roof person places a portable ladder on the left side of the house.

To search all four sides of a private home on the upper levels, firefighters from the second arriving ladder company place portable ladders at the sides of the house not covered by the first ladder company and enter the home through the upper-level windows.

72. The second ladder company to arrive at a fire in a two-story private home sees the aerial ladder being raised to the front porch roof. In this situation, the firefighters should place their portable ladders to the

(A) left and right sides of the house because there is a front porch.

(B) rear and right sides of the house because the aerial ladder is being used.

(C) rear and left sides of the house because there is a front porch.

(D) left and right sides of the house because the aerial ladder is being used.

## QUESTIONS 73 IS BASED ON THE FOLLOWING PASSAGE:

The priority for the removal of a particular victim by aerial ladder depends on the following conditions:

If two victims are at the same window and are not seriously endangered by spreading fire, the victim who is easier to remove is taken down the ladder first and is helped safely to the street. In general, the term "easier to remove" refers to the victim who is more capable of being moved and more able to cooperate. After the easier removal is completed, time can be spent on the more difficult removal.

If there are victims at two different windows, the aerial ladder is first placed to remove the victims who are the most seriously endangered by the fire. The ladder is then placed to remove the victims who are less seriously exposed to the fire.

**73.** Assume that you are working at a fire and that there are a total of three victims at two windows. Victims 1 and 2 are at the same window, which is three floors above the fire and shows no evidence of heat or smoke. Victim 1 is a disabled, 23-year-old male, and Victim 2 is a 40-year-old woman. Victim 3, a 16-year-old male, is at a window of the apartment on fire. From your position on the street, you can see heavy smoke coming from this window and flames coming out of the window next to it. Which one of the following is the proper order for victim removal?

**(A)** Victim 3, Victim 2, Victim 1

**(B)** Victim 1, Victim 2, Victim 3

**(C)** Victim 1, Victim 3, Victim 2

**(D)** Victim 3, Victim 1, Victim 2

**QUESTIONS 74–75 ARE BASED ON THE FOLLOWING PASSAGE:**

The four different types of building collapses are as follows:

1. Building Wall Collapse—An outside wall of the building collapses, but the floors maintain their positions.

2. Lean-To Collapse—One end of the floor collapses onto the floor below it. This leaves a sheltered area on the floor below.

3. Floor Collapse—An entire floor falls to the floor below it, but large pieces of machinery in the floor below provide spaces that can provide shelter.

4. Pancake Collapse—A floor collapses completely onto the floor below it, leaving no spaces. In some cases, the force of this collapse causes successive lower floors to collapse.

74. The most serious injuries are likely to occur in

    (A) pancake collapses.
    (B) lean-to collapses.
    (C) floor collapses.
    (D) building wall collapses.

75. Of the following, a floor collapse is most likely to occur in a

    (A) factory building.
    (B) private home.
    (C) apartment building.
    (D) hotel.

Many subway tunnels contain a set of three rails used for train movement. The subway trains run on two rails. The third rail carries electricity and is the source of power for all trains. Electricity travels from the third rail through metal plates called contact shoes, which are located near the wheels on every train car. Electricity then travels through the contact shoes into the train's motor. Firefighters must be very careful when operating near the third rail because contact with the third rail can result in electrocution.

76. The source of power for subway
    trains is the

    (A) third rail.
    (B) contact shoes.
    (C) motor.
    (D) metal plates.

practice test

**QUESTIONS 77–81 ARE BASED ON THE FOLLOWING PASSAGE:**

Firefighters receive an alarm for an apartment fire on the fourth floor of a 14-story housing project at 1191 Park Place. One firefighter shouts the address as the other firefighters are getting on the fire truck. Knowledge of the address helps the firefighters decide which equipment to pull off the fire truck when they reach the fire scene.

The firefighters know where the water outlets are located in the building on fire. There is an outlet in every hallway. Firefighters always attach the hose at the closest outlet on the floor below the fire.

As they arrive at 1191 Park Place, three firefighters immediately take one length of hose each and go into the building. Because an officer has been told by the dispatcher that two children are trapped in the rear bedroom, the officer and two firefighters begin searching for victims and opening windows immediately upon entering the apartment on fire.

As in all housing project fires, the roof person goes to the apartment above the apartment on fire. From this position, he attaches a tool to a rope in order to break open the windows of the apartment on fire. From this position, the roof person also could make a rope rescue of a victim in the apartment on fire.

77. A firefighter shouted the address of the fire when the alarm was received so that the firefighters would

(A) know which equipment to take from the truck at the fire scene.

(B) be more alert when they arrived at the fire.

(C) be prepared to make a rope rescue.

(D) know that two children were reported to be trapped.

78. The hose should be attached to an outlet on the

(A) floor above the fire.

(B) ground floor.

(C) fire floor.

(D) floor below the fire.

79. Because of the information given to the officer by the dispatcher, the officer and two firefighters

(A) entered the apartment above the fire for a rope rescue.

(B) began immediately to search for victims and to open windows.

(C) opened all the windows before the hose was moved in.

(D) attached a hose and moved to the origin of the fire.

80. The roof person broke the windows of the apartment on fire with a(n)

(A) ax while leaning out the windows.

(B) ax attached to the end of a rope.

(C) tool while standing on the roof.

(D) tool attached to a rope.

81. The proper location for a rescue by the roof person at a fire in a housing project is the

    **(A)** hallway.

    **(B)** apartment below the apartment on fire.

    **(C)** apartment above the apartment on fire.

    **(D)** fire escape.

**QUESTION 82 IS BASED ON THE FOLLOWING PASSAGE:**

You are a firefighter inspecting a building for violations. You must perform the following steps in the order given:

1. Find the manager of the building and introduce yourself.

2. Have the manager accompany you during the inspection.

3. Start the inspection by checking the fire department permits that have been issued to the building. The permits are located in the office of the building.

4. Inspect the building for violations of the Fire Prevention Laws. Begin at the roof and work down to the basement or cellar.

5. As you inspect, write on a piece of paper any violations you find and explain them to the building manager.

82. You are inspecting a supermarket. After entering the building, you identify yourself to the store manager and ask him to come along during the inspection. Which one of the following actions should you take next?

    **(A)** Start inspecting the supermarket, beginning at the basement.

    **(B)** Start inspecting the supermarket, beginning at the roof.

    **(C)** Ask to see the fire department permits that have been issued to the supermarket.

    **(D)** Write down any violations you notice while introducing yourself to the manager.

## QUESTIONS 83–85 ARE BASED ON THE FOLLOWING TABLE:

For a fuel to burn while it is in a gaseous state, it must mix with oxygen. This oxygen usually is found in the air. If there is too much or too little fuel in the mixture for the amount of oxygen, however, the fuel will not burn.

### FLAMMABLE RANGES OF COMMON FUELS

| Fuel | Lower Limit (%) | Upper Limit (%) |
| --- | --- | --- |
| Acetylene | 2.5 | 100 |
| Hydrogen | 4.0 | 75 |
| Propane | 2.2 | 9.5 |
| Natural Gas | 5.0 | 15 |
| Gasoline | 1.4 | 7.6 |

83. When mixed with oxygen, what percentage of propane would be below what is necessary for the mixture to burn?

(A) 1.4%

(B) 2.2%

(C) 2.5%

(D) 4.0%

(E) 5.0%

84. When mixed with oxygen, what percentage of hydrogen would be above what is necessary for the mixture to burn?

(A) 75%

(B) 9.5%

(C) 7.6%

(D) 15%

(E) 100%

85. When mixed with oxygen, what percentage of natural gas would be acceptable for the mixture to burn?

(A) 30%

(B) 2.2%

(C) 10%

(D) 2.5%

(E) 25%

**QUESTION 86 IS BASED ON THE FOLLOWING PASSAGE:**

When administering first aid to a person who is severely bleeding, firefighters should perform the following steps in the order given:

1. Direct pressure

    a. Press a dressing directly on the wound.

2. Elevation

    a. Lift the injured body part above the heart level while continuing direct pressure. This slows down the movement of blood to the wound.

    b. Never elevate a body part if it might be fractured.

    c. If the bleeding continues while the injured body part is elevated and is receiving direct pressure, pressure should be applied to an artery or another pressure point.

86. A firefighter arrives at an automobile accident and finds a woman who is bleeding severely from a cut just above the ankle. She also has a fracture of the upper arm. After placing a dressing on the ankle wound, the firefighter next should

    **(A)** apply pressure to a pressure point to stop the bleeding.

    **(B)** elevate the upper arm and apply direct pressure to it.

    **(C)** apply direct pressure to an artery in the arm.

    **(D)** apply direct pressure to the leg while elevating it.

**QUESTION 87 IS BASED ON THE FOLLOWING PASSAGE:**

One firefighter of the first company to arrive at a fire in a private house is assigned to the roof position. During the beginning stages of a fire, the roof person is part of a team that enters and searches the building for victims.

The roof person's duties are performed in the following order:

1. Climb a portable ladder to the front porch roof, break a window, and enter the building through this opening if there is no victim in immediate danger visible at another window. If there is such a victim, place the ladder at that window.

2. If there is no front porch, use the portable extension ladder at a side of the house.

3. Enter the house after breaking a window.

87. The roof person at a fire in a private house sees a boy at a top floor window on the right side of the house. There is a porch in the front of the house. The roof person should now place the portable ladder at the

(A) front of the porch.

(B) top floor front window.

(C) top floor window on the right side of the house.

(D) top floor window on the left side of the house.

**FOR QUESTIONS 88–89, IDENTIFY WHICH OF THE UNDERLINED WORDS IS SPELLED INCORRECTLY.**

88. The <u>feasibility</u> of opening a floor for <u>ventelation</u> during a firefighting <u>operation</u> <u>obviously</u> depends on how the floor was <u>constructed</u> and with what material.

    (A) feasibility
    (B) ventelation
    (C) operation
    (D) obviously
    (E) constructed

89. Due to a <u>suceptibility</u> to <u>deterioration</u> and wear, fire hoses must be <u>maintained</u> and <u>thoroughly</u> <u>inspected</u> regularly.

    (A) suceptibility
    (B) deterioration
    (C) maintained
    (D) thoroughly
    (E) inspected

90. The decimal .20 is equivalent to what fraction?

    (A) $\frac{1}{4}$

    (B) $\frac{1}{5}$

    (C) $\frac{1}{6}$

    (D) $\frac{1}{7}$

    (E) $\frac{1}{8}$

91. If a fire engine is 8 feet tall and the ladder attached to the top of the engine has three sections that are 18 feet each in length, how high will the ladder reach if all three sections are fully extended?

    (A) 52 feet
    (B) 56 feet
    (C) 62 feet
    (D) 65 feet
    (E) 68 feet

92. If for every mile per hour in speed that a fire engine is traveling it takes $\frac{3}{4}$ of a foot to come to a complete stop, how many feet will it take a fire engine traveling 60 mph to come to a complete stop?

    (A) 60 feet
    (B) 50 feet
    (C) 45 feet
    (D) 30 feet
    (E) 65 feet

93. If water travels through a fire hose at a rate of 5 feet per second, how long will it take a hose that is 125 feet long to fill with water?

    (A) 50 seconds
    (B) 35 seconds
    (C) 25 seconds
    (D) 12 seconds
    (E) 10 seconds

**94.** Fire Station 35 received a call at 4:20 p.m., and the firefighters arrived at the scene at 4:25 p.m. The fire was extinguished at 5:15 p.m., and the firefighters arrived back at the station at 5:30 p.m. How much time (in minutes) elapsed from the time of the call to the firefighters' arrival back at the station?

(A) 60 minutes

(B) 70 minutes

(C) 75 minutes

(D) 80 minutes

(E) 90 minutes

**QUESTIONS 95–97 ARE BASED ON THE FOLLOWING CHART:**

**STATION CALLS BY TYPE AND YEAR**

|                 | 1998 | 1999 | 2000 | 2001 |
|-----------------|------|------|------|------|
| House Fire      | 87   | —    | 68   | 73   |
| Car Accident    | 154  | 167  | 174  | 186  |
| Industrial Fire | 56   | 48   | 59   | 43   |
| **Total**       | 297  | 312  | 301  | 302  |

**95.** How many house-fire calls were received in 1999?

(A) 83

(B) 97

(C) 73

(D) 104

(E) 86

**96.** What was the percent increase in car accidents from 1998 to 2001?

(A) 21%

(B) 22%

(C) 25%

(D) 26%

(E) 28%

**97.** What was the average number of industrial fires from 1999 to 2001?

(A) 47

(B) 48

(C) 49

(D) 50

(E) 51

**98.** A firefighter assigned to the roof position at a fire must notify the officer of dangerous conditions that can be seen from the roof. Which one of the following conditions is the most dangerous?

**(A)** The roof is sagging and might collapse because of the fire on the top floor.

**(B)** Rubbish is visible on the roof of the building next door.

**(C)** The stairway to the roof of the building has poor lighting.

**(D)** An automobile accident in the street is causing a traffic jam.

**99.** Firefighters must often deal with people who need medical assistance. In life-threatening situations, firefighters must perform first aid until an ambulance arrives. In less serious situations, firefighters should make the person comfortable and should wait for the ambulance personnel to give first aid. A firefighter should give first aid until an ambulance arrives when a person

**(A)** appears to have a knee injury.

**(B)** is bleeding heavily from a stomach wound.

**(C)** has bruises on his head.

**(D)** has a broken ankle.

**100.** At a fire on the west side of the third floor of a 10-story office building, a firefighter is responsible for rescuing trapped persons by means of the aerial ladder. The trapped person in the most dangerous location should be removed first. Of the following, the firefighter should first remove the trapped person

**(A)** on the rear fire escape on the east side of the second floor.

**(B)** on the roof.

**(C)** at a window on the west side of the fourth floor.

**(D)** at a window on the tenth floor.

practice test

## ANSWERS AND EXPLANATIONS

| | | | | | | | | | |
|---|---|---|---|---|---|---|---|---|---|
| 1. | B | 22. | B | 43. | D | 64. | A | 83. | A |
| 2. | B | 23. | C | 44. | B | 65. | D | 84. | E |
| 3. | D | 24. | D | 45. | A | 66. | D | 85. | C |
| 4. | B | 25. | A | 46. | C | 67. | B | 86. | D |
| 5. | B | 26. | A | 47. | C | 68. | C | 87. | A |
| 6. | D | 27. | D | 48. | D | 69. | C | 88. | B |
| 7. | A | 28. | C | 49. | C | 70. | C | 89. | A |
| 8. | A | 29. | C | 50. | A | 71. | B | 90. | B |
| 9. | D | 30. | D | 51. | B | 72. | B | 91. | C |
| 10. | B | 31. | D | 52. | C | 73. | A | 92. | C |
| 11. | D | 32. | B | 53. | A | 74. | A | 93. | C |
| 12. | B | 33. | C | 54. | B | 75. | A or B | 94. | B |
| 13. | B | 34. | C | 55. | C | | or C | 95. | B |
| 14. | D | 35. | B | 56. | A | | or D | 96. | A |
| 15. | A | 36. | C | 57. | B | 76. | A | 97. | D |
| 16. | D | 37. | C | 58. | B | 77. | A | 98. | A |
| 17. | D | 38. | C | 59. | A | 78. | D | 99. | B |
| 18. | E | 39. | D | 60. | A | 79. | B | 100. | C |
| 19. | D | 40. | D | 61. | D | 80. | D | | |
| 20. | C | 41. | C | 62. | A | 81. | C | | |
| 21. | B | 42. | E | 63. | B | 82. | C | | |

1. **The correct answer is (B).** The note at the bottom of the picture states that the ground floor is the first floor. Count up from the sidewalk and you will find the fire coming out the fourth-floor windows.

2. **The correct answer is (B).** As you stand in front facing the buildings, the bank is on the left side of the page. The smoke and flames are blowing from the left to the right and up.

3. **The correct answer is (D).** The fire escapes are on the left side of the apartment building and are indicated by the words "fire escape." Count up. The person is on the fire escape at the fifth floor.

4. **The correct answer is (B).** The ground floor is the first floor. Count up again.

5. **The correct answer is (B).** The front window on the fifth floor is directly over the fire and closest to the fire.

6. **The correct answer is (D).** The building on fire is an "isolated" building. This means that there is no direct access to the roof from the roofs of the adjacent buildings.

7. **The correct answer is (A).** As you stand facing the front of the buildings, the fire escapes are on your left-hand side. This is considered the left side of the building for purposes of this exam.

8. **The correct answer is (A).** The fire hydrant is directly in front of the bank.

9. **The correct answer is (D).** The fire escape is in the lower right corner of the diagram. The dining room is in the upper left corner. The dining room is the greatest distance from the fire escape.

10. **The correct answer is (B).** Bedroom 1 has only one door. The living room, dining room, and kitchen each have at least two door openings.

11. **The correct answer is (D).** As indicated in the diagram, the fire escape is directly outside bedroom 3.

12. **The correct answer is (B).** The only door from bathroom 2 leads directly to bedroom 2.

13. **The correct answer is (B).** This is indicated by the door symbol between bathroom 2 and bedroom 2.

14. **The correct answer is (D).** The terrace on the left side of the building has two door symbols to indicate that access from the terrace can be gained to the living room and to the dining room.

15. **The correct answer is (A).** The window symbols are visible in the black outline of the floor plan. This shows that there are two windows in the dining room and three windows in bedroom 2.

16. **The correct answer is (D).** Doorways are shown by the break in the solid lines within the floor plan. The kitchen can be entered using the doorway in the dining room as well as the doorway from the foyer.

17. **The correct answer is (D).** The phrase "should be" is the best phrase to complete the sentence.

18. **The correct answer is (E).** The word "was" is the best word to complete the sentence. Because we are talking about yesterday, the past tense is implied.

19. **The correct answer is (D).** The word "is" is the best word to complete the sentence. Because firefighter Curtis is presently up for promotion, the present tense is used.

20. **The correct answer is (C).** The word "quickly" is the best word to complete the sentence.

21. **The correct answer is (B).** The word "whose" is the best word to complete the sentence. "Whose" refers to firefighter McDaniel.

22. **The correct answer is (B).** As stated in the third sentence of the second paragraph, removal of power from the third rail is the FIRST step firefighters must take when evacuating people through tunnels. By removing the power, the chances of people being electrocuted are removed. This also stops the trains from moving so that people walking on the tracks are not hit by trains.

23. **The correct answer is (C).** This is done to inform the trainmaster of the type of emergency and to ensure that he or she keeps the power turned off.

24. **The correct answer is (D).** The five white lights in a vertical cluster can easily be seen in the darkened tunnel and can be recognized as an exit from a distance.

25. **The correct answer is (A).** The fire cannot be in a factory because all factories are located between 9th Avenue and 12th Avenue. Private houses are all between one and three stories tall.

26. **The correct answer is (A).** There are both stores and private houses on 47th Street between 6th Avenue and 7th Avenue, but the private houses are at most three stories high. Therefore, the fire must be in a one- or two-story store.

27. **The correct answer is (D).** At alarms 1, 3, and 5 (all fires below the third floor of the building) Firefighter Sims, in the control position, assisted the driver in attaching the hose to the hydrant. At the fires on the third floor or above, he helped in getting the hoseline into the building. The control person also is responsible for making sure that enough hose has been taken to reach the fire.

28. **The correct answer is (C).** The compass directional in the upper-left corner of the diagram indicates that travel from 7th Avenue to 6th Avenue is travel in the east-west direction. Corridor D is the east-west hallway in the center of the building. Rooms 107, 109, 125, 126, 127, and 128 are in corridor D.

29. **The correct answer is (C).** The firefighters enter the building on 40th Street, which is the south side of the building. Upon entering the lobby, they would have to turn east in Hallway E, north in Hallway B, and west in hallway C.

30. **The correct answer is (D).** From the diagram, we are able to see that Truck 21 is facing north on 7th Avenue. The nearest hose connection is E, about midway between 39th and 40th Streets.

31. **The correct answer is (D).** The description of row frames and brownstones indicates that either can be three stories high and 25 feet wide. The key is in the fire escape clauses. Fire escapes are required on row frame buildings higher than two stories. Rear fire escapes are required in changed brownstones. This cannot be an old law tenement because it is fewer than five stories high.

32. **The correct answer is (B).** As stated in the description of old law tenements, fire escapes are required on the rear of the building if there are two apartments on a floor, and fire escapes are required on both the front and the rear if there are four apartments on a floor. This is done to provide two means of escape from every apartment. Row frames and brownstones do not have front fire escapes.

33. **The correct answer is (C).** If the building is 30 feet wide, it must be a row frame. A row frame of more than two stories must have a rear fire escape.

34. **The correct answer is (C).** This is stated in example 2 of the "MAY DAY" messages. Depending on the extent of the collapse, it is extremely important that an immediate roll call be taken to make sure all firefighters have escaped injury and none are trapped.

35. **The correct answer is (B).** This is explained in example 3 of the "MAY DAY" messages.

36. **The correct answer is (C).** The probable collapse at this point is more serious than the broken wrist. "MAY DAY" should be used instead of "URGENT, URGENT."

37. **The correct answer is (C).** Stairway B is the dark-colored stairway. It enters the fourth floor on the east side. With the fifth-floor outlet not usable, this is the nearest usable outlet below the fire.

38. **The correct answer is (C).** By using stairway A on the east side of the seventh floor, he would come out on the eighth floor on the west side.

39. **The correct answer is (D).** If the front entrance of the building faces south and the front of the building is Exposure 1, then in clockwise order, Exposure 2 would face west; Exposure 3, north; and Exposure 4, east.

40. **The correct answer is (D).** The first paragraph states that ventilation enables heated air, smoke, and gases to be replaced with fresh air.

41. **The correct answer is (C).** The fourth paragraph states that property also can be damaged by the water that firefighters use to extinguish a fire.

42. **The correct answer is (E).** The second paragraph states that removing dark smoke and gases can improve visibility so that firefighters can locate victims faster.

43. **The correct answer is (D).** The third paragraph states that creating a hole in the roof directly above the fire will keep the fire from spreading.

44. **The correct answer is (B).** The advantages of ventilation are listed at the end of the first paragraph. They are advantages in rescue operations, fire extinguishing, and property conservation.

45. **The correct answer is (A).** Hydrants 3 and 4 are both considerably closer to the fire than hydrants 1 and 7. Hydrant 4 cannot be reached without passing through smoke and fire. Hydrant 3 is in the best position for protecting the warehouse from the fire.

46. **The correct answer is (C).** Try each route and see for yourself. This route also gets them out of the complex and away from danger.

47. **The correct answer is (C).** This is indicated on the diagram for the question.

48. **The correct answer is (D).** Of the choices offered, only hydrants 6 and 7 are outside the fence. Hydrant 7 is closer to the fire.

49. **The correct answer is (C).** Hi-Lo Lane is engulfed in smoke and flame. The route in choice (C) is shorter than the route in choice (D).

50. **The correct answer is (A).** Try each route. This is the shortest way.

51. **The correct answer is (B).** If the door is unlocked, it is the simplest and preferred way to enter an apartment. Sometimes, in the anxious moments of arrival at a fire, the firefighter presumes the door to be locked and immediately starts forcible entry. This is a mistake in procedure.

52. **The correct answer is (C).** Because the aerial ladder is in use, the firefighter must go by Step 1b.

53. **The correct answer is (A).** This is proper procedure as stated in Steps 3b and 4. Opening the skylight will immediately ventilate the interior stairs, causing pent-up smoke and gases to escape. This makes it easier for firefighting interior forces to operate.

54. **The correct answer is (B).** This is proper procedure as stated in Step 2. By doing this, you generally will relieve any fears the people inside may have and might even secure their assistance from their position inside.

55. **The correct answer is (C).** In general, a broken arm is not a life-threatening injury. Rather than trying to remove the person by any means possible, firefighters should adhere to the instruction in Step 5 and send for an elevator mechanic who is better equipped to extricate the victim. The firefighters should, however, maintain verbal contact with the operator to make sure he does not go into shock. If the injury situation appears to worsen, they must move directly to Step 6b and take measures before the elevator mechanic arrives.

56. **The correct answer is (A).** This is the second choice. Victims to be rescued generally are in a very nervous state. Use of the fire escape would be more familiar to them and would be less stressful than climbing over the railing and down a portable ladder.

57. **The correct answer is (B).** Room 304 is on the south side of the building. Room 306 is the nearest room to the west of Room 304.

58. **The correct answer is (B).** Stairway B, as shown in the diagram, is the closest. The stairway used by the firefighter to gain access to the third floor was stairway A.

59. **The correct answer is (A).** Room 303 is directly across the hallway from Room 304. Room 403 is directly above Room 303.

60. **The correct answer is (A).** By checking the compass points in the center of the diagram, we know that Rooms 303 and 305 are on the south side of the courtyard. Rooms 323 and 325 are on the opposite side of the courtyard, which is the north side of the building.

61. **The correct answer is (D).** Stairway D is the correct answer. The north-east stair is stairway C. By coming up stairway D, both hoselines can operate from the same direction to extinguish the fire.

62. **The correct answer is (A).** Diagram A is correct because the rotating base is directly in front of the victim at the window. Ladders generally are positioned with both beams on the sill to prevent twisting of the ladder under the weight of people.

63. **The correct answer is (B).** This is stated in Rule 2. Doors should be locked on the stairway side to force people to exit at the ground level in the case of a fire. In buildings over 100 feet in height, people can seek refuge several floors below the fire floor. Doors on the street level are locked from the street side for security purposes.

64. **The correct answer is (A).** The correct answer is stated in Rules 3 and 4. The reentry signs are posted to safely evacuate occupants to floors below the fire without having occupants walk down many flights of stairs in very tall buildings.

65. **The correct answer is (D).** The answer is directly stated in the third sentence. Removing the victim to street level first would only delay the administration of oxygen.

66. **The correct answer is (D).** The question asks about transmittal of alarms, not about receipt. The fourth and fifth sentences give the answer.

67. **The correct answer is (B).** See the first sentence of the first paragraph. The firefighter on housewatch has other duties such as greeting visitors, answering the telephone, and maintaining security, but his or her primary responsibility is to properly receive alarm information.

68. **The correct answer is (C).** Read carefully. The answer is in the next to last sentence.

69. **The correct answer is (C).** See the third sentence.

70. **The correct answer is (C).** Suspect 3 only differs from the witness's description by 10 lbs. in weight. The other suspects all differ on two or more of the descriptive items.

71. **The correct answer is (B).** Suspect 2 differs from the witness's description by 20 lbs. and also has blond rather than brown hair.

72. **The correct answer is (B).** The aerial ladder is at the front of the house, and the portable ladder is at the left. The second company must now ladder right and rear. Ladder company personnel are assigned specific areas to search so that no area of the building is left unsearched.

73. **The correct answer is (A).** Victim 3 is removed first because he is the most seriously exposed to the fire. Victim 2 is easier to remove because she is not disabled. Victim 1 is the last one removed because this removal will be more difficult and will take more time due to the disability.

74. **The correct answer is (A).** A pancake collapse is not capable of providing refuge for trapped victims because it leaves little or no space.

75. **The correct answer is (A) or (B) or (C) or (D).** After realizing that the wording of this question might lead to different interpretations of its meaning, the Department of Personnel agreed to accept any answer. With the reading originally intended, the answer should be choice (A). The definition of a floor collapse requires that there be large pieces of machinery on the floor below that hold up sections of floor and create spaces that provide shelter. Large pieces of machinery on any or every floor are most likely to be found in a factory building. In other buildings, large pieces of machinery are more likely to be found only in the basement. This limits the number of floor collapses possible.

76. **The correct answer is (A).** The third rail carries the electricity that provides power. The contact shoes and metal plates are one and the same. The shoes or plates carry electricity to the motor.

77. **The correct answer is (A).** The answer is in the third sentence.

78. **The correct answer is (D).** See the second paragraph. The reason is that, if the hoseline is taken from the outlet on the fire floor and the fire gains control of the floor and hall, there is the possibility that the line will have to be abandoned. Using the outlet below the fire enables firefighters to work in a smoke-free atmosphere. Should the fire get out into the hallway, they still can maintain control of the water supply.

79. **The correct answer is (B).** Search and rescue, especially with knowledge of trapped victims as in this case, must be the first order of business.

80. **The correct answer is (D).** See the last paragraph. The tool attached to the rope is likely to be an ax, but this is not specified in the passage. The tool attached to the rope gives the firefighter greater reach when breaking windows from above.

81. **The correct answer is (C).** Almost by definition, the roof person would effect a rescue from above the fire. Fourteen-story buildings do not have fire escapes; therefore, rescue from above, except on the highest floors where it might be made from the roof, must be made from an apartment window above the fire.

82. **The correct answer is (C).** The answer is stated in Step 3. Checking the permits first makes you aware of: (1.) whether the permits are up to date, and (2.) the areas in the building that require special attention such as fuel oil storage, refrigeration equipment, and paint storage. The permit will note information such as the amount to be stored, the number and type of refrigerants being used, and the grade and amount of fuel oil being used and stored.

83. **The correct answer is (A).** The lower limit for propane is listed in the table as 2.2%. Therefore, any percentage less than 2.2% will not burn.

84. **The correct answer is (E).** The upper limit for hydrogen is listed in the table as 75%. Therefore, when the percentage of hydrogen to oxygen is more than 75%, the mixture will not burn.

85. **The correct answer is (C).** In order for natural gas to burn, the percentage must be more than 5.0% and less than 15%. 10% is the only percentage listed that is between these two numbers.

86. **The correct answer is (D).** See Step 2a.

87. **The correct answer is (A).** This question was deleted following candidates' appeal.

88. **The correct answer is (B).** The word that is spelled incorrectly is "ventelation." The correct spelling is "ventilation."

89. **The correct answer is (A).** The word that is spelled incorrectly is "suceptiblility." The correct spelling is "susceptibility."

90. **The correct answer is (B).** If you divide 1 by 5 you get .20.

91. **The correct answer is (C).** If all three sections are fully extended, the ladder will reach $18 \times 3 = 54$ feet. However, the ladders are on top of the truck, which is 8 feet tall, so $54 + 8 = 62$ feet.

92. **The correct answer is (C).** For every mile per hour that a fire engine is traveling, it takes $\frac{3}{4}$ of a foot to come to a complete stop. If an engine is traveling 60 mph, the answer is found by multiplying 60 by $\frac{3}{4}$. In other words, $60 \times 3 = 180$; $180 \div 4 = 45$.

93. **The correct answer is (C).** It takes the water 1 second to travel through 5 feet of hose. Therefore, $125 \div 5 = 25$.

94. **The correct answer is (B).** There are 60 minutes in every hour. 5:30 is 1 hour and 10 minutes after 4:20. Therefore, 60 minutes + 10 minutes = 70 minutes.

95. **The correct answer is (B).** If the total number of fires in 1999 was 312, take $312 - (48 + 167)$. First solve what is in the parentheses. You now have $312 - 215 = 97$.

96. **The correct answer is (A).** Car accidents increased from 154 to 186 from 1998 to 2001. First find the difference between these numbers, $186 - 154 = 32$. Because we want to know the percent increase, we want to know what percentage 32 is of 154. As in problem 90, we take $32 \div 154 = .21$. To convert to percent, take $.21 \times 100\% = 21\%$.

97. **The correct answer is (D).** To find the average, $48 + 59 + 43 = 150$. $150 \div 3 = 50$.

98. **The correct answer is (A).** A firefighter assigned to the roof position who sees the roof sagging and in danger of collapse should immediately give a "MAY DAY" signal. This alerts all firefighters operating on the top floor to evacuate the collapse area at once.

99. **The correct answer is (B).** A person bleeding heavily from a stomach wound is in serious condition. If the bleeding is not halted, the victim can go into shock and can die from loss of blood.

100. **The correct answer is (C).** At this fire, the occupant of the fourth floor should be rescued first. The floor directly above the fire is the most seriously exposed floor. Dangerous products of combustion, mainly carbon monoxide, rise rapidly through convection to fill the area, excluding oxygen and causing death or serious injury to trapped people.

# Practice National Firefighter Selection Inventory (NFSI)™

Fire departments need to hire people who are not only smart, but are also able to work with a diverse group of people, tolerate stress effectively, and have a positive attitude about their jobs.

Developed as an alternative to conventional fire service written entrance examinations, the NFSI™ is a test that combines a traditional cognitive abilities component with a critical attitude/personality, or firefighter orientation, component. The combination of these components help to ensure that candidates are not only mentally fit to handle the rigors of the job, but also have a personality that will be conducive to becoming an excellent firefighter.

## PRACTICE NFSI ANSWER SHEET

1. Ⓐ Ⓑ Ⓒ Ⓓ Ⓔ
2. Ⓐ Ⓑ Ⓒ Ⓓ Ⓔ
3. Ⓐ Ⓑ Ⓒ Ⓓ Ⓔ
4. Ⓐ Ⓑ Ⓒ Ⓓ Ⓔ
5. Ⓐ Ⓑ Ⓒ Ⓓ Ⓔ
6. Ⓐ Ⓑ Ⓒ Ⓓ Ⓔ
7. Ⓐ Ⓑ Ⓒ Ⓓ Ⓔ
8. Ⓐ Ⓑ Ⓒ Ⓓ Ⓔ
9. Ⓐ Ⓑ Ⓒ Ⓓ Ⓔ
10. Ⓐ Ⓑ Ⓒ Ⓓ Ⓔ
11. Ⓐ Ⓑ Ⓒ Ⓓ Ⓔ
12. Ⓐ Ⓑ Ⓒ Ⓓ Ⓔ
13. Ⓐ Ⓑ Ⓒ Ⓓ Ⓔ
14. Ⓐ Ⓑ Ⓒ Ⓓ Ⓔ
15. Ⓐ Ⓑ Ⓒ Ⓓ Ⓔ
16. Ⓐ Ⓑ Ⓒ Ⓓ Ⓔ
17. Ⓐ Ⓑ Ⓒ Ⓓ Ⓔ
18. Ⓐ Ⓑ Ⓒ Ⓓ Ⓔ
19. Ⓐ Ⓑ Ⓒ Ⓓ Ⓔ
20. Ⓐ Ⓑ Ⓒ Ⓓ Ⓔ
21. Ⓐ Ⓑ Ⓒ Ⓓ Ⓔ
22. Ⓐ Ⓑ Ⓒ Ⓓ Ⓔ
23. Ⓐ Ⓑ Ⓒ Ⓓ Ⓔ
24. Ⓐ Ⓑ Ⓒ Ⓓ Ⓔ
25. Ⓐ Ⓑ Ⓒ Ⓓ Ⓔ
26. Ⓐ Ⓑ Ⓒ Ⓓ Ⓔ

27. Ⓐ Ⓑ Ⓒ Ⓓ Ⓔ
28. Ⓐ Ⓑ Ⓒ Ⓓ Ⓔ
29. Ⓐ Ⓑ Ⓒ Ⓓ Ⓔ
30. Ⓐ Ⓑ Ⓒ Ⓓ Ⓔ
31. Ⓐ Ⓑ Ⓒ Ⓓ Ⓔ
32. Ⓐ Ⓑ Ⓒ Ⓓ Ⓔ
33. Ⓐ Ⓑ Ⓒ Ⓓ Ⓔ
34. Ⓐ Ⓑ Ⓒ Ⓓ Ⓔ
35. Ⓐ Ⓑ Ⓒ Ⓓ Ⓔ
36. Ⓐ Ⓑ Ⓒ Ⓓ Ⓔ
37. Ⓐ Ⓑ Ⓒ Ⓓ Ⓔ
38. Ⓐ Ⓑ Ⓒ Ⓓ Ⓔ
39. Ⓐ Ⓑ Ⓒ Ⓓ Ⓔ
40. Ⓐ Ⓑ Ⓒ Ⓓ Ⓔ
41. Ⓐ Ⓑ Ⓒ Ⓓ Ⓔ
42. Ⓐ Ⓑ Ⓒ Ⓓ Ⓔ
43. Ⓐ Ⓑ Ⓒ Ⓓ Ⓔ
44. Ⓐ Ⓑ Ⓒ Ⓓ Ⓔ
45. Ⓐ Ⓑ Ⓒ Ⓓ Ⓔ
46. Ⓐ Ⓑ Ⓒ Ⓓ Ⓔ
47. Ⓐ Ⓑ Ⓒ Ⓓ Ⓔ
48. Ⓐ Ⓑ Ⓒ Ⓓ Ⓔ
49. Ⓐ Ⓑ Ⓒ Ⓓ Ⓔ
50. Ⓐ Ⓑ Ⓒ Ⓓ Ⓔ
51. Ⓐ Ⓑ Ⓒ Ⓓ Ⓔ
52. Ⓐ Ⓑ Ⓒ Ⓓ Ⓔ

53. Ⓐ Ⓑ Ⓒ Ⓓ Ⓔ
54. Ⓐ Ⓑ Ⓒ Ⓓ Ⓔ
55. Ⓐ Ⓑ Ⓒ Ⓓ Ⓔ
56. Ⓐ Ⓑ Ⓒ Ⓓ Ⓔ
57. Ⓐ Ⓑ Ⓒ Ⓓ Ⓔ
58. Ⓐ Ⓑ Ⓒ Ⓓ Ⓔ
59. Ⓐ Ⓑ Ⓒ Ⓓ Ⓔ
60. Ⓐ Ⓑ Ⓒ Ⓓ Ⓔ
61. Ⓐ Ⓑ Ⓒ Ⓓ Ⓔ
62. Ⓐ Ⓑ Ⓒ Ⓓ Ⓔ
63. Ⓐ Ⓑ Ⓒ Ⓓ Ⓔ
64. Ⓐ Ⓑ Ⓒ Ⓓ Ⓔ
65. Ⓐ Ⓑ Ⓒ Ⓓ Ⓔ
66. Ⓐ Ⓑ Ⓒ Ⓓ Ⓔ
67. Ⓐ Ⓑ Ⓒ Ⓓ Ⓔ
68. Ⓐ Ⓑ Ⓒ Ⓓ Ⓔ
69. Ⓐ Ⓑ Ⓒ Ⓓ Ⓔ
70. Ⓐ Ⓑ Ⓒ Ⓓ Ⓔ
71. Ⓐ Ⓑ Ⓒ Ⓓ Ⓔ
72. Ⓐ Ⓑ Ⓒ Ⓓ Ⓔ
73. Ⓐ Ⓑ Ⓒ Ⓓ Ⓔ
74. Ⓐ Ⓑ Ⓒ Ⓓ Ⓔ
75. Ⓐ Ⓑ Ⓒ Ⓓ Ⓔ
76. Ⓐ Ⓑ Ⓒ Ⓓ Ⓔ
77. Ⓐ Ⓑ Ⓒ Ⓓ Ⓔ
78. Ⓐ Ⓑ Ⓒ Ⓓ Ⓔ

79. Ⓐ Ⓑ Ⓒ Ⓓ Ⓔ
80. Ⓐ Ⓑ Ⓒ Ⓓ Ⓔ
81. Ⓐ Ⓑ Ⓒ Ⓓ Ⓔ
82. Ⓐ Ⓑ Ⓒ Ⓓ Ⓔ
83. Ⓐ Ⓑ Ⓒ Ⓓ Ⓔ
84. Ⓐ Ⓑ Ⓒ Ⓓ Ⓔ
85. Ⓐ Ⓑ Ⓒ Ⓓ Ⓔ
86. Ⓐ Ⓑ Ⓒ Ⓓ Ⓔ
87. Ⓐ Ⓑ Ⓒ Ⓓ Ⓔ
88. Ⓐ Ⓑ Ⓒ Ⓓ Ⓔ
89. Ⓐ Ⓑ Ⓒ Ⓓ Ⓔ
90. Ⓐ Ⓑ Ⓒ Ⓓ Ⓔ
91. Ⓐ Ⓑ Ⓒ Ⓓ Ⓔ
92. Ⓐ Ⓑ Ⓒ Ⓓ Ⓔ
93. Ⓐ Ⓑ Ⓒ Ⓓ Ⓔ
94. Ⓐ Ⓑ Ⓒ Ⓓ Ⓔ
95. Ⓐ Ⓑ Ⓒ Ⓓ Ⓔ
96. Ⓐ Ⓑ Ⓒ Ⓓ Ⓔ
97. Ⓐ Ⓑ Ⓒ Ⓓ Ⓔ
98. Ⓐ Ⓑ Ⓒ Ⓓ Ⓔ
99. Ⓐ Ⓑ Ⓒ Ⓓ Ⓔ
100. Ⓐ Ⓑ Ⓒ Ⓓ Ⓔ
101. Ⓐ Ⓑ Ⓒ Ⓓ Ⓔ
102. Ⓐ Ⓑ Ⓒ Ⓓ Ⓔ
103. Ⓐ Ⓑ Ⓒ Ⓓ Ⓔ
104. Ⓐ Ⓑ Ⓒ Ⓓ Ⓔ
105. Ⓐ Ⓑ Ⓒ Ⓓ Ⓔ

## PRACTICE NATIONAL FIREFIGHTER SELECTION INVENTORY™

## 105 Questions • 120 Minutes

**Directions:** Answer questions 1–8 by completing the sentence with the word that is spelled correctly. Mark your answers on the answer sheet.

1. The firefighters were ordered to ____ the burning building because a collapse was imminent.

   (A) evackuate
   (B) evakuate
   (C) evacuate
   (D) evvacuate
   (E) evaquate

2. Firefighter Kingsley opened a hole in the roof in order to _____ the building.

   (A) ventillate
   (B) vintellate
   (C) venntilate
   (D) ventilate
   (E) vintelate

3. The 911 staff was frustrated because they were overwhelmed by phone calls that were not related to true _____.

   (A) emeargencies
   (B) emirgincies
   (C) emergencees
   (D) emerjencies
   (E) emergencies

4. The firefighters were called to the high rise to rescue four people who were trapped in an _____.

   (A) ellevator
   (B) elevaytor
   (C) elevater
   (D) elevator
   (E) elivater

5. The department provided all _____ protective garments such as coats, hats, and boots to new employees.

   (A) essential
   (B) esential
   (C) essentul
   (D) esentiul
   (E) essenteal

6. One ladder company and two _____ companies were sent to the scene.

   (A) enjine
   (B) engin
   (C) engene
   (D) enginn
   (E) engine

7. The victim had severe spinal _____ and had to be extricated from the automobile.

   (A) inguries

   (B) injuries

   (C) injurees

   (D) enjuries

   (E) injeries

8. Because the roof is a dangerous place on which to work, no more firefighters were up there than were absolutely _____.

   (A) necessary

   (B) nessessary

   (C) nesessary

   (D) necesary

   (E) nesessairy

**ANSWER QUESTIONS 9–10 BY IDENTIFYING WHICH ONE OF THE UNDERLINED WORDS IS SPELLED INCORRECTLY.**

9. Firefighter Harris exammined the 1-inch-diameter hose, looking for signs of weakening and corrosion.

   (A) exammined

   (B) diameter

   (C) signs

   (D) weakening

   (E) corrosion

10. Firefighter Perez was trained to recognize the characteristics of a hazardus materials spill.

   (A) trained

   (B) recognize

   (C) characteristics

   (D) hazardus

   (E) materials

**ANSWER QUESTIONS 11–25 BY CHOOSING THE APPROPRIATE WORD OR PHRASE TO COMPLETE THE FOLLOWING SENTENCES.**

11. Firefighters are frequently required _____ basic EMT or paramedic certifications.

   (A) obtain

   (B) obtains

   (C) to obtain

   (D) obtaining

   (E) obtained

12. Before John Winters applied to become a firefighter, he _____ as a police officer for five years.

   (A) serves

   (B) have served

   (C) will serve

   (D) been serving

   (E) had served

13. While in the academy, the new fire-fighters learned about the physical signs that _____ the imminent collapse of a building.

    (A) indicates

    (B) indicate

    (C) are indicated

    (D) is indicating

    (E) to indicate

14. Firefighters' helmets _____ their heads from impact and scalding water.

    (A) protect

    (B) protects

    (C) is protected

    (D) was protecting

    (E) protecting

15. The battalion chief wrote a report about the incident and submitted it _____ the chief.

    (A) from

    (B) at

    (C) to

    (D) on

    (E) of

16. Because Firefighter Davis was obviously overheated, the medical staff _____ him to remain in the rest and rehabilitation sector for another half hour.

    (A) order

    (B) orders

    (C) ordering

    (D) was ordered

    (E) ordered

17. Even when reporting to an emergency, the drivers of fire apparatus _____ ensure that they are driving safely.

    (A) must

    (B) might

    (C) may

    (D) could

    (E) can

18. Firefighter Richards particularly enjoys _____ local classrooms to speak to children about fire safety.

    (A) visits

    (B) visit

    (C) visited

    (D) visiting

    (E) to visit

19. After the inspection, the company officer _____ the results with the owner of the building.

    (A) will review

    (B) review

    (C) been reviewing

    (D) were reviewing

    (E) are reviewing

20. Sandra Marszalek lifted weights six days a week to prepare _____ the physical ability test.

    (A) in

    (B) of

    (C) to

    (D) for

    (E) at

21. Firefighter Johnston and _____ are scheduled to work this Saturday.

    **(A)** me

    **(B)** I

    **(C)** his

    **(D)** him

    **(E)** her

22. Of the three firefighers, Firefighter Randall receives the _____ performance evaluations.

    **(A)** worser

    **(B)** worst

    **(C)** baddest

    **(D)** worse

    **(E)** most worse

23. After the oral interview, the board decided _____ the candidate did not have the right attitudes and motivations to be a firefighter.

    **(A)** because

    **(B)** which

    **(C)** although

    **(D)** what

    **(E)** that

24. Firefighter O'Neill _____ to write a memo explaining why he reported to work late that morning.

    **(A)** instructed

    **(B)** instructs

    **(C)** were instructed

    **(D)** are instructed

    **(E)** was instructed

25. The protection of civilian life is the firefighter's _____ priority.

    **(A)** most high

    **(B)** most highest

    **(C)** highest

    **(D)** high

    **(E)** most higher

practice test

**QUESTIONS 26–30 ARE BASED ON THE FOLLOWING PASSAGE:**

Firefighters who arrive at the scene of a fire must observe the visible smoke conditions because the density and the color of the smoke may help to indicate the type of substance that is being burned. The following are the types of substances that are often associated with a certain type and color of smoke:

**Gray-white smoke of little density:** A fire that is just beginning and is consuming wood, cloth, and other ordinary furnishings.

**Blue-white smoke of little density:** A fire that is just beginning and is consuming wood, cloth, and other ordinary furnishings.

**Black smoke:** A fire that is consuming rubber, tar, or other flammable liquids.

**Brown smoke:** An indication of the presence of the oxides of nitrogen fumes.

**Gray-yellow smoke:** A danger signal of approaching backdraft.

Although the above associations may be helpful in sizing up an incident, it is important to remember that the color of smoke may not always reliably indicate the type of materials that are burning. Only a chemical analysis can truly reveal the materials contained in smoke.

26. Firefighters arrive at the scene of a fire at a strip mall and find gray-yellow smoke leaking out of the door. According to the passage, which of the following conditions is this type of smoke MOST likely to indicate?

    (A) Oxides of nitrogen fumes are in abundance.

    (B) The substance that is burning is likely to be tar or rubber.

    (C) The firefighters should beware of a possible backdraft explosion.

    (D) The fire has just begun and is consuming ordinary furnishings.

    (E) The fire has just begun and is consuming cloth.

27. According to the passage, which of the following substances is MOST likely to produce black smoke?

    (A) Wood

    (B) Nitrogen

    (C) Cloth

    (D) Tar

    (E) Backdraft-producing conditions

28. Which of the following is the MOST accurate way to determine the materials that are present in smoke?

    (A) Observing the color of the smoke

    (B) Observing the density of the smoke

    (C) A chemical analysis

    (D) Observing the smell of the smoke

    (E) There is no way to reliably determine the materials present in smoke

29. Your engine company has just arrived at the scene of a fire. The incident commander tells you that there are nitrogen fumes present. Based on the passage, what color is the smoke at the scene MOST likely to be?

(A) Black

(B) Blue-white

(C) Gray-white

(D) Gray-yellow

(E) Brown

30. Gray-white and _____ smoke can both indicate a fire in its early stages that is consuming wood.

(A) Gray-yellow

(B) Blue-white

(C) Black

(D) Brown

(E) Brown-yellow

## QUESTIONS 31–37 ARE BASED ON THE FOLLOWING PASSAGE:

The following paragraphs list some of the characteristics of common types of hose streams:

Booster: This stream is $\frac{1}{2}$ to 1 inch in size and has a volume of 10 to 40 gallons per minute. It can reach a maximum distance of 25 to 50 feet, and only one person is needed on the nozzle. This type of stream has excellent mobility, control of damage, and control of direction. It is commonly used to fight small exterior fires or chimney fires, as its estimated effective area covers only very small fires.

$1\frac{1}{2}$ inch: This stream has a volume of 40 to 125 gallons per minute. It can reach a maximum distance of 25 to 50 feet, and one or two people may be needed on the nozzle. This type of stream has good mobility and control of damage and an excellent control of direction. It is generally used on developing fires or for quick attack. The estimated effective area covers one to three rooms.

2-inch: This stream has a volume of 100 to 250 gallons per minute. It can reach from 40 to 70 feet, and two or three people are needed on the nozzle. This type of stream has fair mobility and control of damage and good control of direction. It is generally used when there are enough crewmembers and water and when the crew's safety dictates. The estimated effective area covers one fully involved floor or more.

$2\frac{1}{2}$-inch: This stream has a volume of 125 to 350 gallons per minute. It can reach a distance of 50 to 100 feet, and two to four people are generally needed on the nozzle. It has fair to poor mobility, fair control of damage, and good control of direction. It is used when larger volumes or greater reach are required for protection from exposure. The estimated effective area covers one fully involved floor or more.

practice test

Master: This stream has a volume of 350 to 2,000 gallons per minute. The maximum reach is 100 to 200 feet. One person is needed on the nozzle. This type of stream has poor mobility and control of damage but good control of direction. It is generally used when larger volumes or greater reach are required for exposure protection, when massive runoff water can be tolerated, and when interior attack can no longer be maintained. The estimated effective area covers a fully involved large structure.

31. According to the passage, which of the following types of streams has the best mobility?

   (A) Booster
   (B) $1\frac{1}{2}$-inch
   (C) 2-inch
   (D) $2\frac{1}{2}$-inch
   (E) Master

32. You arrive at the scene of a fire that involves an entire six-floor office building. The incident commander informs you that an interior attack is no longer feasible. According to the above passage, which of the following streams would be the MOST effective in this situation?

   (A) Booster
   (B) $1\frac{1}{2}$-inch
   (C) 2-inch
   (D) $2\frac{1}{2}$-inch
   (E) Master

33. The estimated effective area for a $1\frac{1}{2}$-inch stream is _____.

   (A) only very small fires
   (B) one to three rooms
   (C) one fully involved floor
   (D) three fully involved floors
   (E) a fully involved large structure

34. Which of the following streams has a volume of 125 to 350 gallons per minute?

   (A) Booster
   (B) $1\frac{1}{2}$-inch
   (C) 2-inch
   (D) $2\frac{1}{2}$-inch
   (E) Master

35. You arrive at the scene of a fire. You need a stream that can reach at least 76 feet. The smallest type of stream that would be suitable in this situation is a _____ stream.

   (A) booster
   (B) $1\frac{1}{2}$-inch
   (C) 2-inch
   (D) $2\frac{1}{2}$-inch
   (E) master

36. What is the maximum number of people required on the nozzle of the 2-inch stream?

   (A) 1
   (B) 2
   (C) 3
   (D) 4
   (E) 5

**37.** According to the passage, which of the following types of streams has the POOREST control of damage?

(A) Booster

(B) $1\frac{1}{2}$-inch

(C) 2-inch

(D) $2\frac{1}{2}$-inch

(E) Master

### QUESTIONS 38–39 ARE BASED ON THE FOLLOWING PASSAGE.:

Mrs. Jean Norris is explaining to Mr. and Mrs. Hancock, her next-door neighbors, about a small fire that started in her living room two days ago. Her story contains the following five sentences:

1. I fell asleep in the middle of the movie.

2. I made some tea and lit some candles that I had received from my daughter for my birthday before turning on the TV.

3. It had been a long day at work, and I planned to relax by watching *My Fair Lady*, which was going to be on cable that night.

4. I sat up and saw that the coffee table was on fire.

5. I heard a loud "pop" and woke up. The candleholders must have been too hot and shattered.

**38.** Which of the following is the most logical order of the above statements?

(A) 1, 3, 5, 4, 2

(B) 5, 3, 2, 4, 1

(C) 2, 1, 3, 4, 5

(D) 3, 2, 1, 5, 4

(E) 2, 1, 3, 4, 5

**39.** What happened immediately AFTER Mrs. Norris turned on the TV?

(A) She made tea.

(B) She saw that the coffee table was on fire.

(C) She heard a loud "pop".

(D) She lit some candles that her daughter had given to her.

(E) She fell asleep during the movie.

*practice test*

**QUESTIONS 40–45 ARE BASED ON THE FOLLOWING PASSAGE:**

Three firefighters, Frederick Wright, Eric Washington, and Nicholas Freeman, were slightly injured while fighting a fire at a local two-story residence at 2:45 a.m. The home belongs to the Lewis family. Mr. Harrison Lewis and his wife, Jennifer, have three children: Ethan (age 17), Michael (age 13), and Nicole (age 9). Mrs. Lewis remembers waking up at 1:55 a.m. to hear a window break and the curtains catching on fire. The smoke alarm went off, the family dog began barking, and the rest of the family was awakened. The living room had caught on fire, and the fire was rapidly spreading throughout the lower level. The Lewises' next-door neighbors, the Rodriguez family, called the fire department. Jennifer, Ethan, and Michael Lewis all exited the home (along with the family dog) according to the escape plan the family had agreed upon earlier. Harrison Lewis and his daughter, Nicole, remained upstairs; Nicole was too frightened to leave her room, and her father remained with her to try to coax her down the stairs. Eventually, the fire blocked their escape, and they had to be rescued by firefighters. After the remaining family members were rescued, firefighters spent the next few hours extinguishing the blaze. It was at this time that Firefighters Wright, Washington, and Freeman were injured. The firefighters (along with Mr. Lewis and his daughter) were transported to St. Mary Magdalene Hospital to be treated for their injuries and/or smoke inhalation.

While looking into a possible cause for the fire, police officers interviewed the family and neighbors. Mr. and Mrs. Pryce, neighbors who live across the street from the Lewis family, reported seeing teenagers sitting in a car outside of the house before the fire. They noted that shortly afterward they heard the car's tires squealing as it drove away. They did not think anything of this fact until police and firefighters were summoned to the scene. Police officers asked Michael and Ethan Lewis if they knew anyone who had a grudge against them. Ethan looked uncomfortable but was unwilling to speak with the authorities.

**40.** Which of the following is MOST likely the cause of the fire at the Lewis home?

(A) Arson

(B) A defective electronic device

(C) An unattended cigarette

(D) An accident caused by the children's playing with matches

(E) An accident caused by the children's leaving a candle burning all night

**41.** Which of the following people were transported to St. Mary Magdalene Hospital to be treated for their injuries?

(A) Michael Lewis

(B) Firefighter Wright

(C) Jennifer Lewis

(D) Nicole Lewis

(E) Both (B) and (D)

**42.** Who summoned the firefighters to the scene?

**(A)** The Rodriguez family

**(B)** Mr. and Mrs. Pryce

**(C)** Jennifer Lewis

**(D)** Ethan Lewis

**(E)** Teenagers outside of the Lewis home

**43.** If you were investigating the cause of this fire, which of the following would you be MOST likely to do next?

**(A)** Inspect the electronic devices in the home, looking for the defective device that started the fire.

**(B)** Try to get the Lewis children to admit that they carelessly started the fire.

**(C)** Speak with Ethan Lewis and the neighbors across the street to learn more about the teenagers who may have been present outside before the fire.

**(D)** Focus on the Rodriguez family as prime arson suspects.

**(E)** Inspect the living room, looking for the cigarette that may have been left unattended.

**44.** Why didn't Harrison and Nicole Lewis initially leave the house according to the family escape plan?

**(A)** Harrison was trying to coax Nicole out of her bedroom because she was too frightened to leave.

**(B)** The fire was located in Nicole's bedroom.

**(C)** The fire blocked Harrison's escape, and Nicole did not want to leave him by himself.

**(D)** They were asleep and did not know that there was a fire in the living room.

**(E)** Nicole did not want to leave the family dog behind.

**45.** When were the firefighters injured?

**(A)** As they tried to rescue Nicole Lewis

**(B)** As they tried to rescue Harrison Lewis

**(C)** As they attempted to extinguish the blaze

**(D)** As they arrived on the scene

**(E)** As they transported Nicole and Harrison Lewis to St. Mary Magdalene Hospital

practice test

**QUESTIONS 46–52 ARE BASED ON THE FOLLOWING PASSAGE:**

Police are looking for Brian McDowell, a 35-year-old Caucasian male, in connection with two arsons that have occurred in the past three months. The first arson took place at the restaurant owned by Mr. McDowell. The restaurant, a diner featuring American cuisine, was not a profitable business, and investigators were immediately suspicious when it burned to the ground at 3:30 a.m. on January 25, 2002. On February 16, 2002, firefighters were summoned to the home of Mr. McDowell's estranged wife, Diane McDowell, to extinguish a blaze that involved the entire home. Firefighters rescued Diane McDowell and her two children, Robert (age 7) and Elissa (age 13); however, Robert suffered serious burns during the fire, and it eventually took several months for him to recover from his injuries. Evidence at the scene of the second fire suggested arson, and Diane McDowell immediately suspected her husband's involvement. She said that he had been suffering from severe depression due to financial and family problems. She said that he tried to contact her on February 14 regarding his custodial rights, but she refused to speak to him. She claimed that this conversation enraged him, and he threatened her with physical harm. When investigators went to Mr. McDowell's apartment to question him about his whereabouts on the night of February 16, he had apparently fled the area. Neighbors reported him loading some possessions into the back of his dark green 1998 Pontiac Grand Prix, and they had not seen him since.

Brian McDowell is 6′1″ tall with light brown hair and hazel eyes. At the time of the arsons, he had a mustache and a goatee, although he may have shaved off his facial hair to evade discovery. He has a tattoo of a shamrock on his right shoulder, a tattoo of a Celtic cross on his left shoulder, and a long surgical scar on his abdomen. Family members believe that he may be en route to Canada as he has several friends there. Because of his history of depression and mental illness, he is considered dangerous. Extreme care should be taken when approaching or confronting Mr. McDowell. Anyone who sees Brian McDowell should immediately contact the local police.

**46.** On what day was Mr. McDowell's restaurant set on fire?

(A) January 2, 2002

(B) January 5, 2002

(C) January 25, 2002

(D) February 14, 2002

(E) February 16, 2002

**47.** Which of the following individuals was/were seriously injured at the fire that took place at Diane McDowell's home?

(A) Brian McDowell

(B) Diane McDowell

(C) Elissa McDowell

(D) Robert McDowell

(E) Both (C) and (D)

48. Which of the following is a description of Brian McDowell?

    **(A)** A 35-year-old Caucasian male who is 6'1" tall and has light brown hair and hazel eyes.

    **(B)** A 25-year-old Caucasian male who is 6'1" tall and has light brown hair and brown eyes.

    **(C)** A 35-year-old Caucasian male who is 6'1" tall and has dark brown hair and green eyes.

    **(D)** A 25-year-old Caucasian male who is 6'5" tall and has dark blonde hair and brown eyes.

    **(E)** A 35-year-old Caucasian male who is 6'2" tall and has dark brown hair and hazel eyes.

49. Brian McDowell has a tattoo of ____ on his right shoulder and a large surgical scar on his _____.

    **(A)** the Boston Celtics logo; right knee

    **(B)** a shamrock; abdomen

    **(C)** a Celtic cross; left shoulder

    **(D)** a shamrock; left knee

    **(E)** a Celtic cross; chest

50. Why did Diane McDowell believe that her estranged husband was involved with the fire at her home?

    **(A)** A neighbor had witnessed him near her home shortly before the fire.

    **(B)** He confessed to her that he had set the fire before he fled.

    **(C)** He had been suffering from severe depression, and they had an argument two days before the fire.

    **(D)** He had threatened the children with physical harm three days before the fire.

    **(E)** He left a letter for the police confessing to his role in the crime before he fled.

51. Which of the following would be the MOST likely reason that Mr. McDowell would choose to set the restaurant on fire?

    **(A)** The restaurant belonged to his estranged wife.

    **(B)** The restaurant was not profitable, and he wanted to collect the insurance money.

    **(C)** The restaurant belonged to an enemy.

    **(D)** He set someone else's restaurant on fire so that the owner could collect the insurance money.

    **(E)** There seems to be no apparent reason that Mr. McDowell would choose to set the restaurant on fire; the arson demonstrates his mental illness.

52. Which of the following suggests that Mr. McDowell fled the area after the fire at his estranged wife's home?

    **(A)** He shaved his mustache and goatee.

    **(B)** He was not at home when the investigators wanted to question him regarding his whereabouts at the time of the second arson.

    **(C)** He has friends in Canada.

    **(D)** He has a history of severe depression.

    **(E)** His neighbors witnessed him loading his possessions in the back of his car.

**QUESTIONS 53–58 ARE BASED ON THE FOLLOWING PASSAGE:**

Betty Randolph is an EMT. She and her partner, Vince Jackson, have been called to the home of Ms. Michelle Martens. Ms. Martens is a divorced mother who lives with her three children (a 16-year-old girl named Cara and two 5-year-old twin boys named Ben and Jason) and her elderly mother, Mrs. Huber. Mrs. Huber has fallen for the third time that afternoon and is initially somewhat disoriented.

The EMTs notice that Ms. Martens seems quite upset—almost angry—about having to call the paramedics. The EMTs pick up Mrs. Huber, who was sitting on the kitchen floor, and place her in a chair. They examine her and ask her questions about her daily habits and current condition. Mrs. Huber seems hesitant to answer any of these questions. Betty Randolph notices small and medium-sized bruises on Mrs. Huber's arms and legs and asks Mrs. Huber about them. Mrs. Huber first attempts to change the subject and then tells the EMTs to ask her daughter, who is standing behind her mother. Ms. Martens says that her mother must have received the bruises during her recent falls. The EMTs are not entirely convinced by Ms. Martens's explanation. Ms. Martens seems eager to have the EMTs leave. She thanks them for picking up her mother but notifies them that she has to take the children to their various after-school activities.

**53.** Which of the following is MOST likely to be Mrs. Huber's problem?

- **(A)** Mrs. Huber is being abused by her daughter.
- **(B)** Mrs. Huber does not trust doctors or medical personnel.
- **(C)** Mrs. Huber is overwhelmed with the responsibility of caring for her daughter and grandchildren.
- **(D)** Mrs. Huber is not receiving proper nutrition.
- **(E)** Mrs. Huber has financial problems and is worried about having to pay for medical expenses.

**54.** If you were one of the EMTs, which of the following steps should you take NEXT?

- **(A)** Return at a later time in order to interview Ms. Martens's children.
- **(B)** Ensure that Mrs. Huber receives appropriate medical care for her injuries.
- **(C)** Notify your supervisor about your suspicions regarding what is taking place in Ms. Martens's home.
- **(D)** Try to educate Mrs. Huber about the importance of proper nutrition and diet.
- **(E)** Offer Mrs. Huber information about financial resources that may be helpful for her in her situation.

55. Which of the following is the MOST likely reason that the EMTs are unconvinced by Ms. Martens's explanation of her mother's bruises?

   (A) Ms. Marten appears to be too busy with her children to know the true cause of her mother's bruises.

   (B) Ms. Marten does not have enough medical knowledge to know the reason why her mother has so many bruises.

   (C) Ms. Marten appears to have a hostile and dismissive attitude toward the EMTs' concerns.

   (D) Ms. Martens's mother appears to be fearful of answering the EMTs' questions while her daughter is in the room.

   (E) Both (C) and (D)

56. The EMTs arrive at Ms. Martens's home after Mrs. Huber falls for the _____ time.

   (A) first
   (B) second
   (C) third
   (D) fourth
   (E) fifth

57. Where do the paramedics find Mrs. Huber when they arrive?

   (A) In the bathtub
   (B) In a chair in the living room
   (C) On the dining room floor
   (D) On the kitchen floor
   (E) In her bed

58. How many daughters does Ms. Martens have?

   (A) One
   (B) Two
   (C) Three
   (D) Four
   (E) None

practice test

## QUESTIONS 59–64 ARE BASED ON THE FOLLOWING PASSAGE:

The following is a list of some common medications, their actions and side effects, the ways in which the medications are given to patients, and the conditions under which the medications should not be administered.

**Oxygen:** Administering oxygen to a patient increases oxygen pressure and the content of oxygen in the blood. However, a high concentration of oxygen may decrease respiration in patients with chronic obstructive pulmonary disease. Patients inhale oxygen through either a mask or nasal device.

**Glucose:** This medication raises patients' blood glucose levels. However, it can possibly trigger hypoglycemia. An oral paste is rubbed inside the cheeks and mouth and then swallowed; however, unresponsive patients should not be given medication orally.

**Epinephrine:** Epinephrine increases blood pressure by constricting arteries and opens the lung's airways. Side effects include an increased heart rate, arrhythmias, tremor, and hypertension. Epinephrine is commonly administered via injection. However, EMTs who administer epinephrine should note that there is an increased risk of injury in patients with stroke, heart disease, or hypertension.

**Nitroglycerin:** This drug dilates large veins and decreases the workload of the heart. In addition, nitroglycerin dilates large coronary arteries and helps to relieve angina pain. Side effects include hypotension, an increased heart rate, and headache. Nitroglycerin tablets are commonly placed under the tongue or administered via oral spray. This drug should not be given to patients who have an allergy to nitrates, hypotension, or increased pressure within the skull.

**Activated charcoal:** This medication absorbs many poisons from the stomach and intestine and binds them to the surface of the charcoal. If the activated charcoal enters the lungs, it can cause severe pneumonitis. The charcoal will also deactivate any ipecac that has been administered. Activated charcoal is administered orally, but it should never be given to anyone who is semiresponsive, unresponsive, or convulsing.

**59.** You are an EMT, and you respond to a call regarding a child who has swallowed a toxic cleaning product. Which of the following medications would you be MOST likely to administer in this situation?

(A) Oxygen

(B) Glucose

(C) Activated charcoal

(D) Epinephrine

(E) Nitroglycerin

**60.** In order to alleviate some of the symptoms of angina in your patient, you administer nitroglycerin. Which of the following side effects would the patient be MOST likely to experience from this medication?

(A) Severe pneumonitis

(B) Hyperglycemia

(C) Decreased respiratory drive

(D) Headache

(E) None of the above

**61.** Which of the following patients would be MOST likely to receive a dose of glucose?

(A) Patient A, who has a low oxygen content in his blood.

(B) Patient B, whose heart is overworked.

(C) Patient C, whose airways are closed.

(D) Patient D, who has a large concentration of toxins in her small intestine.

(E) Patient E, whose blood sugar levels are abnormally low.

**62.** According to the passage, how is epinephrine administered?

(A) Tablets placed under the tongue

(B) Injection

(C) Oral paste rubbed inside the cheeks

(D) Inhaled via mask or nasal device

(E) Oral spray

**63.** Based on the information provided in the passage, under which of the following circumstances should you NOT administer glucose via an oral paste?

(A) When the patient has increased pressure inside the skull

(B) When the patient has heart disease

(C) When the patient is allergic to nitrates

(D) When the patient is unresponsive

(E) When the patient has hypertension

**64.** According to the passage, _____ is a common side effect of oxygen.

(A) decreased respiratory drive in patients with chronic obstructive pulmonary disease

(B) hypotension

(C) arrhythmias

(D) tremor

(E) hyperglycemia

practice test

## QUESTIONS 65–68 ARE BASED ON THE FOLLOWING PASSAGE:

A nasogastric tube (NG tube) is a device used to decompress the stomach and bowels, administer medication, provide nutrition, and flush and empty the stomach (a process known as gastric lavage). NG tubes should never be placed inside the nose when a patient has experienced facial and skull fractures. In this case, the NG tube should be placed through the mouth. Nasogastric tubes are available in several sizes. The following sizes are appropriate for the following age groups:

**Adult/Adolescent:** 14–16

**School-Age Child:** 12

**Preschool-Age Child/Toddler:** 10

**Infant/Newborn:** 8

65. A 10-year-old child is involved in a car accident and is severely injured. In order to ease artificial ventilation efforts, paramedics will use a nasogastric tube in order to decompress the child's stomach. According to the information provided, which size of nasogastric tube should be used in this situation?

(A) 16

(B) 14

(C) 12

(D) 10

(E) 8

66. For which of the following patients would it be appropriate to use a size-10 nasogastric tube?

(A) A 3-year-old girl

(B) A 24-year-old woman

(C) A 6-month-old boy

(D) A 13-year-old boy

(E) A 10-year-old boy

67. Under which of the following circumstances should an NG tube be placed in a patient's mouth?

(A) When the NG tube is used to administer medication

(B) When the patient's stomach must be decompressed

(C) When the patient has facial or skull fractures

(D) When the patient's stomach must be flushed

(E) When the patient is unresponsive

68. "Gastric lavage" is the process of _____.

(A) using an NG tube to decompress the bowels

(B) using an NG tube to provide nutrition

(C) using an NG tube to administer medication

(D) placing an NG tube in a patient's mouth instead of in his/her nose

(E) using an NG tube to flush and empty a patient's stomach

**QUESTIONS 69–73 ARE BASED ON THE FOLLOWING PASSAGE:**

Firefighter Linda Rollins has been out of the academy for six months. One night, her immediate supervisor asked her to get a drink with him after work. She refused, primarily because her supervisor was married. She noticed shortly afterward that even though she believed that she was working as hard as ever, she began to receive poor performance evaluations, and she often felt that she was singled out for criticism in front of other firefighters. A friend who also works at the department told her that he hadn't noticed her being treated differently from anyone else and pointed out instances in which he had been mildly reprimanded in front of others. Firefighter Rollins is not sure what, if anything, should be done about this problem.

**69.** Which of the following is MOST likely to be Firefighter Rollins's problem?

(A) She is too sensitive to criticism.

(B) She is being penalized because she rejected her supervisor's advances.

(C) She is not so competent a firefighter as her coworkers.

(D) She is receiving poor evaluations because she does not have enough firefighting experience.

(E) Both (A) and (D)

**70.** If you were Firefighter Rollins, which of the following steps would you take NEXT?

(A) Try to ignore the criticism and poor performance evaluations.

(B) Try to be friendlier toward your supervisor.

(C) Accept responsibility for your own poor performance.

(D) Express your concerns to your supervisor's immediate supervisor.

(E) Look for a position in another department.

**71.** How long has Firefighter Rollins been out of the academy?

(A) Six weeks

(B) Three months

(C) Six months

(D) One year

(E) Six years

**72.** Why did Firefighter Rollins primarily refuse to get a drink with her supervisor?

(A) He had given her a poor performance evaluation.

(B) He criticized her often in front of her peers.

(C) She did not feel it was appropriate to get a drink with a supervisor.

(D) He was married.

(E) She did not want the other firefighters to think that she received special privileges.

**73.** How did her friend in the department react to Firefighter Rollins's complaints?

**(A)** He agreed that she was often treated unfairly.

**(B)** He offered to write a letter to the deputy chief on her behalf.

**(C)** He offered to talk to her supervisor for her.

**(D)** He dismissed her concerns and pointed out instances when he had been similarly criticized.

**(E)** He accused her of lying.

### QUESTIONS 74–75 ARE BASED ON THE FOLLOWING PASSAGE:

At 1:22 a.m. on a Thursday evening, a fire alarm was activated in a college dormitory. Resident assistants had to ensure that each and every resident on their floors evacuated the building, even though most students were convinced that the fire alarm was merely an intoxicated student's prank. The students stood outside of the dorm; the weather was cold that night, and many students were improperly dressed for the weather. They were irritated because final exams were about to begin the following week, and many students had been trying to rest and study.

Firefighters were summoned to the dorm. Before the students were allowed indoors, firefighters had to inspect each floor for signs of a fire. Firefighters found a small fire inside a student's room on the tenth floor. The student had left a candle burning unattended while she visited a friend on the fourth floor. Her curtains, which were too close to the candle, caught on fire. The fire spread to her bedspread. A neighbor who smelled the smoke coming from the student's room activated the fire alarm. The firefighters easily extinguished the flames. The students were allowed to return to their rooms. The student who had accidentally caused the fire was distraught to find her room damaged. A university disciplinary committee later decided that she would have to pay for the damages to the room.

74. Why had the fire alarm been activated?

(A) Intoxicated students were playing a prank.

(B) An unattended candle had set a student's clothes and carpet on fire.

(C) The fire alarm was accidentally activated.

(D) A student who had mistakenly believed there was a fire on his floor activated the fire alarm.

(E) An unattended candle had set a student's curtains and bedspread on fire.

75. What decision did the university disciplinary committee make regarding this incident?

(A) The student who had purposely caused a false alarm would be expelled from the dormitory.

(B) The student who had left a candle unattended in her room would be suspended.

(C) The student who had left a candle unattended in her room would have to pay for the damages to the room.

(D) No one would be disciplined because the incident was clearly an accident.

(E) No one would be disciplined because committee members could not prove which students caused the false alarm.

76. You read a report that states that 28 percent of elderly people suffer from some kind of hearing loss. According to the report, if there are 3,500 elderly residents in a small community, how many of these residents will suffer from some kind of hearing loss?

(A) 28

(B) 98

(C) 980

(D) 1,120

(E) 1,400

*practice test*

**QUESTIONS 77–80 ARE BASED ON THE FOLLOWING TABLE:**

**Estimated Number of Arsons per year from 1995–2000**

| Year | Number of Arsons |
|------|------------------|
| 1995 | 2,634 |
| 1996 | 3,332 |
| 1997 | 2,735 |
| 1998 | 1,302 |
| 1999 | 2,988 |
| 2000 | 3,219 |

**77.** The total number of arsons that took place from 1995–2000 is _____.

**(A)** 10,620

**(B)** 16,210

**(C)** 17,512

**(D)** 18,844

**(E)** 26,160

**78.** What is the average number of arsons per year from 1995–2000?

**(A)** 1,770

**(B)** 2,702

**(C)** 2,842

**(D)** 3,140

**(E)** 3,242

**79.** How many more arsons occurred in 1999 than in 1998?

**(A)** 1,686

**(B)** 1,299

**(C)** 698

**(D)** 231

**(E)** 168

**80.** If there was a 3 percent decrease in arsons from 2000–2001, how many arsons took place in 2001?

**(A)** 3,315

**(B)** 3,221

**(C)** 3,216

**(D)** 3,122

**(E)** 2,642

**QUESTIONS 81–85 ARE BASED ON THE FOLLOWING TABLE AND PASSAGE:**

A homeowner who recently lost many of her possessions in a fire is trying to estimate the value of the lost items. Below is a list of some of the lost items and their approximated values.

| Item | Approximate Value |
| --- | --- |
| Computer | $3,000 |
| Wardrobe | $2,200 |
| Jewelry | ? |
| Stereo system | $1,500 |
| Oriental rug | $500 |

81. If the total value of the lost items was equal to $10,378, what was the approximate value of the lost jewelry?

(A) $6,178

(B) $6,678

(C) $3,178

(D) $2,678

(E) $1,178

82. If the jewelry were worth $\frac{1}{3}$ of the value of the wardrobe, how much would the jewelry be worth?

(A) $763.66

(B) $733.33

(C) $933.33

(D) $1,113.13

(E) $6,600.00

83. As she replaces the items in her home, the homeowner buys a dining room table that is four times as expensive as the Oriental rug. How much does the dining room table cost?

(A) $1,000

(B) $1,500

(C) $2,000

(D) $2,500

(E) $3,000

84. If the homeowner were to buy a stereo system that is 6 percent more expensive than the original, how much would it cost?

(A) $900

(B) $1,410

(C) $1,590

(D) $2,400

(E) $3,300

85. Assuming that the jewelry was worth $1,600, what is the average value of the lost items?

    (A) $880
    (B) $1,760
    (C) $2,200
    (D) $2,760
    (E) $3,000

86. Joe Davis is applying to become a firefighter. To prepare himself for the physical ability test, Joe runs every morning. If Joe can run 1.5 miles in 9 minutes, how long would it take him to run 5 miles, assuming he runs at a steady rate?

    (A) 15 minutes
    (B) 25 minutes
    (C) 30 minutes
    (D) 45 minutes
    (E) 60 minutes

87. Amounts of heat transfer are measured in terms of British thermal units (Btu) or in joules. One British thermal unit is equal to 1,055 joules. Seven Btu would then be equal to _____ joules.

    (A) 150
    (B) 989
    (C) 6,330
    (D) 7,385
    (E) 8,440

88. The square root of 121 is _____.

    (A) 9
    (B) 10
    (C) 11
    (D) 12
    (E) 13

89. In order to ventilate a room, Firefighter Moore breaks a window that is 6 feet in length and 4 feet in width. What is the total area of the window in square feet?

    (A) 10 sq. ft.
    (B) 12 sq. ft.
    (C) 14 sq. ft.
    (D) 24 sq. ft.
    (E) 32 sq. ft.

90. Firefighters have attached three lengths of hose together to fight a fire. The first length is 200 feet. The second length is 150 feet. If the total length of the attached hoses is 575 feet, how long is the third length of hose?

    (A) 100 feet
    (B) 150 feet
    (C) 175 feet
    (D) 200 feet
    (E) 225 feet

91. For a training session, a group of 623 firefighters are broken up into groups of 6. How many groups will there be?

    (A) 102 groups of 6 and one group of 4
    (B) 103 groups of 6 and one group of 5
    (C) 104 groups of 6 and one group of 3
    (D) 105 groups of 6
    (E) 106 groups of 6

**92.** A firefighter has to climb from the ground floor to the fifth floor of an eight-story building using the stairs. Each set of stairs has eighteen steps. How many total steps does the firefighter have to climb to reach the fifth floor?

**(A)** 144

**(B)** 108

**(C)** 90

**(D)** 88

**(E)** 72

**93.** 0.2 is equivalent to what fraction?

**(A)** $\dfrac{1}{2}$

**(B)** $\dfrac{1}{3}$

**(C)** $\dfrac{1}{4}$

**(D)** $\dfrac{1}{5}$

**(E)** $\dfrac{1}{6}$

**94.** The total weight load on a roof ladder must not exceed 750 pounds. Two firefighters are currently on a roof ladder while fighting a fire. Firefighter Colby weighs 185 pounds and is currently carrying 50 pounds of equipment. Firefighter Dunn weighs 220 pounds and is currently carrying 50 pounds of equipment. How much more weight can the roof ladder withstand?

**(A)** 505 pounds

**(B)** 345 pounds

**(C)** 245 pounds

**(D)** 195 pounds

**(E)** 145 pounds

**95.** Since instituting a new training program in 1999, there were 28 percent fewer on-the-job injuries in 2000. If there were 98 on-the-job injuries in 1999, approximately how many injuries took place in 2000?

**(A)** 71

**(B)** 64

**(C)** 58

**(D)** 39

**(E)** 27

**96.** Polyester rope begins to lose strength at 300°F. If the temperature of a room is 86°F, how many times hotter must the room become before the polyester rope begins to lose strength?

**(A)** 2.6

**(B)** 3.5

**(C)** 4.1

**(D)** 5.9

**(E)** 6.8

**97.** If $12x + 64 = 325$, then $x$ is equal to _____.

**(A)** 32.42

**(B)** 31.64

**(C)** 28.92

**(D)** 21.75

**(E)** 15.68

**QUESTIONS 98–103 ARE BASED ON THE THE FOLLOWING TABLE:**

The following is a list of firefighters and their respective heights and weights.

| Name | Height | Weight |
|------|--------|--------|
| Firefighter Kurz | 6'2" | 168 pounds |
| Firefighter Johansen | 5'7" | 140 pounds |
| Firefighter Lao | 5'7" | 145 pounds |
| Firefighter Nagle | 6'5" | 200 pounds |
| Firefighter Lonergan | 5'11" | 185 pounds |

**98.** What is the average height of the firefighters listed above (in feet and inches)?

(A) 5'7"

(B) 5'10"

(C) 5'11"

(D) 6'1"

(E) 6'5"

**99.** Firefighter Nagle is _____ taller than Firefighter Lao.

(A) 1 foot 3 inches

(B) 1 foot

(C) 11 inches

(D) 10 inches

(E) 7 inches

**100.** If the second-tallest firefighter were to gain 6 pounds each year for the next three years, how much would he or she weigh at the end of the third year?

(A) 158 pounds

(B) 163 pounds

(C) 186 pounds

(D) 203 pounds

(E) 218 pounds

**101.** The average weight of the firefighters listed is _____ pounds.

(A) 155.7

(B) 167.6

(C) 172.9

(D) 184.4

(E) 198.4

**102.** If each of the firefighters listed gained 10 pounds, how much would the tallest firefighter weigh?

(A) 210 pounds

(B) 195 pounds

(C) 178 pounds

(D) 150 pounds

(E) 155 pounds

**103.** Firefighter Kurz is _____ pounds heavier than Firefighter Lao.

(A) 10

(B) 15

(C) 23

(D) 28

(E) 32

**104.** A group of firefighters at one department take an anonymous survey that is being conducted by the city's human resources department. On this survey, two out of every three firefighters claims that they use less than their allotted sick days per year. This means that _____ percent of the department's firefighters do not use all of their sick days every year.

(A) 33

(B) 50

(C) 66

(D) 75

(E) 90

**105.** What is the answer to the equation $D(A + B) - \sqrt{C}$ when $D = 6$, $A = 4$, $B = 8$, $C = 81$?

(A) 104

(B) 90

(C) 84

(D) 71

(E) 63

## ANSWERS AND EXPLANATIONS

| | | | | | | | | | |
|---|---|---|---|---|---|---|---|---|---|
| 1. | C | 22. | B | 43. | C | 64. | A | 85. | B |
| 2. | D | 23. | E | 44. | A | 65. | C | 86. | C |
| 3. | E | 24. | E | 45. | C | 66. | A | 87. | D |
| 4. | D | 25. | C | 46. | C | 67. | C | 88. | C |
| 5. | A | 26. | C | 47. | D | 68. | E | 89. | D |
| 6. | E | 27. | D | 48. | A | 69. | B | 90. | E |
| 7. | B | 28. | C | 49. | B | 70. | D | 91. | B |
| 8. | A | 29. | E | 50. | C | 71. | C | 92. | C |
| 9. | A | 30. | B | 51. | B | 72. | D | 93. | D |
| 10. | D | 31. | A | 52. | E | 73. | D | 94. | C |
| 11. | C | 32. | E | 53. | A | 74. | E | 95. | A |
| 12. | E | 33. | B | 54. | B | 75. | C | 96. | B |
| 13. | B | 34. | D | 55. | E | 76. | C | 97. | D |
| 14. | A | 35. | D | 56. | C | 77. | B | 98. | C |
| 15. | C | 36. | C | 57. | D | 78. | B | 99. | D |
| 16. | E | 37. | E | 58. | A | 79. | A | 100. | C |
| 17. | A | 38. | D | 59. | C | 80. | D | 101. | B |
| 18. | D | 39. | E | 60. | D | 81. | C | 102. | A |
| 19. | A | 40. | A | 61. | E | 82. | B | 103. | C |
| 20. | D | 41. | E | 62. | B | 83. | C | 104. | C |
| 21. | B | 42. | A | 63. | D | 84. | C | 105. | E |

1. **The correct answer is (C),** "evacuate."

2. **The correct answer is (D),** "ventilate."

3. **The correct answer is (E),** "emergencies."

4. **The correct answer is (D),** "elevator."

5. **The correct answer is (A),** "essential."

6. **The correct answer is (E),** "engine."

7. **The correct answer is (B),** "injuries."

8. **The correct answer is (A),** "necessary."

9. **The correct answer is (A),** "examined."

10. **The correct answer is (D),** "hazardous."

11. **The correct answer is (C).** The phrase "to obtain" is the best phrase to complete this sentence.

12. **The correct answer is (E).** The phrase "had served" is the best phrase to complete this sentence. Note that the use of the past perfect tense is required. John Winters had worked as a police officer *before* he applied to become a firefighter. A singular auxiliary verb (or "helping verb"), "had," is used with the past participle "served" in this case. You would not say that *John Winters have served as a police officer* because John Winters is only one person. Therefore, you would not use a plural auxiliary verb in this instance.

13. **The correct answer is (B).** The word "indicate" is the best word to complete this sentence. The plural noun "signs" requires a plural verb, "indicate." The sentence also requires a verb in the present tense and active voice.

14. **The correct answer is (A).** The word "protect" is the best word to complete this sentence. The sentence requires a verb in the present tense and active voice. The plural noun "helmets" also requires a plural verb, "protect."

15. **The correct answer is (C).** The preposition "to" is the most appropriate preposition to use in this sentence.

16. **The correct answer is (E).** The word "ordered" is the best word to complete this sentence. The sentence requires the use of a verb in the simple past tense and the active voice.

17. **The correct answer is (A).** The word "must" is the best word to complete this sentence. Because driving safely is a requirement, "must" is the most appropriate word to indicate the necessity of driving safely.

18. **The correct answer is (D).** The word "visiting" is the best word to complete this sentence.

19. **The correct answer is (A).** The phrase "will review" is the best phrase to complete this sentence. The sentence requires the use of a singular verb in the future tense.

20. **The correct answer is (D).** The word "for" is the most appropriate preposition to complete this sentence.

21. **The correct answer is (B).** The word "I" is the most appropriate preposition to complete this sentence. The sentence requires a pronoun in the nominative case.

22. **The correct answer is (B).** The word "worst" is the best word to complete this sentence. The appropriate way to indicate the superlative form of "bad" is to use "worst."

23. **The correct answer is (E).** The word "that" is the best word to complete this sentence. "That" is the most appropriate conjunction to introduce the

clause "the candidate did not have the right motivations and attitudes to be a firefighter."

24. **The correct answer is (E).** The phrase "was instructed" is the best phrase to complete this sentence. Firefighter O'Neill was ordered *by someone else* to write the memo. Although Firefighter O'Neill is the subject of the sentence, he did not initiate the action "to instruct"; therefore, the sentence requires the use of a verb in the passive voice. Because Firefighter O'Neill is a singular subject, a singular auxiliary verb, "was," is required in this sentence.

25. **The correct answer is (C).** The word "highest" is the most appropriate word to use in this sentence. The correct way to write the superlative form of "high" is "highest."

26. **The correct answer is (C).** As stated in the sixth paragraph, gray-yellow smoke is a sign of backdraft conditions.

27. **The correct answer is (D).** As stated in the fourth paragraph, burning tar could produce black smoke.

28. **The correct answer is (C).** The last paragraph of the accompanying passage indicates that a chemical analysis is the most accurate way to determine the type of substance that is being consumed in a fire.

29. **The correct answer is (E).** The fifth paragraph of the accompanying passage indicates that the presence of the oxides of nitrogen fumes can produce brown smoke.

30. **The correct answer is (B).** The second and third paragraphs of the accompanying passage reveal that both gray-white and blue-white smoke can indicate a fire in its early stages that is consuming wood.

31. **The correct answer is (A).** The second paragraph of the accompanying passage reveals that booster streams have the best mobility.

32. **The correct answer is (E).** The last paragraph of the accompanying passage reveals that a master stream would be most appropriate to use in the situation presented in the question.

33. **The correct answer is (B).** The third paragraph of the accompanying passage reveals that the estimated effective area for a $1\frac{1}{2}$-inch stream is one to three rooms.

34. **The correct answer is (D).** The fifth paragraph of the accompanying passage states that the $2\frac{1}{2}$-inch stream has a volume of 125–350 gallons per minute.

35. **The correct answer is (D).** The fifth paragraph of the accompanying passage states that the $2\frac{1}{2}$-inch stream can reach 50–100 feet and is thus the smallest stream that can reach the required distance of 76 feet.

36. **The correct answer is (C).** The fourth paragraph of the accompanying passage reveals that three is the maximum number of people needed on the nozzle of the 2-inch stream.

37. **The correct answer is (E).** The last paragraph explains that the master stream has the poorest control of damage.

38. **The correct answer is (D).** The passage should read as follows: It had been a long day at work, and I planned to relax by watching *My Fair Lady*, which was going to be on cable that night. I made some tea and lit some candles that I had received from my daughter for my birthday before turning on the TV. I fell asleep in the middle of the movie. I heard a loud 'pop' and woke up. The candleholders must have been too hot and shattered. I sat up and saw that the coffee table was on fire.

39. **The correct answer is (E).** According to the most logical arrangement of the sentences, after Mrs. Norris turned on the television, she fell asleep during the movie.

40. **The correct answer is (A).** The presence of the teenagers, the broken window, and the evident discomfort of the Lewis' oldest son, arc all indicators that arson was the cause of the fire.

41. **The correct answer is (E).** According to the first paragraph, both Firefighter Wright and Nicole Lewis were transported to the hospital.

42. **The correct answer is (A).** According to the first paragraph, the Rodriguez family summoned firefighters to the scene.

43. **The correct answer is (C).** Because there is evidence of arson and because there were reports of teenagers acting suspiciously before the fire and because Ethan Lewis appears to know more than he admits, interviewing the neighbors across the street and Ethan Lewis about the teenagers would be the most sensible way to pursue an investigation of this case.

44. **The correct answer is (A).** According to the first paragraph, Harrison and Nicole Lewis were unable to escape according to the preplanned family escape plan because Nicole Lewis was too frightened to leave her bedroom and because Harrison Lewis was attempting to persuade her to escape.

45. **The correct answer is (C).** According to the first paragraph, the firefighters were injured as they attempted to extinguish the blaze.

46. **The correct answer is (C).** According to the first paragraph, Mr. McDowell's restaurant was set on fire on January 25, 2002.

47. **The correct answer is (D).** According to the first paragraph, Robert McDowell was seriously injured at the fire that took place in Diane McDowell's home.

answers

48. **The correct answer is (A).** According to the second paragraph, Brian Mc-Dowell is a 35-year-old Caucasian male who is 6'1" tall and has light brown hair and hazel eyes.

49. **The correct answer is (B).** According to the second paragraph, Brian Mc-Dowell has a tattoo of a shamrock on his right shoulder and a large surgical scar on his abdomen.

50. **The correct answer is (C).** According to the first paragraph, Diane Mc-Dowell suspected her husband's involvement in the fire at her home due to his depression and their recent disputes.

51. **The correct answer is (B).** According to the first paragraph, Brian Mc-Dowell's restaurant was not profitable. The most logical reason that the restaurant was set on fire was that Brian McDowell wanted to collect the insurance money.

52. **The correct answer is (E).** Of the choices available, the strongest piece of evidence that Brian McDowell fled the area is that neighbors witnessed him loading his possessions into his car.

53. **The correct answer is (A).** Because of the attitudes displayed by Mrs. Huber and her daughter and because of the bruises found on Mrs. Huber's limbs, the most logical conclusion is that Mrs. Huber is being abused by her daughter.

54. **The correct answer is (B).** The EMTs should immediately ensure that Mrs. Huber's injuries are cared for before taking any further steps.

55. **The correct answer is (E).** The EMTs likely find the situation to be suspicious because Mrs. Huber seems fearful and because Ms. Martens appears hostile.

56. **The correct answer is (C).** According to the first paragraph, the EMTs arrive at Ms. Martens's home after Mrs. Huber falls for the third time.

57. **The correct answer is (D).** According to the second paragraph, the EMTs find Mrs. Huber on the kitchen floor.

58. **The correct answer is (A).** According to the first paragraph, Ms. Martens has one daughter.

59. **The correct answer is (C).** According to the last paragraph, activated charcoal would be the most appropriate medicine to administer in the case of the ingestion of a toxic substance.

60. **The correct answer is (D).** According to the fifth paragraph, patients taking nitroglycerin would be most likely to experience headaches as a side effect.

**answers**

61. **The correct answer is (E).** According to the third paragraph, glucose is administered to those patients with low blood-sugar levels.

62. **The correct answer is (B).** According to the fourth paragraph, epinephrine is administered via injection.

63. **The correct answer is (D).** According to the third paragraph, glucose should not be administered to an unresponsive patient via an oral paste.

64. **The correct answer is (A).** According to the second paragraph, a decreased respiratory drive in patients with chronic obstructive pulmonary disease can be a side effect of oxygen.

65. **The correct answer is (C).** Because a size-12 NG tube is appropriate for school-age children, a size-12 NG tube should be used on a 10-year-old child.

66. **The correct answer is (A).** Size-10 NG tubes are suitable for preschool-age children or toddlers. Therefore, a size-10 NG tube would be most appropriately given to a 3-year-old girl.

67. **The correct answer is (C).** According to the passage, NG tubes should be placed in the mouth when the patient has facial or skull fractures.

68. **The correct answer is (E).** According to the passage, gastric lavage is the process of using an NG tube to flush and empty a patient's stomach.

69. **The correct answer is (B).** Because Firefighter Rollins's problems seemed to begin after she had rejected her supervisor's advances, the most likely problem is that Firefighter Rollins is being penalized for refusing her supervisor's invitation.

70. **The correct answer is (D).** The most productive way to deal with discrimination on the job is to try to work with the established chain of command to find a solution. Therefore, the best solution would be to speak with the immediate supervisor of the offending supervisor.

71. **The correct answer is (C).** According to the passage, Firefighter Rollins has been out of the academy for six months.

72. **The correct answer is (D).** According to the passage, the primary reason that Firefighter Rollins refused to get a drink with her supervisor was that he was married.

73. **The correct answer is (D).** According to the passage, her friend pointed out that he had not noticed her being treated differently from anyone else and noted occasions when he too faced criticism in front of his coworkers.

74. **The correct answer is (E).** According to the second paragraph, the fire alarm was activated because an unattended candle caught a student's curtains and bedspread on fire.

75. **The correct answer is (C).** According to the second paragraph, the university disciplinary committee decided that the student would have to pay for the damages to her room.

76. **The correct answer is (C).** In order to determine the number of elderly residents with hearing loss, multiply 3,500 by .28. The answer is 980.

77. **The correct answer is (B).** In order to determine the total number of arsons from 1995–2000, add the total number of arsons for each year. The answer is 16,210.

78. **The correct answer is (B).** In order to determine the average number of arsons from 1995–2000, add the total number of arsons for each year (16,210), then divide this total by 6. The answer is 2,702 (rounded to the nearest whole number).

79. **The correct answer is (A).** In order to determine how many more arsons occurred in 1999 than in 1998, subtract 1,302 from 2,988. The answer is 1,686.

80. **The correct answer is (D).** To learn how many arsons took place in 2001, multiply the number of arsons in 2000 (3,219) by 3 percent, or .03. 3,219 × .03 = 97 (rounded to the nearest whole number). Because the number of arsons *decreased* by 3 percent, you will then subtract 97 from 3,219. The answer is 3,122.

81. **The correct answer is (C).** To determine the value of the jewelry, add together the values of the computer, wardrobe, stereo system, and Oriental rug. Then subtract this total ($7,200) from $10,378. The answer is $3,178.

82. **The correct answer is (B).** To calculate the value of the jewelry, divide the value of the wardrobe ($2,200) by 3. The answer is $733.33.

83. **The correct answer is (C).** To calculate the value of the dining room table, multiply the value of the Oriental rug ($500) by 4. The answer is $2,000.

84. **The correct answer is (C).** To determine the value of the new stereo system, multiply the value of the old stereo system by 6 percent, or .06. $1,500 × .06 = 90. Then add 90 to 1,500. The answer is $1,590.

85. **The correct answer is (B).** To determine the average value of the lost items, add together the total value of the items, assuming that the jewelry was worth $1,600. The total is $8,800. Then divide the total by 5. The average value is $1,760.

86. **The correct answer is (C).** To learn how long it would take Joe Davis to run 5 miles, multiply 9 (minutes) by 5 (miles) to obtain 45. Then divide 45 by 1.5 (miles). The answer is 30 minutes.

87. **The correct answer is (D).** To calculate how many joules constitute 7 Btu, multiply 1,055 by 7. The answer is 7,385 joules.

88. **The correct answer is (C).** To square a number, simply multiply a number by itself. $11 \times 11 = 121$. Therefore, the square root of 121 is 11.

89. **The correct answer is (D).** To determine the area of a window, multiply the length by the width of the window. $6 \times 4 = 24$; therefore, the area of the window is 24 sq. ft.

90. **The correct answer is (E).** To determine the length of the third hose, add together the lengths of the first two lengths of hose. $200 + 150 = 350$. Then subtract 350 from the total length of the hose (575). $575 - 350 = 225$ feet.

91. **The correct answer is (B).** To determine the number of groups in the training session, divide 623 by 6. There are 103 groups of 6 with 5 firefighters remaining.

92. **The correct answer is (C).** In this question, the ground floor is equivalent to the first floor. To calculate the total number of steps, multiply 18 by 5. The answer is 90.

93. **The correct answer is (D).** 0.2 is equivalent to $\frac{2}{10}$. $\frac{2}{10}$ can be further reduced to $\frac{1}{5}$.

94. **The correct answer is (C).** In order to figure out how much more weight the ladder can withstand, first add together the total weight of both firefighters and their equipment ($185 + 50 + 220 + 50 = 505$). Then subtract 505 from 750. The answer is 245 pounds.

95. **The correct answer is (A).** To determine the number of injuries in 2000, multiply 98 by 28 percent. $98 \times .28 = 27$ (rounded to the nearest whole number). Then subtract 27 from 98. The answer is 71.

96. **The correct answer is (B).** To learn how much hotter the room must become before the rope loses strength, divide 300 by 86. The answer is 3.5 (rounded to the nearest tenth).

97. **The correct answer is (D).** To determine the value of $x$, subtract 64 from 325. $325 - 64 = 261$. Then divide 261 by 12. The answer is 21.75.

98. **The correct answer is (C).** The average height of the firefighters can be calculated by adding together the total heights of each firefighter in inches. To convert each firefighter's height into inches, multiply the number of feet by 12 and then add the remaining inches. The total heights are equal to 356 inches. Then divide the total by 5. The average height is 71 inches (rounded to the nearest whole number). 71 inches converted into feet and inches is 5'11".

99. **The correct answer is (D).** Firefighter Nagle is 6'5" tall (77 inches), and Firefighter Lao is 5'7" tall (67 inches). Firefighter Nagle is therefore 10 inches taller than Firefighter Lao.

100. **The correct answer is (C).** Firefighter Kurz is the second-tallest firefighter. Were Kurz to gain 6 pounds every year for three years, the firefighter would gain a total of 18 pounds by the end of the third year. Firefighter Kurz weighs 168 pounds. $168 + 18 = 186$ pounds.

101. **The correct answer is (B).** To calculate the average weight of the firefighters, add together the weights of all of the firefighters. The firefighters' total weight is 838 pounds. Divide 838 by 5. The average weight is 167.6 pounds.

102. **The correct answer is (A).** Firefighter Nagle is the tallest firefighter and weighs 200 pounds. Were Nagle to gain 10 pounds, the firefighter would then weigh 210 pounds.

103. **The correct answer is (C).** To determine how much heavier Firefighter Kurz is than Firefighter Lao, subtract Firefighter Lao's weight from Firefighter Kurz's. $168 - 145 = 23$ pounds.

104. **The correct answer is (C).** Two out of three, or $\frac{2}{3}$, is equivalent to .66, or 66 percent.

105. **The correct answer is (E).** $D(A + B) - \sqrt{C}$ is equal to $6(4 + 8) - \sqrt{81}$. First, add together 4 and 8 in the parentheses. $4 + 8 = 12$. Then multiply 12 by 6. $12 \times 6 = 72$. The square root of 81 is 9. So the answer is $72 - 9 = 63$.